Mechanisms of Brain Function and Adaptive Potential

Mechanisms of Brain Function and Adaptive Potential

Liana Spytska

JENNY STANFORD
PUBLISHING

Published by

Jenny Stanford Publishing Pte. Ltd.
101 Thomson Road
#06-01, United Square
Singapore 307591

Email: editorial@jennystanford.com
Web: www.jennystanford.com

British Library Cataloguing-in-Publication Data
A catalogue record for this book is available from the British Library.

Mechanisms of Brain Function and Adaptive Potential

ISBN 978-981-5129-91-5 (Hardcover)
ISBN 978-1-003-71697-6 (eBook)
DOI: 10.1201/9781003716976

Contents

Preface

Traditional neuroscience has long relied on structural-anatomical approaches to studying brain function, focusing on unchanging neural networks and static cerebral organization, leaving the phenomena of consciousness understudied. Beginning in the 1980s, pioneering studies, including the Silver Spring monkey experiments, initiated a paradigm shift toward neuroplasticity, demonstrating that the brain is not a rigid structure with fixed contours but an organ capable of dynamic reorganization in response to experiential factors and traumatic influences. As the discipline evolved and integrated research on hypnosis, alternative states of consciousness, and neural network modeling, researchers have faced complex methodological challenges related to modified states of consciousness, subjective experience, and complex cognitive interactions between biological and psychological determinants.

Modern neuroimaging techniques, including functional magnetic resonance imaging (fMRI), electroencephalography (EEG), positron emission tomography (PET), and magnetoencephalography (MEG), have enabled researchers to directly observe neuronal activity in cortical and subcortical structures involved in attention, sensory perception, nociceptive modulation, and emotional response. Studies of the phenomenon of neuroplasticity have radically transformed scientific understanding of the adaptive potential of the mature brain, establishing a conceptual understanding of cerebral structures as dynamic systems that demonstrate the ability to undergo morphofunctional reorganizations during ontogenesis in response to pathological processes, cognitive load, and environmental factors.

This book is a scientific analysis of the convergence of the phenomena of hypnosis, neuroplasticity, and neural networks to reveal the dynamic properties of the human brain. The book is structured to systematically study cerebral functions, hypnotic states, and neural network systems. The first section is devoted to fundamental neuroscientific concepts that form the theoretical basis for further analysis. The second section presents a comprehensive overview of hypnosis through the prism of historical retrospectives and modern scientific approaches. The third section researches artificial and biological neural networks, analyzing their architectural organization, synaptic dynamics, and implications for perception and cognition. The fourth section integrates the previously presented content to study the specific relationship between hypnotic states and cerebral functions, and also analyzes the impact of hypnosis on brain activity. The final section focuses on practical applications and promising directions in the medical field (pain therapy, psychiatric intervention), the psychological field (psychotherapeutic methods, cognitive enhancement), technological innovations (brain–computer interfaces, artificial intelligence development), and educational processes (learning optimization, competence development). The conclusions summarize the results on the interdependencies between cerebral processes, hypnotic states, and neural networks; consider ethical aspects of hypnotherapy and neural network research; and identify promising directions for further scientific research and technological progress.

The interdisciplinary research methodology, based on the principles of evidence-based medicine, gives this work special relevance in the context of modern scientific discourse, where solving complex problems requires the synthesis of achievements of multiple fields of knowledge. The target audience are scientists and PhD candidates who study cerebral functions, the phenomenology of consciousness, and altered states of the psyche or focus on cognitive mechanisms, conscious processes, and therapeutic interventions; psychiatric and psychological clinicians who are interested in the implementation of hypnotherapeutic techniques; neuroscientists who study cerebral functionality and neuroplastic processes. The book may also be of interest to computer scientists who are engaged in the development of artificial neural networks and brain–computer interfaces.

The book represents an interdisciplinary approach that integrates the achievements of neuroscience, hypnological studies, and neural network theory. The conceptual basis of the work is based on the hypothesis that hypnotic phenomena are realized through specific neurophysiological mechanisms that are amenable to empirical study, theoretical understanding, and practical optimization in the context of therapeutic application. The book emphasizes the need to synthesize methodological approaches of psychological science, neurobiological research, and computational technologies to fully reveal the potential of neuromodulatory therapeutic interventions. The prospects for further scientific exploration in this problem area will contribute to the deepening of scientific understanding of the cognitive and neuroplastic capabilities of the human psyche.

Liana Spytska
July 2025

Introduction

Introduction to the Subject

In the world of science and psychology, research on the brain, its functions, and complex relationships with mental phenomena has always attracted a lot of attention. The brain, as the central organ of the nervous system, remains one of the most mysterious structures in the body. One of the interesting aspects is its interaction with the phenomena of hypnosis and the functioning of neural networks. Hypnosis, as it is a mysterious tool for influencing human consciousness and behaviour, has long attracted the attention of researchers and doctors. Neural networks, in turn, become the basis for our understanding of how the brain processes information and performs complex functions. In this book, the relationship between the brain, hypnosis, and neural networks is considered, and important aspects of this subject are explored in an attempt to solve some of its mysteries.

Important insights into the evolution of brain function research can be gained from a historical perspective on the field's development. Over time, the idea of brain plasticity—which is essential to contemporary neuroscience—has changed significantly. The Silver Spring monkeys in the early 1980s were the subject of a crucial study that influenced this change [1]. These experiments provided novel insights into the concept of cortical reorganisation by demonstrating that the brain is not a rigid structure with fixed circuits, but an organ capable of undergoing changes in shape in response to injuries. The notion that adult neurons are static was challenged by the observation that monkeys with severed sensory nerves exhibited functional adjustments in the associated cortical areas. The results obtained in the present study provided a foundation for current neurorehabilitation methodologies and

DOI: 10.1201/9781003716976-1

further refined our understanding of brain plasticity. This concept is highly relevant to the domain of hypnosis, as alterations in brain connections during hypnosis may bear resemblance to the processes of cortical reorganisation observed in studies on neuroplasticity.

The brain, a complex and incredibly powerful organ, has always attracted the attention of researchers in the field of psychology and neurobiology. It performs important functions that determine our thinking, feelings, reactions, and behaviour. The brain is made up of billions of neurons that interact with each other via synapses [2]. This complex system is called a neural network, and it forms a fundamental structural component of the brain. The brain is divided into several main parts, including the cerebral cortex, brainstem, and limbic system (Fig. 1). Each of these parts is responsible for certain functions, and their interaction determines a person's ability to learn, perceive information, solve problems, and respond to the environment. In psychology, the brain plays a key role in examining cognitive processes that include perception, thinking, memory, and decision-making. Learning cognitive functions helps discover and understand how humans process information, solve problems, and make decisions. Functional techniques such as functional magnetic resonance imaging (fMRI) allow visualising brain activity during tasks by examining the responses of various brain regions to stimuli and tasks [3]. In neurobiology, the brain is an object of study from a biological standpoint. Research on the physiology of the brain and its structure and functions helps identify the mechanisms by which the nervous system works. Neuroscientists examine neurons, synapses, neurotransmitters, and other biological aspects that underlie information perception and signal transmission in the brain. The brain is a common research subject in both psychology and neurobiology [4]. Understanding psychological processes such as attention, memory, and decision-making requires knowledge of the physiological mechanisms that underlie them. Neurobiology, on the other hand, benefits from understanding how these physiological processes affect behaviour and mental disorders. The conclusion is that the role of the brain in psychology and neurobiology is extremely important. Researching this organ helps to discover and understand the internal mechanisms of the human mind and behaviour. The integration of data from both fields contributes to the further development of scientific research and the application of the acquired knowledge in important areas such as medicine, education, and psychological practice.

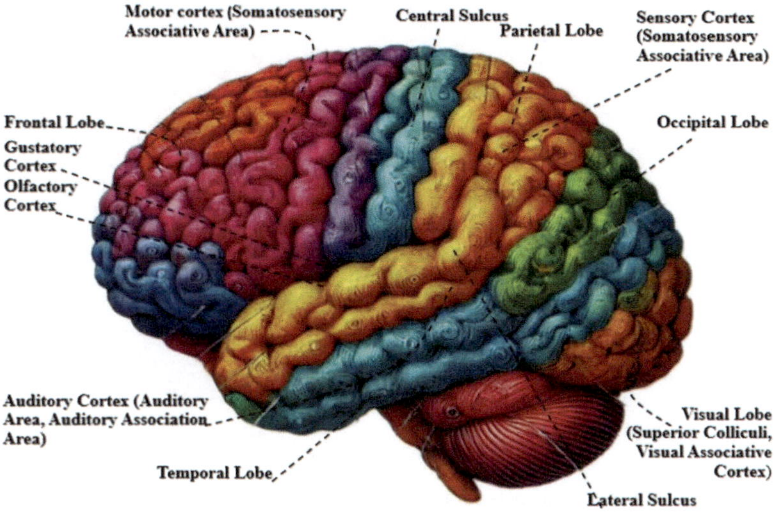

Figure 1 Regions of the brain.

Hypnosis has always been a mysterious and interesting phenomenon for science and society. It evokes associations with mysterious states of consciousness and control over conscious processes. Recently, researchers have increasingly turned their attention to hypnosis as a tool for examining the functioning of the brain and identifying its secrets. Hypnosis is interesting from the standpoint of brain research because it can serve as a kind of window into a person's consciousness. During the hypnotic state, changes occur in perception, imagination, memory, and control. Scientists use hypnosis to investigate how the brain responds to changes in the state of consciousness and how this affects the functioning of neural networks (Fig. 2). They use various techniques to do this, such as fMRI, to visualise brain activity during hypnotic states [5]. Hypnosis research provides important information about brain function and neurophysiological mechanisms. Researchers are examining how hypnosis affects the activity of different parts of the brain and the connections between them. For example, some studies show that during hypnosis, the activity of the frontal lobes of the brain, which are responsible for executive functions and behaviour control, can change [6].

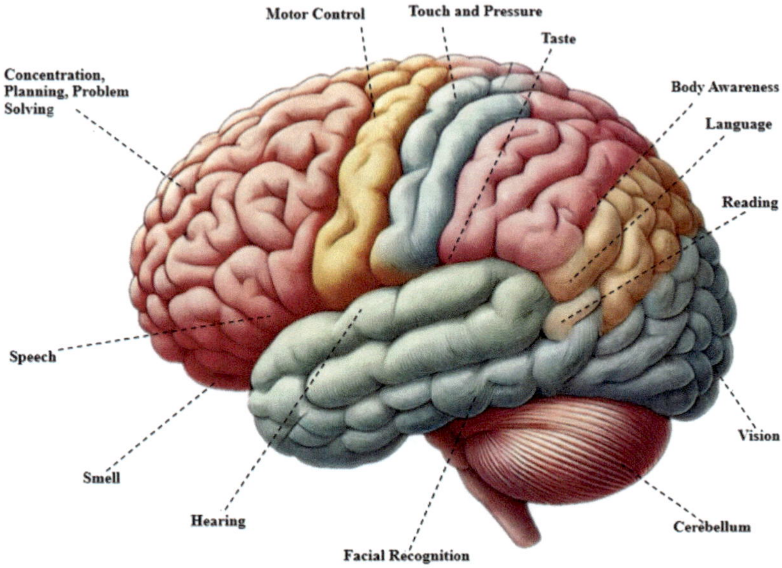

Figure 2 How hypnosis affects the human brain and what parts of the brain are involved in hypnosis.

Other papers point to the effects of hypnosis on the limbic system, which is responsible for emotional responses. Hypnosis also helps researchers understand the nature of consciousness. Hypnosis is used to investigate changes in perception of time, identity, and other aspects of consciousness. Examining hypnosis can show how different parts of the brain interact to create people's perception of the world and themselves [7]. Hypnosis also has practical applications in medicine and psychotherapy. It can be used to reduce pain, manage stress, and treat phobias and depressive disorders. Research on hypnosis helps to establish its mechanisms of action and develop effective treatments.

Hypnosis plays an important role in brain research, helping to identify its secrets and understand neurophysiology, consciousness, and mental processes. This unique state of consciousness serves not only as an object of research but also as a tool for developing new examination methods and practical applications in medicine and psychology. Research on hypnosis continues to discover the secrets of the brain that remain at the heart of human consciousness.

Mental processes that encompass thinking, perception, memory, decisions, and other aspects of cognition have always interested researchers in psychology and neurobiology. These processes determine our understanding of the world and our interaction with it. Recent scientific advances in neural networks, or artificial neural networks (ANNs), have opened up new opportunities for understanding and modelling mental processes [8]. ANNs are mathematical models that mimic the structure and function of biological neural networks in the brain. These models consist of a large number of artificial neurons that interact with each other through weights and signal transmission. They are able to "learn" from input data and solve complex tasks such as image classification, speech recognition, and predicting future events. Neural networks can be used to model cognitive processes such as perception and attention. They help identify how the brain processes information and determine what a person pays attention to. For example, neural networks can mimic how the brain responds to stimuli and thus determine their importance.

Memory and learning are other important aspects of mental processes that can be examined and modelled using neural networks. They help to identify the mechanisms of formation of short-term and long-term memory, and the processes of learning and memorising information. Neural networks have many practical applications in the modern world. They are used in the fields of artificial intelligence, automated control systems, image recognition, speech processing, and many other areas. The examination of mental processes using neural networks is also important for the development of psychotherapy, rehabilitation, and mental medicine. Neural networks open up new horizons in the research on mental processes [9]. They help to understand how the brain works and identify the mechanisms underlying behaviour and understanding of the world. The growing interest in neural networks opens up new prospects for further research and development of modern science.

During the recent decades, a growing interest in the examination of hypnosis by neuroscientists and psychologists has been detected. Current studies focus on examining the effect of hypnosis on the activity of various brain regions using functional methods of imaginative thinking and fMRI. Researchers are trying to determine what mechanisms are triggered by the hypnotic state and how it can affect cognitive processes and consciousness [10]. Current research also focuses on the interaction between hypnosis and neural networks. Scientists use ANNs to simulate the brain's response to hypnotic influences. This allows identifying which parts of the brain are responsible for hypnotic effects and why they can occur. Researchers are also investigating the role of neural networks in understanding the phenomenon of hypnosis [11]. They examine how neural networks can be used to expand the understanding of hypnotic phenomena such as amnesia, hypnotic suggestion, and hypnotherapy. Despite the growing interest in this area of research, scientists face some challenges. Firstly, hypnosis is a complex phenomenon, and its exact mechanism is still not fully understood. Researchers should continue to develop research methods that allow them to better understand this process. In addition, it is important to consider the ethical aspects of the use of hypnosis in research and practice, in particular, hypnotherapy. All of these challenges open up new possibilities for understanding the interactions between the brain, hypnosis, and neural networks. This field of science remains fascinating and important for developing an understanding of mental processes and their application in practice.

The brain has always remained one of the most mysterious and complex structures in the human body. It controls all aspects of our lives, from physical functions to mental activity and internal processes. For centuries, scientists have tried to unravel and understand the functioning of the brain, and their studies have led to numerous discoveries. However, many aspects of brain function remain unexplored, and one such aspect is hypnosis and its relationship to neural networks. Hypnosis is a phenomenon that has always attracted the attention of researchers and specialists in psychology and neuroscience. It is known for its ability to influence a person's consciousness and subconscious mind, contributing to changes in perception and behaviour. There are many questions related to the mechanisms of hypnosis that remain open and require further research.

In recent years, there has been a period of intense neuroscience development, where researchers have focused on examining neural networks and their impact on various aspects of brain function. Neural networks are becoming an increasingly important tool for understanding brain function and solving various tasks related to information processing and artificial intelligence. There are many studies that combine aspects of the brain, hypnosis, and neural networks. For example, researchers use hypnosis combined with fMRI and electroencephalography (EEG) to investigate how the brain responds to hypnotic states and changes in consciousness. Research on memory and hypnosis helps to identify how hypnosis can affect the recall of events and why, under hypnosis, people may have altered susceptibility to influences on their memory. In addition, researchers use neural network models to understand which parts of the brain are responsible for hypnosis states and altered consciousness [6, 12, 13]. Examining the effects of hypnosis on pain sensations and how the brain responds to hypnotic, analgesic techniques is an important area of scientific research. Understanding how hypnosis affects the sensation of pain can lead to the development of new, effective treatments. This can be especially useful for patients who are unable or unwilling to take medication due to side effects or other reasons. Hypnotherapy can be useful for reducing pain and discomfort in patients during medical procedures or in the postoperative period. This helps to improve the quality of life and reduce patient stress. Pain often has a psychological component, and hypnosis can help deal with these aspects. The examination of hypnosis can separate the mechanisms by which it is possible to change the perception of pain and the stress response to it. Hypnotherapy can be useful for treating various psychosomatic diseases and pain-related disorders [14]. Understanding the mechanisms of hypnotherapy can improve the effectiveness of psychotherapy techniques.

Investigating the effects of hypnosis on biological processes in the brain can help identify mysterious aspects of brain function and contribute to the development of neuroscience. Overall, the analysis of the effects of hypnosis on pain sensation is of great practical and scientific importance, which can contribute to improving medical practice and the quality of life of patients. Examining the interaction between hypnosis and brain neuroplasticity can show how hypnosis can influence changes in brain structure and function. These studies contribute to understanding how the brain functions during hypnotic states and what important practical effects this can have, including the use of hypnosis in medicine, psychotherapy, and other fields.

Purpose and Justification of the Importance of Brain Research, Hypnosis, and Neural Networks

The purpose of this book is to consider and analyse the relationship between the brain, hypnosis, and neural networks. For many years, researchers have sought to identify the nature of hypnosis, its effects on brain function and information perception, and the role of neural networks in this context. The scientists' tasks included examining the history of hypnosis research, analysing current theories and approaches to explaining hypnotic phenomena, and investigating the neural mechanisms underlying hypnotic states.

Defining the purpose and objectives of research is an important step in any scientific work or project. This defines the main goals of the paper and reflects what exactly needs to be achieved. The importance of defining the purpose and objectives of research is manifested in the fact that the goal and objectives indicate the direction of research. They help determine exactly what information is needed and how to collect it. This embeds a logical structure into the study and avoids ambiguity. Defining goals and objectives helps justify why research is important. Goals and objectives help determine how ambitious or realistic research plans are. Defining goals and objectives helps allocate time, money, and other resources efficiently to achieve the goal and helps understand in advance what specific expected results are expected. Thus, determining the purpose and objectives of research is a fundamental stage in the scientific process, which determines its success and importance.

In this section of the study, the overall goal was to establish the relationship between the phenomenon of hypnosis, brain activity, and neural networks to identify and understand the mechanisms of hypnotic phenomena and their impact on mental processes. This purpose defined the main direction of the study and indicated that it was aimed at identifying the essence of hypnosis, determining its physiological and psychological correlations with brain activity, and examining the role of neural networks in this context. This goal is important because research on the brain, hypnosis, and neural networks can identify new aspects and mechanisms that can influence the understanding of brain function and consciousness. Understanding the relationship between hypnosis, brain activity, and neural networks may make practical sense for psychotherapists, physicians, and psychologists to develop new treatments and therapies. Research on this subject contributes to the development of brain science and neurobiology in general, which can have important implications for many areas and scientific fields. Identifying the mechanisms of hypnotic phenomena and their impact on mental processes can have practical implications for society and individuals, for example, to improve psychological health and better understand human consciousness. Thus, the definition of the overall purpose of the study indicates its importance for the development of scientific knowledge and possible practical applications in medicine, psychology, and neurobiology. An important stage in examining the relationship between the research goal and its relevance is to understand exactly how this goal meets the pressing issues and needs in modern science and practice.

Early philosophers conjectured about the role of the brain in perception and consciousness in ancient civilisations, which is where a historical understanding of neuroscience originates. By dissecting human cadavers throughout the Renaissance, Andreas Vesalius transformed neuroanatomy and refuted Galen's antiquated views about the brain [15]. During this period, evidence-based anatomical research replaced speculative hypotheses, establishing the foundation for modern neuroscience. The development of neurobiology as a separate field enabled the systematic study of brain activity, which paved the way for more recent investigations into the neurological mechanisms behind cognitive processes like altered states of consciousness and hypnosis.

Examining the brain has always been one of the main goals of science. Identifying how hypnosis affects brain activity and neural networks helps identify new aspects of the functioning of this complex organ, which is of general importance for neurobiology. Hypnosis has the potential to be a tool for improving mental health and treating certain diseases. Determining its relationship with brain activity and neural networks identifies the possibilities of using hypnosis in medicine and psychology. Understanding the mechanisms of hypnotic phenomena and modelling them using neural networks contributes to the development of artificial intelligence and the creation of more intelligent systems. Knowledge of the relationship between the brain, hypnosis, and neural networks may make practical sense for developing hypnotherapy, psychotherapy, and other forms of psychological care. The examination of hypnosis and its effect on brain activity discovers the secrets of consciousness and its work, which is one of the key scientific problems. Therefore, the specified purpose of the book is to answer the relevant questions and challenges in the fields of neurobiology, psychology, and medicine. The study establishes a link between the relevance of this subject and its scientific purpose, which makes it important and justified in the modern research context.

The investigation of the effects of hypnosis on neural networks is a fascinating field that combines psychology, neurobiology, and neurophysiology. Some scientists have addressed this subject, directing their research to understand how hypnosis can affect the activity of neural networks in the brain. The history of the investigation of the effects of hypnosis on neural networks covers a number of important discoveries and conclusions that contribute to understanding this interesting phenomenon. Already in the Middle Ages, scientists and doctors began to investigate hypnosis and its effect on human physiology. However, the real breakthrough in this field came in the 19th century due to the work of scientists such as J.-M. Charcot and H. Bernhard.

Jean-Martin Charcot was a prominent French physician and neurologist who lived from 1825 to 1893. He is known for his contributions to the fields of medicine, especially in neuroscience and psychiatry, and is considered one of the founders of neuroscience. He examined and detailed the symptoms and characteristics of Parkinson's disease, now known as Parkinsonism. His work has contributed to understanding this neurodegenerative disease. Charcot was also known for his research on hysteria and hypnosis. The scientist examined cases of hysteria and believed that hypnosis could be a useful tool in treating this mental illness. He founded the Pitié-Salpêtrière Hospital in Paris, where his research was conducted and patients were treated. This hospital has become an important centre for the study of neuroscience and psychiatry. Charcot has made substantial contributions to the classification and description of various neurological disorders and diseases. He helped formulate concepts that influenced the further development of neuroscience. He was elected a member of the French Academy of Medicine and was an important figure in the research and treatment of neurological diseases of his time. J.-M. Charcot left a substantial mark on medical science and is considered one of the fathers of modern neuroscience and psychiatry. His research and work continue to influence medical practice and research to this day [16, 17]. However, despite substantial achievements and contributions to medicine, J.-M. Charcot faced some limitations and failures in his research. Many of Charcot's limitations were due to the fact that in the 19th century, the available research tools and methods were limited compared to modern standards. He was unable to use modern technologies such as MRI or EEG. Charcot investigated cases of hysteria and used hypnosis, but this caused criticism and controversy. Some of his methods may have been unethical by modern research standards. At that time, there was a widespread belief in "hypnotic effects", and some of his research may have distorted the results due to this belief. Compared to today, when medical information is available to the general public, the sources and exchange of information at that time were much more limited. Given the limitations of the time, Charcot still made substantial contributions to medicine and neuroscience. His work became the basis for further research and development of these branches of science, and many of his discoveries remain relevant to this day.

Hippolyte Bernheim was a French physician and psychologist known as one of the key figures in the development of hypnotherapy and the examination of hypnosis in the late 19th and early 20th centuries. He was born on January 7, 1840, and died on February 2, 1919. Bernhard made an important contribution to the field of psychotherapy and became a prominent representative of the School of "Nancy Hypnosis" or "Nancy Hypnotism School", named after the city of Nancy in France. Bernhard investigated and improved hypnosis techniques as a means of influencing the human psyche and physical well-being. He believed that hypnosis could be useful for treating various functional and psychosomatic diseases. Bernhard actively influenced the spread and examination of hypnosis in the scientific and medical environment. He published numerous papers and studies on this subject and attracted the attention of scientists and doctors to hypnotherapy. Bernhard developed a theory known as "monotheism of hypnosis", in which he argued that hypnosis is the only system of influence on the patient's psyche. He believed that hypnosis is not divided into different states or phases. Bernhard used hypnotherapy to treat a variety of mental and physical illnesses, including headaches, neuroses, insomnia, and other functional disorders. Bernhard's contributions to hypnotherapy and the investigation of the psyche substantially influenced the development of modern psychotherapy and psychiatry. His methods and approaches to hypnosis have remained relevant in modern methods of psychotherapy. Bernhard was an important figure in the history of medicine and psychology, and his contribution to the development of hypnosis and psychotherapy cannot be overemphasised. The scientist contributed to the development of scientific research in these areas and helped introduce hypnotherapy as one of the methods of treatment and psychotherapy [18]. Despite his substantial contribution to the development of hypnotherapy and psychology, Hippolytus Bernhard failed to achieve some results, and his work also faced certain limitations and criticism. Bernhard often received approval and criticism from colleagues as well as from other scientists and doctors. Due to the fact that he was a representative of the Nancy School of Hypnosis, which used hypnosis as a method of treatment and psychotherapy, some scientists and doctors viewed this approach with doubts and criticised the effectiveness and ethics of hypnotherapy. At the time Bernhard was working, evidence-based medicine was less developed, and standards for scientific research and clinical trials were less stringent. Therefore, his work may not meet modern standards of evidence-based medicine. As a result of his research and methods, Bernhard sometimes found himself in disputes and conflicts with other scientists and specialists in the field of psychiatry and psychotherapy. Despite the limitations and difficulties, H. Bernhard managed to make a substantial contribution to the examination of hypnotherapy and the use of hypnosis in treatment and psychotherapy. He helped spread this method and raise interest in it in the scientific community, which influenced the further development of this field of medicine and psychology.

In the 20th century, interest in brain activity research increased substantially. Scientists have started using new technologies, such as EEG, to investigate brain activity during hypnosis. Due to the development of neuroscience, it became clear that many processes in the brain are associated with the work of neural networks. Research on the effects of hypnosis on these networks has become increasingly interesting. Currently, there are various methods of examining the brain during hypnosis, including fMRI and magnetoencephalography (MEG). These techniques allow scientists to more accurately assess the activity of different parts of the brain and their interactions during a hypnotic state. The examination of the effects of hypnosis on neural networks provided insights for improving psychotherapy methods and treating various psychosomatic diseases. All of these studies contributed to a deeper understanding of how hypnosis affects neural networks and brain function in general. This is an important area for further research, which may have practical applications in medicine, psychotherapy, and other fields. Other well-known scientists have also made a substantial contribution to this industry. For example, Dr. D. Spiegel (1947) is a well-known psychiatrist and researcher who has made substantial contributions to the examination of the effects of hypnosis on neural processes. He has conducted research on the use of hypnosis in the treatment of mental conditions and investigated its effect on brain function, including the activity of neural networks [19]. M. Erickson (1901–1980) was an outstanding psychotherapist and pioneer of clinical hypnosis. He studied how hypnosis can affect psychological and physiological processes in the brain and developed treatments based on hypnosis [20]. M. Posner (1936) is a prominent neuroscientist who has studied the relationship between attention, meditation, and hypnosis. His research is aimed at understanding how these states of consciousness can affect the functioning of neural networks [21]. E. Rossi (1933) is known for his research in clinical hypnosis and psychophysiology. He investigated the interaction between hypnosis and neural networks in the context of therapy and self-affirmation. These scientists and other researchers are directing their efforts to investigate the effects of hypnosis on the activity of neural networks in the brain to identify new knowledge about the functioning of consciousness and develop treatments and psychotherapy [22]. Psychotherapist and writer M. Newton (1931–2016) examined the effects of hypnosis and regression on patients' memory and consciousness. His research in regression hypnotherapy helps patients discover their previous lives and explore soul experiences [14]. Neuroscientist and states of consciousness researcher S. Laureys (1968) conducted studies on hypnosis states using fMRI to investigate their effects on the brain [23]. He investigated the physiological changes that occur during hypnotic states.

These studies may have important implications for the development of clinical psychology and neurobiology, and for understanding the mind-brain relationship. The problem of the effect of hypnosis on the brain has attracted the attention of many scientists and doctors. Research helps to identify how hypnosis can affect brain activity and consciousness. The examination of hypnosis can contribute to the development of effective treatments, such as hypnotherapy, to relieve physical and psychological symptoms. Hypnosis research helps to investigate various aspects of brain function and its response to mental states. However, hypnosis is a complex condition, and examining it in the laboratory can be difficult and not always accurate. The response to hypnosis can vary from person to person, making it difficult to draw general conclusions. In some studies, there is a possibility of placebo exposure, when patients perceive an improvement in the condition through waiting for the effect and not through hypnosis itself. Despite these limitations, research on the effects of hypnosis on the brain and consciousness continues to evolve and bring new knowledge that can be useful both in clinical practice and in science.

The analysis of the effects of hypnosis on the brain and neural networks can be conducted using various methods and technologies. The fMRI method allows measuring the activity of various areas of the brain and tracking changes in brain activity during hypnotic states. During this examination, patients are asked to perform tasks in a hypnotic state, and X-rays record the activity of various parts of the brain [6]. EEG is used to measure the electrical activity of the brain. It allows recording changes in brain activity that may occur during hypnosis. Changes in the frequency and amplitude of electric waves can be indicators of hypnotic states [24, 25]. The positron emission tomography (PET) method allows measuring the volume of blood circulation and glucose intake in different areas of the brain (Fig. 3). It can be useful for determining which parts of the brain become more active during hypnosis [26]. MEG allows measuring magnetic fields caused by the activity of neurons in the brain. This method can record rapid changes in brain activity that occur during hypnosis [27]. Also, doctors use neurophysiological and psychophysiological measurements, which cover the determination of physiological indicators, such as heartbeat, blood pressure, respiration, etc., during hypnosis. Changes in these indicators may indicate the effect of hypnosis on the autonomic nervous system.

These methods allow scientists to investigate the effects of hypnosis on the brain and neural networks and obtain objective data to better understand this phenomenon. However, it is important to keep in mind that hypnosis is a complex state of consciousness, and its study can be difficult due to individual variations and methodological problems.

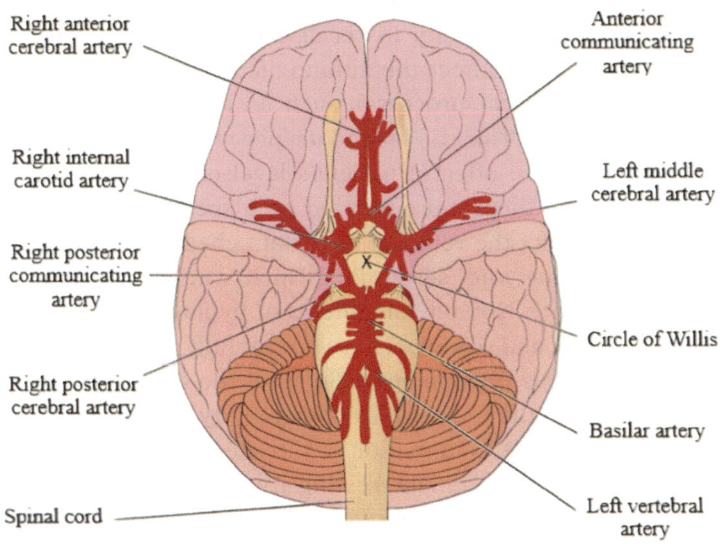

Figure 3 Cerebral circulation: Diagram of brain vessels and their role in oxygen supply.

The effects of hypnosis on the brain and neural networks are a complex and not fully understood process, but researchers have identified several possible mechanisms. During hypnosis, changes in the activity of different parts of the brain can be observed. Hypnosis can lead to a decrease in the activity of the anterior part of the brain, known as the frontal region. This can promote relaxation and reduce the analytical functions of the brain. Hypnosis can also affect the back of the brain, which is responsible for processing sensory information and body awareness. Hypnosis can help activate the limbic system, which is responsible for emotional responses and motivation. Some studies indicate changes in the sensory cortex during hypnosis that may affect the perception of sensory stimuli [7, 24]. Notably, the exact changes in brain activity during hypnosis may vary depending on individual characteristics and the hypnosis methods used in a particular study. Hypnosis can affect pain perception and pain signals by activating the brain's opioid system. This can lead to a decrease in the sensation of pain and an increase in the pain threshold while the patient is in a hypnotic state. Hypnosis can cause changes in consciousness and imagination. This may include more immersion in a state of hypnosis, an imaginary transition to another place or time, or the ability to recall events that were forgotten earlier. Hypnosis can affect the autonomic nervous system, which regulates physiological processes such as heartbeat, respiration, and blood pressure. This can lead to deeper relaxation and reduced stress. Hypnosis can affect a patient's level of consciousness and attention, making them more focused on a specific object or task. Hypnosis can affect the removal of certain chemicals in the brain, such as dopamine and serotonin, which can affect mood and psychological state. Notably, the effects of hypnosis can vary from person to person, and scientists have not yet fully understood all the mechanisms of this phenomenon. Some of the effects of hypnosis may be related to the individual's vulnerability and imagination of the patient. Thus, the examination of the effects of hypnosis on the brain remains an active and interesting area of scientific research.

This book is divided into sections and subsections to better identify various aspects of the subject. Each chapter covers issues related to the brain, hypnosis, and neural networks, considering current research and theories. This study can be an important contribution to understanding hypnosis, the brain, and the role of neural networks in these processes. Research on these issues can have important implications for modern science and medicine, the development of new methods for influencing brain activity, and improving a person's quality of life. The purpose of the examination of the brain, hypnosis, and neural networks is to identify and understand a number of key aspects that relate to the functioning of the human brain, the phenomenon of hypnosis, and the role of neural networks in these processes. The rationale for the importance of this study is that the brain is the central nervous system and the leader of all physiological and psychological processes in the human body. Its functioning covers information perception, decision-making, memory, emotions, and other key aspects. Understanding the brain helps solve problems related to neurological disorders, improve cognitive abilities, and develop new treatments. Hypnosis is a state of consciousness that opens up new opportunities for examining and influencing the human psyche and physiology. Understanding the nature of hypnosis can lead to the development of new treatments, pain control, and better functioning of the subconscious mind.

Neural networks mimic the structure and function of the brain and are used to solve complex tasks in artificial intelligence. The investigation of neural networks can contribute to the development of more powerful and intelligent artificial systems that can perform many functions similar to those performed by the human brain. The examination of the brain, hypnosis, and neural networks is important in psychology and psychiatry. Information about these aspects helps to understand mental disorders, depression, anxiety, and other conditions, and develop more effective methods of psychotherapy and psychiatric treatment. These studies address questions about the existence of consciousness, volitional influence, and limitations of the human mind. They can also influence cultural attitudes and public perceptions of how our mental sphere functions. As a result, the investigation of the brain, hypnosis, and neural networks is an important task that has huge potential for the development of science, medicine, and technology, improving the quality of human life, and understanding the fundamental aspects of human nature.

Basic Concepts Related to Neuroscience

Neuroscience is an interdisciplinary field of science that considers the nervous system, brain, and related processes. Neuroscience uses numerous terms and concepts to describe and understand the functions of the nervous system and brain. Together with the brain, many concepts and terms are connected in various scientific fields.

The outer layer of the brain, i.e., the cerebral cortex, is responsible for higher functions such as thinking, perception, speech, memory, and decision-making. The main functional units of the brain are neurons that are responsible for transmitting and processing information in the brain. The points of contact between neurons where neurochemical information signals are transmitted are called synapses. Groups of interconnected neurons that work together to perform specific functions are combined into neural networks. The mechanisms and processes that allow neurons to communicate with each other and transmit signals are a neural transmission system, and the chemicals that transmit signals between neurons are called neurotransmitters. The biochemical processes of nerve impulse transmission in the brain and nervous system are complex and involve several key stages. The main stages of biochemical impulse transmission involve the synthesis of neurotransmitters, chemicals that are used to transmit signals between neurons. Neurotransmitters are synthesised in nerve cells in different parts of the brain. Synthesised neurotransmitters are usually stored in vesicles at the ends of neurons, ready for release. When a nerve impulse reaches the end of a neuron, it stimulates the release of neurotransmitters from the vesicles into the space between neurons, known as the synaptic gap. Neurotransmitters pass through the synaptic cleft and bind to receptors on the membrane of the next neuron. This leads to changes in the electrical potential of the neuron membrane. The binding of neurotransmitters to receptors initiates an electrical signal in the next neuron. This can lead to an increase or decrease in the electrical potential of the membrane, depending on the type of receptor and neurotransmitter. An electrical signal triggered by neurotransmitters can be transmitted along a neural network, affecting the activity of subsequent neurons and determining the system's response to stimuli. After signal transmission, neurotransmitters must be redistributed or removed from the synapse. This process involves the return of neurotransmitters back to the neuron that synthesised them, or their cleavage by enzymes [28]. Neurotransmitters are chemicals used to transmit signals between neurons in the nervous system. These substances play a key role in regulating brain and body functions. Thus, acetylcholine is responsible for transmitting signals between nerves and muscles. Acetylcholine is important for controlling movement, memory, and attention. Serotonin affects mood, sleep, appetite, and sexual activity. Low serotonin levels are linked to depression and other psychiatric disorders. Dopamine is responsible for feelings of pleasure and also regulates movement and coordination. Norepinephrine (noradrenaline) is responsible for the feeling of cheerfulness and readiness for action. It regulates heart rate, blood pressure, and stress response. Gamma-aminobutyric acid (GABA) is an inhibitory neurotransmitter that reduces neural activity and regulates feelings of anxiety and stress. The excitatory neurotransmitter glutamate increases neural activity and is used to transmit rapid signals to the brain. Endorphin is responsible for the feeling of euphoria and reducing the feeling of pain. Each of these neurotransmitters has its own important functions in regulating various aspects of human physiology and psychology. These biochemical processes allow neurons to transmit information and control body functions by creating electrical signals that are transmitted throughout the nervous system. Precise regulation of these processes is crucial for the normal functioning of the brain and nervous system [29]. Disequilibrium in the levels of these neurotransmitters can cause a variety of mental and physical disorders, so they are the object of research and treatment in medicine and psychiatry.

The two main hemispheres of the brain, the left hemisphere and the right hemisphere, are responsible for various aspects of thinking and function. The structure of the brain that is important for the formation of new memories and learning is called the hippocampus. The group of structures, including the hippocampus, that is responsible for emotional responses and memory, is the limbic system. One of the most compelling case studies in neuroscience involves Clive Wearing, a musician who suffered severe damage to the hippocampus due to herpes encephalitis, leading to profound anterograde and retrograde amnesia [30]. His procedural memory was unaffected, which allowed him to play the piano flawlessly even if he was unable to create new memories or remember the majority of his past. This scenario highlights the difference between declarative and procedural memory and demonstrates the many memory systems in the brain. While it shows that procedural memory, which is connected to several brain regions, including the cerebellum and basal ganglia, can survive even in cases of severe memory impairment, it also emphasises the function of the hippocampus in encoding and recovering explicit memories. This scenario strongly supports the idea that the brain has various memory systems and functional specialisation.

There is a concept that indicates that different parts of the brain have specialised functions and are responsible for performing specific tasks or processing certain types of information—a functional specialisation in the brain. This concept is based on the fact that the brain is made up of different structures, and each of them plays an important role in regulating various aspects of the body's functioning. Different parts of the cerebral cortex are responsible for processing information from different sensory systems, such as visual, auditory, somatic (body sensation), taste, and olfactory. Each of these areas specialises in processing specific types of sensory impressions. The brain has specialised areas responsible for controlling the body's movements. For example, the motor cortex of the brain coordinates volitional movements, and the brainstem regulates autonomous functions such as breathing and heartbeat. The brain has specialised areas responsible for understanding and producing speech (Fig. 4). The limbic system, which includes the hypothalamus, hippocampus, amygdala, and other structures, is responsible for regulating emotions, memory, and motivation. Different parts of the hippocampus and cerebral cortex are responsible for different aspects of memory. Memory is a complex process that involves varieties and phases. For example, sensory memory is the ability to capture impressions from the environment through the senses (visual, auditory, tactile, etc.). Sensory memory is very short-lived and allows temporarily saving impressions for further processing. The type of memory that allows keeping a small amount of information in the mind for a few seconds or minutes is called short-term memory. This memory is used, for example, to temporarily save a phone number before dialling it. Long-term memory stores information for a long time, perhaps even for life. Long-term memory includes facts, skills, experiences, events, and more. Episodic memory is a type of long-term memory that includes the ability to reproduce specific events, scenes, or "episodes" from the past. This allows remembering personal experiences and stories. Another type of long-term memory is semantic memory, which contains knowledge about facts, concepts, and meanings. For example, knowledge of dictionaries, mathematics, or history. Procedural memory encompasses skills and procedures. For example, how to ride a bicycle, drive a car, or play a musical instrument. The emergence of memory varieties is associated with different mechanisms in the brain and specific areas responsible for their functioning. For example, the hippocampus plays a key role in the formation of episodic and spatial memory, while the cerebral cortex is responsible for semantic and procedural memory. Varieties of memory interact with each other and allow storing and using information in daily lives.

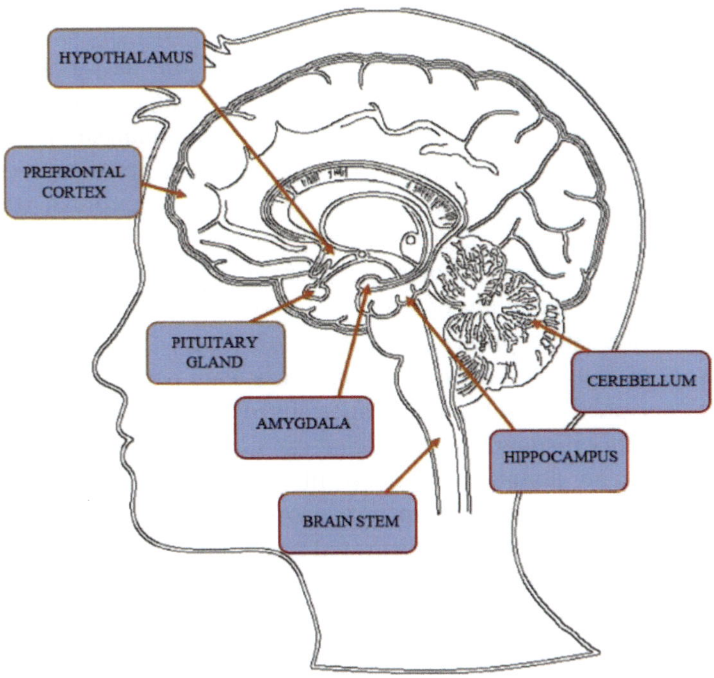

Figure 4 Brain cross-section: Diagram of internal structures.

The brain also has specialised areas for performing specific tasks, such as regulating breathing, processing pain sensations, distinguishing faces and emotions, and many others. The process of recognising faces and understanding other people's emotions and feelings in the human brain is associated with complex neural mechanisms and includes several key areas. In the visual cortical region, the main visual impressions from the retina of the eyes are processed. The hippocampus plays a role in the formation of memory and associations. When people see faces, the hippocampus helps them preserve the memory of those faces and the details associated with them. The area in the brain that is responsible for analysing faces and recognising emotions is also of great importance. It responds to subconscious impressions associated with faces and helps determine the emotional state of another person. The prefrontal cortex plays a vital role in higher cognitive functions such as reasoning, decision-making, and understanding social situations. It helps analyse information about the perception of faces and facial expressions, understand social contexts, and identify the emotions of others. The limbic system, which includes the amygdala and hypothalamus, plays an important role in detecting and regulating emotions. When a person sees a face, their brain analyses various details, such as eye shape, facial expressions, and lip movements, which help determine a person's personality and emotional state. These information signals are processed and interpreted by different areas of the brain, which allows recognising faces and understanding the emotions and feelings of others.

The brain's perception and response to pain are due to a complex network of neurons and neurotransmitters. Pain perception begins in the peripheral nerves, where specialised pain receptors known as nociceptors respond to stimuli that can damage the tissue. These receptors send signals along nerve fibres to the spinal cord. The spinal cord processes and transmits pain signals to higher levels of the brain. Here, the pain is interpreted and modulated. For example, reflexes such as pulling one's hand away from a hot object can occur without brain involvement. Pain signals are sent to various areas of the cerebral cortex, including the somatosensory cortex. This gives humans a conscious perception of pain and the ability to respond to it. Pain can also trigger an emotional response through interaction with the brain's limbic system, which is responsible for emotions and motivation. The brain can modulate pain by releasing neurotransmitters such as endorphins, which can block pain or reduce its intensity. The brain also considers context and the individual when perceiving pain [31]. This means that the feeling of pain can vary depending on the situation, emotional state, and other factors. In general, pain perception is a complex process that involves many different parts of the brain and can be modulated by various factors. This concept of functional specialisation helps to better understand how different parts of the brain interact and interact to ensure the normal functioning of the body. The brain's ability to adapt and change, especially when influenced by learning and experience, confers neuroplasticity. Disorders of brain function, i.e., brain diseases such as Alzheimer's disease and Parkinsonism, and others that can affect brain function in which neurodegeneration occurs, i.e., the process of losing neurons and connections in the brain, which can also occur with age. These concepts are just a few of many that are used to investigate and understand the brain, its functions, and its role in our lives and health.

Regarding hypnosis, it is a complex state of consciousness that is associated with numerous concepts and aspects. A form of psychotherapy in which hypnosis is used to treat psychological and physical problems, such as stress, anxiety, phobias, chronic pain, etc., is called hypnotherapy. During a hypnotherapy session, the patient is in a hypnotic state of deep relaxation, i.e., in a trance. A specialist who uses hypnosis for therapeutic or entertainment purposes is commonly referred to as a hypnotist. During hypnosis, accessing a part of a person's consciousness (subconscious) to change beliefs, habits, and opinions is possible. These changes in perception, memory, and awareness that can occur during hypnosis can include amnesia, hypnotic communication, and other phenomena, i.e., hypnotic effects. During a hypnotherapy session, the doctor may provide instructions or hints (suggestions) that may affect the patient. A person can use autohypnosis on their own to achieve certain goals, such as reducing stress or improving productivity. There is evidence for the ability of hypnosis to reduce the perception of pain and provide relief from pain symptoms, i.e., hypnotic pain control. These concepts provide only a partial overview of the complex and interesting field of hypnosis that is being researched in psychology, medicine, and brain science.

The use of hypnosis raises various ethical questions and debates among professionals and specialists in the field of psychology and medicine. The main ethical issue is obtaining the patient's consent to conduct a hypnosis session. Patients should be aware of the procedure, its possible risks and side effects, and be able to stop hypnosis at any time. A doctor or psychologist who uses hypnosis must maintain strict confidentiality. Information obtained during hypnosis should remain confidential and not be disclosed without the patient's consent. Hypnosis can be a useful tool in psychotherapy practice, but it is important to use it with caution and ethics. Therapists should have appropriate training and experience in hypnosis [7]. Hypnosis is also used in entertainment shows and performances. In such cases, it is important to adhere to ethical standards and not harm the mental state of the participants. Hypnotherapists should use hypnosis within the framework of laws and regulations that relate to the practice of medicine or psychotherapy. It is also important to take care of the rights and safety of patients. The use of hypnosis in medical practice should be based on scientific research and evidence of effectiveness. A scientific approach helps reduce risks and provide better outcomes for patients. Individuals who use hypnosis in their practice should have appropriate training and licenses, especially in areas where it is required. Education and training help ensure a high level of professionalism and ethics. In general, the ethics of using hypnosis require specialists to adhere to high standards and ensure the safety and well-being of patients.

As for neural networks, this is a complex and multidimensional field of science that covers numerous concepts and terms. Electrical impulses propagate along the axons of neurons and play an important role in transmitting information in the brain. These are just a few key concepts in neuroscience, and this field of science continues to evolve, making important contributions to understanding the functioning of the brain and nervous system. These concepts form a fundamental component of neural networks and help you understand how they work and how they are used in various industries.

As a result, there is a general idea of the importance of examining the effects of hypnosis on the brain and neural networks. Hypnosis remains one of the most interesting and mysterious areas of psychology and neurobiology, and examining its effects on brain activity can identify new insights into the nature of consciousness, imagination, and neurophysiology. In this book, the effect of hypnosis on brain function and neural networks is considered in detail, and possible mechanisms of this influence and current research in this area are investigated. Understanding these aspects can be important for clinical research, developing new hypnotherapy techniques, and unlocking the potential of neural networks in the context of cognition and changing consciousness. This subject is becoming increasingly relevant as it discovers the secrets of the functioning of the human mind and opens up new opportunities for applying this knowledge in many areas.

Chapter 1

Brain Basics

1.1 Brain Anatomy

Brain anatomy research is an in-depth examination of the structure and complexity of the central nervous system (CNS), which plays a crucial role in controlling various aspects of our lives. The brain is an extraordinary organ that has not only extreme complexity in structure but also incredible power in functioning. It is the central command post responsible for regulating movements, thinking, and emotions and for coordinating the work of other body systems. When examining the anatomy of the brain, researchers dive into a world of complex architecture of neurons and their connections that form an amazing network. Understanding this anatomy helps solve mysterious questions about how we perceive the world, how our thinking and memory work, and how the brain responds to external influences. This discipline allows close examination of the structure of the brain, its various parts, and the connections between them. Brain anatomy identifies and explains how different areas of the brain are responsible for different functions and how changes in its structure can affect mental and physical health.

The importance of investigating brain anatomy cannot be overemphasised as it has a profound impact on various areas of

Mechanisms of Brain Function and Adaptive Potential
Liana Spytska

ISBN 978-981-5129-91-5 (Hardcover), 978-1-003-71697-6 (eBook)
www.jennystanford.com
DOI: 10.1201/9781003716976-2

human life and the development of modern science and medicine. Examining brain anatomy helps to understand how different parts of the brain work and how they interact to provide functions such as thinking, memory, movement, and emotions. Knowledge of brain anatomy helps doctors diagnose and treat various neurological and mental illnesses, such as stroke, epilepsy, schizophrenia, and head injuries. Brain anatomy is the foundation for understanding psychology and mental processes, helping to investigate intelligence, memory, behaviour, and emotions. Knowledge of brain anatomy can improve the learning and development methods of children and adults, contributing to the optimal development of neural networks. Brain anatomy is an important component of artificial intelligence research and the development of brain computer interfaces. Therefore, understanding brain anatomy is an important prerequisite for further research and advances in neuroscience, medicine, psychology, and other fields, and for improving people's quality of life and health.

The investigation of the brain is a long and fascinating story that includes the contributions of many outstanding scientists and researchers [32]. The first attempts to understand the nature of the brain belong to ancient civilisations. The Egyptians, Greeks, and other ancient peoples were engaged in medicine and anatomy, including brain research. Their knowledge and brain research were limited compared to modern standards, but they made a substantial contribution to the development of these industries.

Significant advances in the localisation of brain activities were made during the 19th century. The identification of the brain regions in charge of particular cognitive capacities was one of the most important discoveries made during this time. Our knowledge of how various brain regions contribute to intricate cognitive processes was greatly enhanced by Karl Wernicke's (1874) discovery of Wernicke's area, a region in the brain associated with language comprehension [33] (Fig. 5). His study expanded on previous research by Paul Broca, who had determined that Broca's region was essential for speech production. Wernicke's research laid the foundation for future studies of neurological illnesses that impact language and communication by confirming the idea that different parts of the brain are specialised for particular functions.

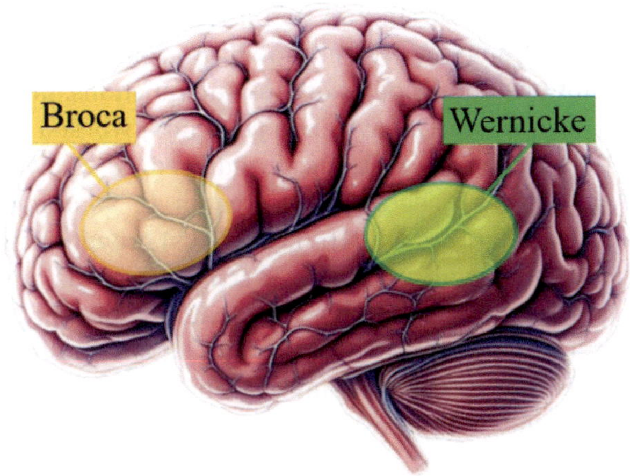

Figure 5 Broca's and Wernicke's areas.

Egypt, for example, was one of the first civilisations to consider medicine and have its own doctors. Egyptian mummies show that they had knowledge of some aspects of anatomy. Their medical practices included procedures such as corn removal and surgery. Greek philosophers and doctors such as Hippocrates, Galen, and Aristotle made substantial contributions to the development of medicine and anatomy. Hippocrates defined the principles of ethics in medicine, and his oath is the basis of modern medical ethics. Galen investigated anatomy and physiology, including the brain, and his work was important for medieval and Renaissance medicine. The ancient Greek physician Galen made a great contribution to the early anatomy and physiology of the brain, making the first attempts to understand its role in the functioning of the body. He was one of the most famous doctors of his time and examined various aspects of medicine, performing anatomical studies. Galen is known for his research on the structure of the brain and the functions of its various parts. He believed that the brain is made up of four basic fluids. He also examined the connections between the brain and the nervous system. Although many of his theories proved inaccurate over time and with the development of science, Galen's contribution to early brain anatomy and physiology was substantial, and his work influenced further scientists and their medical research.

In China, medical teaching was also developed, and some Chinese doctors looked at the brain and its functions. In Indian medicine, some aspects of anatomy were known. Ancient Indian texts had some descriptions of the body's organs and functions. In the Middle Ages, Arab doctors such as Avicenna (Ibn Sina) worked on medical treatises that contained anatomical and medical knowledge. The Middle Ages led to a decline in research in anatomy, but the Renaissance period (15th and 16th centuries) brought new interest to the field. Renowned anatomists such as Vesalius and Harvey have made substantial discoveries in anatomy and physiology. In general, although ancient civilisations had limited opportunities to investigate the brain and anatomy in general, their curiosity and knowledge provided the basis for further development of these sciences in the future.

During the Renaissance, great scientists such as Leonardo da Vinci and Vesalius made substantial contributions to understanding the structure and function of the brain through anatomical research. Leonardo was a versatile genius and artist who also studied anatomy. He produced the most famous anatomical drawing of his time, known as the "Vitruvian Man". In his drawings, there was an attempt to present the structure of the human body, including the brain. Leonardo made an important contribution to the investigation of anatomy and contributed to the development of modern body science. Andreas Vesalius was a Belgian physician and anatomist who lived in the 16th century. He published his famous work "Factory of the male body" (lat. "De humani corporis fabrica"), where for the first time, a detailed illustration of the structure of the human brain and other organs was provided. Vesalius revolutionised approaches to anatomy, recognised the importance of direct observations, and developed an anatomical method based on realistic research rather than beliefs or perceptions. These outstanding scientists used their anatomical knowledge and drawings to better understand the structure of the brain and contribute to further research in anatomy and neurobiology. They played a key role in setting the stage for further brain research.

The 19th century was marked by the discovery of neurons, the main structural and functional unit of the brain. Scientists such as C. Golgi and S. Ramón y Cajal were outstanding researchers who made substantial contributions to the examination of the structure of nerve cells and synapses. C. Golgi was an Italian physician and neuroscientist who developed a technique known as "Golgi staining". This method allowed for the first time in the world to investigate in detail the structure of neurons and the synapses through which they interact. Using his staining method, Golgi created remarkable drawings where individual components of nerve cells, including dendrites, axons, and synapses, could be seen. S. Ramón y Cajal (1852–1934) was a Spanish neuroscientist and Nobel laureate, known for his important contributions to brain anatomy and physiology. He was a pioneer in the field of neuron research and made substantial contributions to understanding the structure of the nervous system. He first identified and described synapses, that is, the contact points between neurons where nerve impulses are transmitted. This discovery was a step towards understanding how the nervous system works. Ramón y Cajal created a detailed atlas of brain structures, including descriptions of many types of neurons and their structure. This atlas has become an important resource for scientists examining the structure and function of the brain. He also used a microscope for the first time to reconstruct nerve pathways in the brain, which allowed him to better understand their structure and connections. Ramón y Cajal's contribution was very important for the development of neuroscience and brain science, and their research remains relevant and important to this day. He worked on the work of Golgi and other scientists who used Golgi colouring. He carefully examined neurons and, for the first time, determined their structure and functions, in particular, the structure of synapses. His works and drawings became the foundation for the development of modern neurobiology. Due to the contributions of Golgi and Ramón y Cajal, the first detailed knowledge was obtained about the structure of neurons, their connectivity, and synapse function. The use of microscopy allowed scientists to investigate brain structures in detail at the micro level. The discovery of neurons and their processes has been the result of microscopic research and observations by many scientists over the years.

Examining the chemical composition of the brain, including neurotransmitters and receptors, helps in understanding various aspects of the nervous system and brain functions. The examination of brain chemistry is an important branch of neurobiology and neurochemistry that investigates the chemical components and processes that occur in the brain. The history of research on the chemical composition of the brain contains the work of many scientists and substantial discoveries. Franz A. Mesner (1772–1857) was the first German anatomist and physiologist to identify isolated structures in the brain that later became known as neurons. He was also one of the first to discover that the brain contains a large amount of protein. W. Kuhn (1839–1902) was a German chemist and first described simple chemical compounds that perform important functions in the brain, such as neurotransmitters. He examined the work of choline and acetylcholine. J. Wilhelms (1870–1948), a Belgian physiologist, made substantial contributions to understanding glutamate, one of the most important neurotransmitters in the brain. American neuroscientist E. Kandel was awarded the Nobel Prize in Physiology or Medicine in 2000 for his research on the molecular mechanisms of learning and memory, which include chemical processes in the brain [34]. Research on brain chemistry allowed a better understanding of how the brain functions and what chemical reactions and compounds play a key role in regulating various aspects of the nervous system, including signalling between neurons, controlling emotions, and interacting with other body systems.

These important discoveries served as the basis for further research in the field of neurobiology and the examination of the functioning of the nervous system. A. Alzheimer (1864–1915) was a German physician and pathologist who made substantial contributions to the investigation of brain pathology and nervous diseases. He became best known for his contribution to understanding and describing the disease, which was later called Alzheimer's disease. He described in detail the clinical symptoms and pathological changes in the brains of patients suffering from a certain type of dementia. His work, published in 1907, was the first description of the pathological changes that cause this disease. Alzheimer's disease has shown brain cell damage in patients with the disease. He first described specific inclusions in neurons, which he called "neurofibrillary vesicles" and "plates". Alzheimer's work served as the basis for further scientific research and the development of Alzheimer's disease science [35]. It has had a substantial impact on understanding neurodegenerative diseases and brain pathology. Alzheimer's disease is now known as one of the most common forms of dementia and memory disorders. Alois's research on Alzheimer's disease was a step towards understanding the disease and developing treatments and diagnostics.

R. Feynman (1918–1988), an outstanding physicist and Nobel Prize winner, made an important contribution to understanding quantum physics. However, his contribution to the nervous system is more general and based on the principles of quantum mechanics. One of the important concepts Feynman developed and used is "interference", a phenomenon in which quantum particles (such as photons) can interact and create phenomena similar to wave vibrations. In the context of the nervous system, this may be important for understanding how neurons exchange information through electrochemical signals. Feynman also made a substantial contribution to the development of quantum field theory, which describes the interaction of elementary particles. This may be important in examining the physiological aspects of neural communication and how signals are processed in the nervous system. In general, Richard Feynman's contributions to physics and quantum mechanics provide a basis for understanding the nature of interactions in the nervous system, although specific research in this area is usually conducted by neuroscientists and physiologists.

The development of electrophysiology allowed scientists to investigate the electrical activity of the brain. I. Pavlov and E. Adrian were among the first to use an electroencephalogram (EEG) to record brain activity. I. Pavlov and E. Adrian were two outstanding scientists who made a great contribution to the development of the study of brain activity using an EEG. Each of them had their own unique contributions to this industry. I. Pavlov, known for his work in the field of physiology and conditioned reflex, was one of the first scientists to use the recording of brain activity in his experiments [36]. I. Pavlov conducted experiments on dogs and studied their reactions to various stimuli. He identified that dogs can learn to associate certain stimuli with food, and this association leads to conditioned reflexes, such as drooling after hearing a call. Pavlov actively measured physiological processes in dogs' bodies during conditioned reflexes, including salivation, gastrointestinal activity, and other functions. I. Pavlov took a huge step by applying his discoveries regarding conditioned reflexes to the psychology of human behaviour. He noted that people also form conditioned reflexes that affect their behaviour and reactions to stimuli. Pavlov's work opened a new path for research in psychology, especially in the field of conditioned reflexes, and sparked interest in psychological theories of behaviour. His studies gave rise to a new direction in psychology, which became known as "Pavlov's psychology" and helped to understand how organisms form conditioned reflexes and how this affects behaviour. The impact of Pavlov's work has proven important in both physiology and psychology, and his research continues to be investigated and used to this day. He used simple electrophysiological techniques to examine the activity of the nervous system and the functioning of the brain. His investigation focused, in particular, on the functioning of the cerebral cortex in relation to conditioned reflexes and the relationship between stimulus and response. E. Adrian was a British physiologist and neuroscientist who was known for his research on the electrophysiology of neurons and the introduction of methods for recording brain activity. In 1920, he received the Nobel Prize in Physiology or Medicine for developing EEG. E. Adrian designed a special electrode that could be used to record electrical signals generated by the brain. He, along with Charles Sherrington, won the Nobel Prize in Physiology or Medicine in 1932 for his research on how the nervous system works, including synaptic transmission. This method allowed researchers to investigate the electrical activity of the brain in detail and observe waves known as "alpha rhythms" displayed on the EEG for the first time. The introduction of techniques such as electrophysiology allowed researchers to understand what functions the brain performs and how it transmits signals between neurons. The work of scientists examining the spontaneous activity of neurons and synaptic connections was very important.

Many scientists have conducted studies on spontaneous neuronal activity and synaptic connections. J. Ackles, an Australian physiologist, also examined synaptic connections and their role in signalling between neurons. He co-won the Nobel Prize in Physiology or Medicine in 1963 for his research on synapse function. D. Huberman, an American neuroscientist, investigated the spontaneous activity of neurons and synaptic connections, in particular, their role in brain development and learning. J. LeDoux was an Italian physiologist who examined the mechanisms of synaptic transmission and spontaneous neuronal activity. These are just a few examples of scientists who worked on spontaneous neuronal activity and synaptic connections [37]. Their research was important for understanding the physiology of the nervous system and brain. Together, their contributions to the field of physiology and neurophysiology have contributed to the development of methods for recording brain activity and opening up new opportunities for examining the functions of the nervous system. The results of their studies have become important for further research on brain function and the development of modern neurophysiology.

Examining the genetic aspects associated with brain development provides an opportunity to understand the genetic factors that influence how understanding is related to brain functions. The examination of genetic aspects related to brain development is an important area of research in genetics and neurobiology. R. Levy-Montalcini, an Italian neuroscientist and Nobel Prize winner, is recognised for her great achievements in the investigation of nerve growth factor (NGF), which is a key component of neuronal growth and development. Her research played an important role in our understanding of how neurons form and perceive signals. In the 1950s, R. Levy-Montalcini and her colleague Stanley Cohen conducted a series of experiments in which they identified and isolated NGF, which is a protein that helps in the growth, development, and differentiation of neurons in the nervous system. By discovering NGF, Levi-Montalcini made a huge contribution to the development of neurobiology. She showed that NGF plays a key role in the survival and growth of neurons and the development of the nervous system. Her studies helped improve knowledge about how neurons interact and form networks. Levi-Montalcini's work is also of great importance for medicine [38]. It has helped to better understand neural diseases and develop new treatment approaches, including neuroregeneration and the treatment of neurodegenerative diseases such as Alzheimer's disease. For her outstanding achievements, R. Levi-Montalcini has been awarded numerous prestigious awards, including the Nobel Prize in Physiology or Medicine in 1986. This award highlighted the importance of her research for science and medicine. R. Levy-Montalcini is a prominent figure in neurobiology, and her work has generated considerable interest in the world of science and has contributed to further research on the nervous system and neurodegenerative diseases.

Neuroscientist E. Kandel, winner of the 2000 Nobel Prize in Physiology or Medicine, investigated genetic aspects related to learning and memory in the neurons of sea slugs. He investigated the molecular mechanisms underlying learning and memory. British biologist S. Brener won the Nobel Prize in Physiology or Medicine in 2002 for his work with a gene that encodes neural growth factors and affects the development of the nervous system. Many other researchers are examining the genetic aspects of brain development, including working on genes that regulate the growth and development of neurons and genes associated with various aspects of neuroplasticity. Research on the genetic aspects of brain development helps to better understand how genes influence the formation of neurons, their migration, relationships, and synapse development, and what genetic factors may be associated with neurological and psychiatric disorders [39].

American neurologist and psychiatrist O. Sacks published several popular books about various neurological conditions. He made substantial contributions to brain research and understanding neurological and psychiatric disorders. O. Sacks was known for his deep compassion for his patients and interest in their unique stories. He made an effort to communicate with patients and understand their problems, which helped to investigate various neurological conditions and disorders with a deeper understanding. O. Sacks conducted scientific research in the field of neuroscience, in particular, the examination of neurodegenerative diseases such as Parkinsonism and Huntington's chorea. His work emphasised the importance of a humanitarian approach to medicine. He developed the idea that brain disease does not exclude a person from a number of humane values and individual preferences. O. Sacks became known not only in the scientific community but also among the general public for his ability to teach complex scientific concepts in an accessible and engaging way, and this contributed to the spread of understanding of neurological and mental states among the general public [31].

British biologist R. Dawkins investigated the evolution and understanding of the spread of genes that affect brain activity. R. Dawkins, a biologist and evolutionist, has made substantial contributions to understanding the evolution and distribution of genes that influence brain activity, especially in the context of evolutionary psychology and behaviour. Dawkins first introduced the idea of "gene centrism". He argued that genes are the main drivers of evolution and saw them as "self-centred" structures that work for their own reproduction. In the context of the brain and behaviour, this helped to understand how genes can influence the development of various aspects of the brain and behaviour in animals and humans. Dawkins was one of the pioneers of evolutionary psychology, a field that explores how the evolutionary process shaped the psychological structures and behaviour of humans. He helped show how psychological aspects such as social interactions, partner choices, and cultural practices can be justified and understood from an evolutionary perspective. Richard Dawkins actively promoted the dissemination of scientific ideas, especially in the field of evolutionary biology and psychology, to the public. He was known for his clarity of presentation and ability to explain complex concepts in an accessible way [37].

Nowadays, it is known that many disorders in the nervous system are associated with abnormalities in genes, and their names may vary depending on the specific disorder or disease. Some examples of gene disorders that are known to be associated with neurological diseases are presented below [40]:

1. Huntington's disease is a deadly hereditary disease associated with the expansion of CAG repeats in the HTT gene (Huntington's gene). This leads to a decrease in the functionality of neurons and severe neurodegeneration.
2. Down syndrome: This is caused by an additional copy of chromosome 21 and can lead to the development of various neurological problems in children with Down syndrome, including developmental disorders.
3. Amyotrophic lateral sclerosis (Lou Gehrig's disease). Some cases of amyotrophic lateral sclerosis (ALS) are associated with genetic mutations, in particular, mutations in the C9ORF72, SOD1, and other genes.
4. Autism. There are genetic factors that are associated with the risk of developing autism. For example, mutations in genes that affect neuronal function and synapses may play a role in the development of this disorder.

5. Schizophrenia also has a genetic component, and many genes have been identified as potentially associated with the risk of developing this mental illness.

These are just some examples. A large number of studies are aimed at elucidating the role of specific genetic mutations in the development of neurological and psychiatric disorders. Scientists are examining genetic features that can affect the function of the nervous system and developing new treatments based on genetic information.

In the 20th century, many new methods were created, such as functional magnetic resonance imaging (fMRI) and positron emission tomography (PET), which allowed improving our understanding of brain functions to a great extent. In the 20th century, there was a substantial development in the field of brain research due to the invention and improvement of various methods for recording and visualising brain activity [39]. fMRI was developed in the 1990s. This method allows visualising the activity of different parts of the brain while performing certain tasks or activities. Peter Mansfield, who made an important contribution to the development of fMRI, deserves mention. PET uses radioactive markers to visualise functional processes in the brain. Invented by David E. Kahn and George F. Pealy, PET has become an important method for examining brain activity and understanding nervous system functions. Magnetoencephalography (MEG) uses magnetic signals generated by the brain to record the activity of neurons. This method allows for investigating brain activity with high accuracy over time. Y. Zikila, who made an important contribution to the development of MEG, is worth mentioning. These methods and scientists have contributed to the development of brain science, allowing researchers to better understand the structure and function of the brain and investigate the relationships between its various parts and processes. They have become important tools for research and building a more complete understanding of how the human brain works.

The growing interest in examining various neurological and psychiatric disorders, such as Alzheimer's disease and schizophrenia, has led to further discoveries at the functional and molecular levels of the brain. Research on various neurological and mental disorders is conducted by scientists, doctors, and researchers in the field of neurology and psychology. These studies began many decades ago and continue to be conducted all over the world. Some prominent scientists and organisations are working in this field. The World Health Organisation (WHO) conducts research and contributes to the development of strategies for mental health and neurological disorders and international standards in this area. The American Psychiatric Association (APA) is engaged in research and development of clinical guidelines for mental disorders and is an important organisation for psychiatrists and psychologists. The World Bank for Mental Health (World Psychiatric Association) is an international organisation dedicated to promoting mental health and research on mental disorders. In many countries, some national organisations and foundations fund research in the field of neuroscience and psychiatry. For example, in the United States of America, these are the National Institute of Mental Health and the National Institute of Neurological Disorders and Stroke. Many universities and research centres have departments that specialise in neuroscience and psychology research. Research on neurological and psychiatric disorders is very important for improving the treatment, diagnosis, and understanding of these conditions. Scientists continue to work to increase knowledge in these areas to improve the quality of life of people who experience such disorders.

In today's interdisciplinary field of research, scientists are making efforts to elucidate the brain map and examine the connections between neurons. This interdisciplinary field of research is called connectomics and deals with the examination of brain structure and function by mapping connections between neurons and different parts of the brain. Researchers use a variety of techniques and technologies to do this, including functional magnetic resonance imaging; diffusion tensor magnetic resonance imaging, which allows investigating the structure and pathways of white matter in the brain and connections between different areas; electroencephalography; and MEG. Scientists also use laser light to visualise the activity of neurons and the connections between them. There are neural mapping techniques that include techniques for tracking the activity of individual neurons and their connections. Researchers in this field use these techniques to create a brain map and understand how different parts of the brain interact with each other. It can help to investigate various aspects of brain function, including various neurological and psychiatric disorders, and the development of new treatments and diagnostics. This story points to continuous progress in brain research, which contributes to the development of medicine, psychology, and neurobiology, and identifies many more unknowns that await further research.

Scientists have learned about the structure of the brain and its functions through a combination of many research methods that have continuously evolved over the years. Due to these methods and many others, scientists have gained a vast amount of knowledge about the structure of the brain and its functions. Brain research is an ever-evolving industry, and new methods and technologies help us better understand this complex organisation. A long history of research on the brain, its structures, and functions has resulted in the confidence that the brain is the most important organ in the CNS and is responsible for controlling many vital functions. Brain anatomy can be classified according to various criteria, including its functional structure, areas, and important structures.

Some basic aspects of classifying brain anatomy are provided below:

1. Structural anatomy. The brain can be classified according to its structural components. This classification includes the main division of the brain into the brainstem, cerebellum, and large hemispheres. Individual structures, such as the hypothalamus, thalamus, hippocampus, and amygdala, can also be distinguished.
2. Functional anatomy. The brain can also be classified by functional areas (Fig. 6). For example, areas responsible for motor skills, sensations, speech, visual processing, memory, emotions, and other functions can be highlighted.
3. Lateralisation. There is a concept of the lateralisation of the brain, when the hemispheres of the brain perform different functions. For example, the left hemisphere usually contains areas responsible for speech, while the right hemisphere is responsible for pattern recognition and spatial imagination.
4. Physiological anatomy. The brain can be classified according to physiological characteristics, such as areas with the highest neuronal activity, regulation of chemical levels, signalling systems, and other aspects.

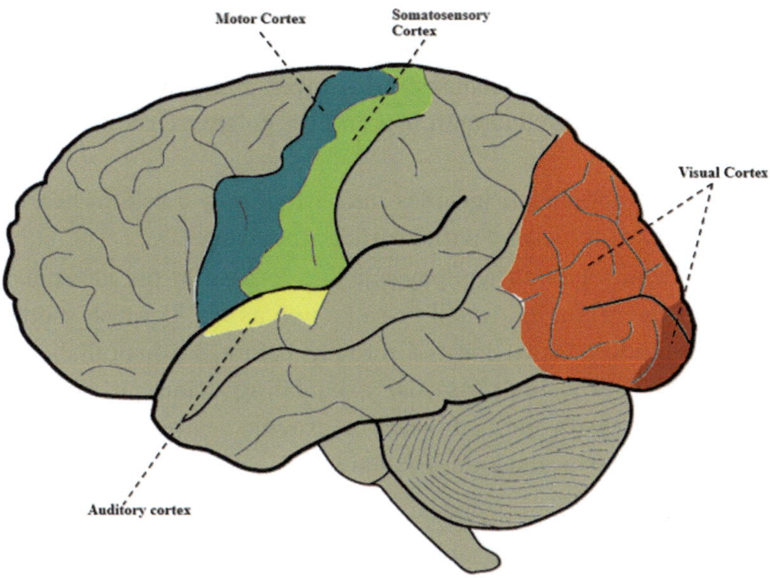

Figure 6 Functional areas of the brain.

These classifications are not exhaustive, and researchers continue to improve their understanding of brain anatomy using state-of-the-art research techniques such as MRI and connectomics. In general, the brain consists of different parts, each of which is responsible for its own functions [41]. The main structures of the brain are:

1. Brainstem: It connects the brain and spinal cord, is responsible for basic vital functions such as breathing, heartbeat, and coordination of movements, and consists of:

- *Medulla oblongata*: Its main function is to control the main automatic functions such as breathing, heartbeat, and blood circulation.
- *Bridge* (pons): It is responsible for functions related to sleep and the coordination of movements.
- *Midbrain*: It contains neurons that control responses to external stimuli and regulate vision.

2. Brain department (diencephalon): It comprises the:

- *Thalamus*: It shares information between different parts of the brain and is responsible for processing sensory information.

- *Hypothalamus*: It controls hunger, thirst, body temperature, sleep, and hormone production by the pituitary gland.
- *Epithalamus*: It includes the grooved gland, which produces the hormone melatonin. Melatonin regulates circadian rhythm and sleep.
- *Metathalamus*: It includes nuclei that play a role in hearing and are associated with hearing processing.
- *Pituitary gland* (hypophysis): It is located in the lower part of the brain, in a small fossa known as the "Turkish saddle" (pituitary fossa) located directly below the hypothalamus, with which it interacts closely, regulating various physiological processes in the body such as growth, sexual development, metabolism, and kidney function.

3. **Limbic system:** It includes:
    - *Hippocampus*: It plays a key role in forming new memories and learning.
    - *Amygdala*: It is responsible for emotion and fear processing.

4. **Cerebrum:** It consists of two hemispheres, left and right, and is responsible for higher mental functions, such as thinking, memory, problem-solving, and perception. It also has:

- *Cerebral cortex*: It regulates functions related to consciousness, language, thought, and decisions.
- *Frontal lobe*: It is responsible for thinking, motor skills, and personality.
- *Parietal lobe*: It processes sensory impressions and information about touch and spatial orientation.
- *Occipital lobe*: It is responsible for processing visual information.
- *Temporal lobe*: It is responsible for auditory processing and learning ability.

5. **Cerebellum:** It regulates movement coordination, balance, and posture.

6. **The spinal cord (medulla spinalis):** It is located in the spine and is responsible for transmitting signals between the brain and the rest of the body and performing certain reflex actions.

These parts of the brain work together to provide a variety of functions, from basic vital responses to complex mental processes such as understanding language and thinking. Understanding the structure and functions of different parts of the brain helps science and medicine justify disorders and develop treatments for various neurological and mental illnesses. This is the general structure of the brain, which is a very complex organisation with many other structures and regions that perform different functions.

1.2 Explanation of the Structure and Functions of Different Parts of the Brain

External anatomical features of the brain include its shape and structures. The brain consists of two cerebral hemispheres—the left and right, which are separated by the lumbar beam. Each brain hemisphere is responsible for the functions of the opposite side of the body and has its own characteristics. The upper parts of the brain hemispheres have a folded surface, which is known as the cerebral cortex, where a large number of neurons are concentrated, and many higher cognitive functions are performed. The brainstem is located at the bottom of the brain and includes the medulla oblongata, the bridge, and the midbrain, which then goes directly to the spinal cord. It provides basic functions such as breathing, heartbeat, and blood circulation. The brainstem also contains important nuclei responsible for regulating various functions. The cerebellum is located behind the brainstem and is responsible for coordinating movements and maintaining balance. It has a complex texture that contains numerous components.

The cerebral cortex has a characteristic appearance with furrows and protrusions. Deep protrusions are the convex parts of the cerebral cortex, and furrows are the pits between these protrusions. Furrows and deep protrusions help to increase the area of the cerebral cortex, which allows more space for neural connections and functions. The brain also has chambers or cavities known as the anterior and posterior chambers. They contain CSF, the fluid that surrounds the brain and spinal cord and provides protection and nutrition for these structures. External anatomical features of the brain allow scientists to examine and understand various aspects of its functioning and structure [42]. This information is important for the diagnosis and treatment of various neurological conditions, diseases, and further research in the field of neuroscience.

The corpus callosum, a network of nerve fibres that connects the left and right hemispheres and permits interhemispheric communication, is an essential part of the brain's anatomy. Particularly in the case of split-brain patients, research on the corpus callosum has yielded important insights into hemisphere specialisation. There were noticeable cognitive alterations in patients who had corpus callosotomy, which involves cutting the corpus callosum to treat epilepsy. Their left hand could sketch an image, but their right hemisphere could not describe it. This occurrence showed strong evidence of lateralisation, showing that speech is primarily controlled by the left hemisphere and spatial processing is handled by the right [43]. These results support the notion that cognitive processes may be lateralised towards particular areas but are dispersed throughout both hemispheres.

The brainstem is responsible for performing vital functions. It is one of the main parts of the CNS that provides communication between the brain and the spinal cord. The brainstem plays an important role in survival and provides functions necessary for life, such as respiration, metabolism, and blood circulation regulation [44].

The medulla oblongata is one of the three main parts of the brainstem, and it performs several important functions necessary for life. It is located at the lowest level and is directly adjacent to the spinal cord. It contains a respiratory centre that controls the rhythm and depth of breathing. It responds to the level of carbon dioxide (CO_2) in the blood and affects the frequency and volume of breathing to ensure normal oxygenation of the body. The medulla contains a heart centre that controls the frequency and strength of heart contractions. It regulates blood circulation and provides blood throughout the body. The medulla is also responsible for regulating blood pressure and responding to changes in blood pressure. It also contains nuclei that control reflexes such as swallowing and coughing, allowing humans to take food and drink fluids. It contains white and grey matter and specialised neurons. White matter contains fibres that carry information up and down the brainstem and spinal cord, while grey matter contains cells that process this information and coordinate the functions of the medulla. It is very important for life, and any damage or disruption of its functions can be very serious [45].

The bridge is an important structure located above the meninges in the brainstem. It is used to transmit information between different parts of the brain and between the brain and the spinal cord. This area of the brain provides a connection between the CNS and the rest of the body. The bridge plays an important role in coordinating movements and contains many neurons and important pathways that regulate complex movements. The bridge has an effect on the regulation of sleep and the rhythm of sleep and wakefulness. It contains structures that affect the biological clock and sleep phases. The bridge contains neurons that analyse information received from various sensory systems, including visual and auditory information. It consists of various cores and paths that interact to perform important specialised functions. An important part of the bridge is the various nuclei that affect facial movements, and respiratory and circulatory functions [46].

The midbrain is an important part of the brainstem, located between the bridge and the diencephalon. It includes several key structures and performs a number of important functions. The midbrain contains four colliculi: the lower colliculi are responsible for auditory perception and the upper colliculi regulate the visual process. These structures are involved in coordinating movements and responses to external stimuli. The black substance in the midbrain produces dopamine, an important neurotransmitter. Dopamine deficiency is associated with diseases such as Parkinsonism, so the functioning of this structure is crucial. The red core plays a key role in movement coordination and muscle activity. It interacts with motor pathways to regulate body movements. The tegmentum is the upper layer of the midbrain that includes various neurons responsible for controlling various aspects of movement and sensation. The midbrain performs functions related to auditory and visual perception, coordination of movements, and responses to environmental stimuli. It is an important part of the brainstem and plays a role in ensuring the normal functioning of the body [47, 48].

The brainstem, including its components, interacts at various levels, including chemical, electrical, and molecular ones. The brainstem is made up of a large number of neurons that interact with each other through electrical impulses. Neurons transmit information through areas called axons, the impulses of which are electrically transmitted along nerve fibres. All elements of the brainstem interact through synapses. Synapses are places where information is transmitted between neurons. In synaptic transmission, neurons use chemical signals in the form of neurotransmitters, such as acetylcholine, serotonin, gamma-aminobutyric acid (GABA), glutamate, etc. These chemicals help transmit signals from one neuron to another through the space between them. At the molecular level, intracellular processes regulate the functioning of neurons and other brainstem cells. These include the synthesis and breakdown of neurotransmitters, the operation of ion channels to transmit impulses, and the production of proteins and other molecules that are necessary for normal functioning. The interaction between all these levels—electrical, chemical, and molecular—allows the brainstem to control various vital functions, such as breathing, heartbeat, movement coordination, and many other processes. In the case of disorders in these processes, various neurological and physiological problems may occur [49, 50]. Notably, any injury or disease of the brainstem can lead to serious disorders in the functioning of these important functions and endanger a person's life.

The diencephalon brain division is one of the five main parts of the brain and is located between the brain and the brainstem. It is of vital importance for many physiological and neurological functions. The midbrain consists of several structures, including the thalamus, hypothalamus, epithalamus, metathalamus, and pituitary gland [51].

The thalamus plays a central role in relaying sensory signals. It processes information from various senses, such as vision, hearing, touch, and temperature, and sends it to the appropriate areas of the brain for further processing. The thalamus also plays a role in regulating sleep and consciousness. The thalamus is formed from paired structures and includes a number of nuclei and nuclear groups that interact with other parts of the brain and transmit sensory signals. It processes information received from various sensory organs, such as the eye, ear, and skin. It acts as a "sorter" of sensory information and sends it to the appropriate areas of the brain for further processing. The thalamus plays a role in the emergence of consciousness, as it transmits information about the outside world to the cerebral cortex, where this information is analysed and perceived. Its interaction with other structures, such as the hypothalamus, regulates circadian rhythm and sleep, controlling the transition from sleep to wakefulness. The thalamus also plays a role in evaluating emotions and influencing the formation of internal experiences and emotional responses. The thalamus interacts with other parts of the brain, such as the cerebral cortex, hypothalamus, and brainstem, to ensure proper processing and integration of sensory information. This interaction is important for our perception of the world and response to it [52].

The hypothalamus is an important part of the brain, which plays a critical role in providing various physiological functions and regulating many body systems. It performs numerous functions and is built from different cores and structures. The hypothalamus is located below the thalamus, in the centre of the brain, near the midbrain. It is responsible for regulating many important physiological processes, such as regulating body temperature, controlling appetite and food intake, regulating water balance, regulating sleep and wake cycles, and controlling important hormonal systems through the pituitary gland. The hypothalamus interacts with other parts of the brain, such as the brainstem and limbic system, to ensure internal harmony and regulate body functions. The hypothalamus consists of numerous nuclei and subsystems that are responsible for various functions. These structures interact with each other and with other parts of the brain. In general, the hypothalamus plays a critical role in maintaining homeostasis (the stability of the internal environment) and regulating many aspects of physiology [47].

The Epithalamus is a small structure in the brain located at the posterior end of the third layer of the brain, known as the midbrain. It consists of several main parts, including several important structures, such as the pineal gland and the choroid layer. The pineal gland produces the hormone melatonin, which regulates the sleep and wake cycle and also affects the regulation of sleep rhythms. The choroid layer produces CSF. Melatonin, produced by the pineal gland, affects the regulation of sleep and wake cycles. It helps the body respond to changes in the light environment by regulating sleep and wake rhythms depending on the time of day. The epithalamus also interacts with other brain structures, such as the hypothalamus and large hemispheres of the brain, to regulate the emotional state and body functions. For example, melatonin affects mood and may be associated with seasonal affective disorder (a condition that occurs as a result of changes in the amount of light available throughout the year). The epithalamus is also able to influence hormonal processes through interaction with the hypothalamus and pituitary gland [44].

The metathalamus is one of the parts of the midbrain located above the thalamus and connected to the limbic system. It performs several important functions in the body. The metathalamus includes part of the limbic system that deals with the processing of odours. This helps in evaluating and interacting with the environment. This part of the brain, connected to the limbic system, plays a role in emotional responses and emotion formation. The metathalamus interacts with other parts of the brain, including the hypothalamus, thalamus, and brainstem, enabling the regulation of various aspects of physiology and behaviour. In general, the metathalamus plays an important role in ensuring the normal functioning of the body, and its interaction with other brain structures helps to ensure harmony of physiology and behaviour [49].

The pituitary gland is located in the brain, in a special depression known as the Turkish saddle. This small depression is on the lower surface of the brain that surrounds and protects the pituitary gland. The location of the pituitary gland in the Turkish saddle allows it to interact with the hypothalamus, which regulates the production of various hormones that affect the functions of various organs and systems of the body. The pituitary gland, also known as the hypophysis, plays a key role in regulating numerous body functions. It produces and releases various hormones into the bloodstream that affect other endocrine glands in the body. Specifically, the pituitary gland produces hormones such as adrenocorticotropic hormone (ACTH), which affects the adrenal cortex and thyroid-stimulating hormone (TSH), which regulates thyroid function. The pituitary gland also controls the release of neurohormones produced by the hypothalamus, such as oxytocin and vasopressin. These hormones regulate water-salt metabolism and kidney function. The pituitary gland produces growth hormone (somatotropic hormone), which affects the growth and development of the body. The pituitary gland consists of two parts: anterior and posterior. Each of these parts produces different hormones and performs different functions. Pituitary activity is controlled by the hypothalamus, which produces neurohormones that stimulate or inhibit the pituitary gland's production of hormones. Pituitary hormones affect the functioning of other endocrine glands, such as the kidneys, thyroid, adrenal glands, etc., regulating their function and hormonal balance in the body. All these functions make the pituitary gland a very important organ that regulates many aspects of physiology and hormonal balance in the body [28].

The midbrain interacts with the brainstem through many important pathways and neural connections. It plays a key role in regulating health-related functions, internal regulation, and responses to external stimuli.

The limbic system is a complex group of structures in the deep parts of the brain that plays a critical role in regulating various aspects of our behaviour and emotions. The limbic system includes the hypothalamus, which controls important physiological processes such as hunger, thirst, temperature regulation, and hormone balance. It also plays a role in the emergence of emotions and is responsible for regulating aggression and sexual behaviour. The hippocampus is mainly responsible for the formation of new memories and learning [52].

The hippocampus is a small structure in the brain that plays a vital role in the formation and processing of information and the storage of certain types of memory. It is located in the limbic system of the brain and is shaped like a curved roller. The hippocampus plays a vital role in the formation of new episodic memories, i.e., memories of specific events and experiences. It is also responsible for spatial memory, which allows navigating in space and even forming maps of the environment. The hippocampus helps with orientation in the world around people and reorientation. The hippocampus interacts with other parts of the brain, such as the cortex and other limbic structures, to process and integrate information. For example, it collaborates with the cortex to store memory and learn. Damage to the hippocampus can lead to loss of the ability to form new memories and problems with navigation. The hippocampus is also examined in the context of various mental disorders, such as schizophrenia and depression, as it plays an important role in regulating mood and emotions [26].

The amygdala is a pair of small structures located deep in the brain's nape, in each hemisphere. It consists of several sub-regions, and each of them is responsible for different functions. The main function of the amygdala is to process and respond to emotionally charged stimuli. It is important for responses to fear and anxiety and for the formation and memory of emotions. In addition, the amygdala plays a role in determining the importance of a stimulus and deciding whether it is a threat. It also interacts with other brain structures, such as the hypothalamus (to regulate physiological responses to emotions), the cerebral cortex (to regulate rational thinking and responses), and the hippocampus (to create and preserve memory). In general, the amygdala plays a crucial role in regulating emotions and is important for the body's survival and adaptation to its environment [17].

The limbic cortex is an important part of the brain responsible for processing emotions and important functions related to memory creation and storage. This structure includes various parts of the cerebral cortex that play a role in controlling emotions and expressing feelings. This part of the brain includes various areas that control emotions, feelings of pleasure, and motivation. The limbic system interacts with other parts of the brain, including the cerebral cortex, the pituitary gland, which is responsible for hormone production, and other structures that regulate physiological and emotional responses. It helps determine how we respond to stimuli and perform different behaviours. Overall, the limbic system plays an important role in regulating emotions, memory expression, and regulating important aspects of physiology and behaviour [35].

The brain consists of two hemispheres—the left and right. These hemispheres are separated by a vertical barrier structure known as the corpus callosum, which allows them to interact and exchange information. Each hemisphere has a large number of components, including the cerebral cortex, which is responsible for higher cognitive functions such as thinking, speech, memory, and decision-making. Usually, the left hemisphere is responsible for speech and language abilities. It controls speech comprehension, word pronunciation, reading, and writing. The left hemisphere plays a role in logical thinking, analysis, and problem-solving. The left hemisphere helps process symbols such as numbers and mathematical operations. The right hemisphere is responsible for the perception of spatial relationships, drawing, pattern recognition, and processing of non-spatial information. The right hemisphere is more involved in regulating and expressing emotions through gestures, facial expressions, and vocal intonation. The right hemisphere is important for musical abilities and understanding of music. Notably, many functions require the cooperation of both brain hemispheres [53].

For example, when reading, if the left hemisphere is responsible for understanding words, then the right hemisphere can play a role in interpreting context and emotionalising the text. Each person has their own balance in the functional use of the left and right brain hemispheres, and this can affect their individual abilities and responses to various activities. The left hemisphere of the brain is responsible for the right side of the body, while the right hemisphere is responsible for the left side of the body.

The cross-cutting principle of defining body control means that control over the movements and sensations of the body passes from one hemisphere to the other. The left-brain hemisphere controls the motor functions of the right side of the body, and the right hemisphere controls the left side of the body. This applies not only to limb movements but also to motor functions related to the face, throat, and other parts of the body. Sensory input from the body is also shared between the left and right hemispheres. Information from the right side of the body goes to the left hemisphere and vice versa. This means that the left hemisphere perceives sensory information from the right side of the body and vice versa. The interaction between the left and right hemispheres plays an important role in cognitive function. Brain plasticity refers to the brain's ability to switch functions between hemispheres in response to brain development or damage. For example, if one of the hemispheres of the brain is damaged, the other may try to compensate for this loss. The division into left and right hemispheres in the brain is a complex and important aspect of brain organisation. Interaction and collaboration between these hemispheres allow for the deployment of various functions and abilities [54]. Notably, these are common patterns, and there may be individual differences in the distribution of functions between hemispheres. In addition, the interaction and collaboration between the hemispheres allow the implementation of complex functions and tasks that require the joint work of both hemispheres.

The brain's ability to redistribute functions between the hemispheres (neuroplasticity) allows it to adapt to changes, including restoring functions after brain damage or adapting to new conditions. Brain lesions, such as injuries, strokes, or tumours, can affect the function of the cerebral hemispheres, and this effect depends on the location and extent of the lesion. Damage to the left hemisphere can lead to aphasia, which is a loss of speech skills. Patients may lose the ability to speak, understand speech, read, or write. If damage to the left hemisphere affects motor areas, it can lead to problems in motor coordination on the right side of the body. Damage can affect logical thinking and the ability to analyse and solve problems. Damage to the right hemisphere can affect the perception of spatial relationships and observation. Patients may have trouble recognising images, drawing, or even perceiving their body in space. Damage to the right hemisphere can cause changes in emotional state and expression. Patients may become less emotionally stable or expressive when expressing emotions. Damage to the right hemisphere can affect musical abilities, the ability to draw art, and musical interpretation [55–57]. Notably, the response to brain damage is individual and depends on many factors, including the size of the lesion, brain plasticity, and the general condition of the patient. Rehabilitation and therapy can help patients restore some functions and learn how to compensate for losses. In most people, some functions may be more strongly associated with one of the hemispheres, but it is important to remember that brain activity always involves the cooperation of both hemispheres [45]. Understanding the role of brain hemispheres helps explain how the brain performs various functions and how this organisation helps with understanding and interaction with the world around us.

The brain is made up of two main types of tissue: grey and white matter. The white and grey matter of the brain are the two main types of tissue that make up the structure of the brain and spinal cord. They perform various functions and are essential for the functioning of the CNS. The grey matter has a grey colour and consists mainly of neurons and their appendages. Grey matter is responsible for processing information, creating integrated responses to stimuli, and performing higher functions such as thinking, memory, understanding, and sensation. White matter is white in colour and consists mainly of nerve fibres (axons) and myelin, which is insulation for axons. White matter is responsible for transmitting signals between different parts of the brain and between the brain and other parts of the nervous system. It helps to coordinate movements, transmit information between the brain and spinal cord, and control organ functions. These two types of tissues interact as part of the nervous system, and their proper functioning is necessary to ensure the normal functioning of the body. The grey matter of the brain is located in different parts of the brain and spinal cord, and it performs different functions in the CNS. It forms the outer layer of the cerebral cortex. The cerebral cortex is responsible for higher cognitive functions such as thinking, speech, memory, information awareness, decisions, and other cognitive processes. Grey matter is located in the hypothalamus, which is a key part of the limbic system and is responsible for regulating many important physiological functions, such as temperature, hunger, thirst, sleep, endocrine regulation, and pleasure. Grey matter also forms part of the basal ganglia, which play an essential role in regulating movement and participating in decision-making [58, 59]. In the spinal cord, grey matter is in the shape of the letter "H" and is responsible for processing signals coming from different parts of the body and leading to movements and other reactions. The grey matter of the midbrain regulates functions such as controlling audition and vision and regulating pulse and blood pressure. Grey matter is located in various other parts of the brain and spinal cord and is involved in regulating various functions, including sensitivity, movement, internal organisation, emotions, and other processes. In general, grey matter is key to processing information in the CNS and performing various functions that allow normal functioning and interaction with the world around us.

The white matter of the brain plays a key role in signalling between neurons and various parts of the CNS. The main function of white matter is to provide communication between neurons and coordinate various brain functions. White matter has its colour due to the myelin surrounding the axons of neurons. Myelin is a fatty membrane that isolates axons and helps store and transmit signals with greater speed and efficiency. White matter contains the axons of neurons, which are long twists and turns and are responsible for transmitting electrical and chemical signals along them. Myelin helps keep axons isolated, providing speed for signal transmission.

The white matter, which is located in the upper part of the spinal cord and cerebellum, is responsible for coordinating muscle contraction and controlling body movements. White matter provides communication and transmission of information between different parts of the brain and spinal cord. It allows the brain to function as a single organ, coordinating various aspects of perception, processing, and response to information. White matter in the spinal cord is responsible for regulating the functions of internal organs such as the heart, lungs, stomach, etc. It allows transmitting signals from the brain to these organs and back, monitoring their work. Therefore, the white matter of the brain is important for the proper functioning of the CNS and the transmission of signals between neurons, helping to provide coordination of movements, perception, and processing of information, and control over the functions of internal organs. The grey matter of the brain has various anatomical connections to other body systems. Grey matter is the main component of the CNS, which contains the brain and spinal cord. It forms neurons that process and transmit information in these structures. Grey matter in the brain interacts with the peripheral nervous system via the spinal cord and nerves. It helps guide signals from the peripheral nervous system to the CNS and vice versa, providing control over movement, perception, other functions, and responses to stimuli. Grey matter is responsible for controlling the somatic nervous system, which contains conscious movement and the perception of sensory information such as touch, pain, temperature, and other sensations. Grey matter also has connections to the autonomic nervous system, which controls the functions of internal organs such as the heart, stomach, lungs, blood vessels, etc. It regulates the body's automatic functions, such as respiration and blood circulation. Grey matter interacts with the endocrine system, including the hypothalamus and pituitary gland, which regulate endocrine glands such as the thyroid, adrenal glands, ovaries, etc., through hormones. Grey matter is surrounded by vessels that deliver blood and nutrients to the brain and spinal cord. Ensuring a good blood supply is important for the normal function of the CNS [60, 61]. The spinal cord, which contains grey matter, is surrounded by vertebrae that fix it in the body and provide support. These connections help ensure the proper functioning of the CNS and coordinate the actions of the brain and spinal cord with other body systems to ensure normal functioning. Notably, grey and white matter interact to ensure proper brain function [46]. Grey matter processes information, while white matter transmits signals and allows different parts of the brain to communicate with each other. Understanding this structure is important for research.

The brain chamber is a system of cavities or chambers inside the brain that is filled with cerebrospinal fluid, a special liquid that performs the function of protecting and nourishing the brain. The brain chamber and CSF spaces play an important role in the functioning of the CNS and the brain's interaction with the environment. This fluid performs several functions, including brain shock absorption and protection from damage. In addition, CSF is important for nutrition and removing waste products from the brain. CSF also contains essential substances for the provision of nutrients to the neurons and for removing waste products. CSF is produced in special structures known as choroid plates and then circulates through a system of brain chambers and CSF spaces. Excess CSF can be removed from the brain and spinal cord through the vesicular spaces and enter the bloodstream. CSF and the brain chamber system are an important part of the physiology of the CNS and help ensure optimal conditions for the functioning of the brain and spinal cord. Any disruption in the production, circulation, or elimination of CSF can cause severe medical conditions and require appropriate treatment. The overall structure of the brain is an extremely complex and well-organised system that allows us to perform a wide variety of functions necessary for our physical and cognitive functioning [47]. Understanding this structure is important for further research and applications in neuroscience and medicine. The internal structures of the brain include various components that play an important role in the functioning of the CNS and the implementation of various cognitive and motor functions.

The cerebral cortex, or the neocortex, is the outer layer of the brain, which is made up of groups of cells known as neurons and performs important functions related to perception, thinking, understanding, decision-making, memory, and many other aspects of cognitive and motor function. The cerebral cortex is a key structure in the brains of humans and other higher mammals. The cerebral cortex is divided into different areas, each of which is responsible for specific functions. These areas include the area responsible for visual perception, auditory perception, speech, voluntary movement, etc.

The cerebral cortex contains many neurons that communicate with each other via synapses (connections). This allows transmitting and processing information in the form of electrical signals. The cerebral cortex helps in implementing various aspects of cognitive function, such as perception, memory, speech, thinking, decision-making, and movement regulation. The cerebral cortex is the largest part of the brain and provides a higher level of information processing. Individual differences in the cerebral cortex can affect various aspects of personality, intelligence, and behaviour. The cerebral cortex performs functions important for the functioning of the human brain and regulates a large number of processes that help us adapt to the world around us and make decisions.

Deep structures include structures such as the hypothalamus, thalamus, basal nuclei, and hippocampus. These structures regulate various functions such as emotions, autonomic functions, sensory information processing, and involvement in memory formation and movement coordination. The brainstem regulates important functions such as respiration, heartbeat, and blood circulation. It also contains nuclei responsible for somatic and vegetative control. The spinal cord carries signals between the brain and the rest of the body and regulates reflexes. The limbic system includes structures such as the hypothalamus, hippocampus, and amygdala. It is responsible for regulating emotions, forming and preserving memory, and for learning and motivation. The pituitary gland is a small structure located at the base of the brain that produces and secretes hormones that regulate the activity of other endocrine glands. It plays a key role in regulating the functions of the endocrine system. The cerebellum plays an important role in regulating movement and motor skills. It is involved in the formation and coordination of movements and has an impact on the training of new motor skills [48]. Understanding both the internal and external structures of the brain helps scientists and doctors better apprehend the functioning of the CNS, diagnose and treat various brain disorders and diseases, and develop new treatments and research in neuroscience.

The cerebellum is an important part of the brain located behind the brainstem and the upper part of the cerebral cortex. It is responsible for the coordination of movements and maintains balance. The cerebellum plays a vital role in performing precise movements and learning new motor skills. The cerebellum has a complex anatomy and consists of two main hemispheres—the left and right. It has deep furrows and a folded surface with a wood-like texture. The cerebellum connects to other parts of the brain through the white matter, which contains numerous neural connections. The main function of the cerebellum is the coordination of movements. It helps ensure precision and consistency of movements, enabling the performance of complex movements such as writing, walking, dancing, etc. The cerebellum is responsible for maintaining the balance of the body. It measures head and body movements and then sends signals to the muscles to adjust movements and maintain balance. The cerebellum plays an important role in learning and improving new motor skills. It affects the learning and development of motor circuits, which become automatic after training. In addition to motor functions, some studies suggest that the cerebellum may affect cognitive functions such as thinking, attention, and working with short-term memory. Damage to the cerebellum can lead to a variety of problems, including motor disorders and balance and coordination problems [44]. Understanding the role of the cerebellum in the functioning of the human body helps in the treatment and rehabilitation of patients with motor disorders and other disorders associated with this part of the brain. The cerebellum also regulates muscle tone, helping to keep the muscles ready for movement and in a certain state of tension. This is important for ensuring stability and control of the body. The cerebellum plays a role in learning new motor skills and improving existing ones. It tracks movements and makes adjustments to improve the accuracy and efficiency of movements. The cerebellum affects the activity of the heart muscle and regulates blood pressure, helping to ensure adequate blood supply to the brain and other organs. The cerebellum interacts with the cerebral cortex, basal ganglia, and other brain structures to regulate movement, memory, thinking, and other cognitive functions. Damage to the cerebellum can lead to various motor disorders, such as dysarthria (speech disorder), ataxia (impaired coordination of movements), increased reflexes, and disorders in various activities. The cerebellum is important for ensuring the proper functioning of the nervous system and coordination of movements and other body functions [51]. The cerebellum interacts with other parts of the brain, receiving sensory information and sending instructions for movement and coordination of motor functions. This is vital for ensuring precise and smooth movements and consistent functioning of the body. The cerebellum plays an extremely important role in coordinating movements and ensuring balance in the body. It receives sensory information from various parts of the body about the position of the body, the location of the limbs and feet, and the degree of muscle tension. Based on this information, the cerebellum makes corrections in the work of the muscles to ensure accuracy and smoothness of movements. It helps to avoid loose or ragged movements that can occur with improper coordination. For example, when walking or performing complex movements, the cerebellum allows different muscle groups to work together so that a human can perform these actions with precision and smoothness. It receives information from the vestibular system (which is responsible for the sense of movement and balance) and helps regulate the muscles and body position to avoid falling and maintain stability. Damage to the cerebellum can lead to motor disorders and impaired balance and coordination. The cerebellum also affects cognitive function, although its role in this aspect is not its primary function. The cerebellum helps to learn movements and improve them [52]. This is crucial for the development of muscle memory, which allows performing movements automatically and without conscious effort. For example, when learning a new sport or game, the cerebellum helps optimise movements and preserve them in memory. The cerebellum affects the ability to plan and perform consecutive movements. The cerebellum is important for developing cognitive functions such as concentration, attention, memory, and problem-solving in children. The development of a child's cerebellum can affect their learning and ability to solve tasks. Although the cerebellum is not the centre of cognitive function, its role in learning movement and coordination has an important impact on various aspects of activity and learning, where accuracy and coordination are important.

The spinal cord is one of the main parts of the CNS, which is located in the spinal column. It plays a critical role in transmitting information between the brain and other parts of the body and coordinates many different functions. It is important to understand its structure, functions, and role in regulating the body. The spinal cord is protected by the spine, which gives it physical protection from damage. It is responsible for transmitting information from other parts of the body to the brain and back. Sensory signals that come from the senses and motor commands are transmitted through the spinal cord. The spinal cord regulates reflexes—rapid, automatic movements or responses to stimuli that do not require conscious brain involvement. The spinal cord is responsible for performing movements, coordinating the activity of various muscles, and executing commands coming from the brain. It contains switching points where sensory information can be processed and forwarded to other parts of the brain. The spinal cord is an important component of the CNS and interacts with the brain to ensure normal body function. It performs many automatic functions that allow the body to adapt to changes in the environment and perform various actions, such as walking, responding to stimuli, etc.

The connections between the brain and the nervous system are important for ensuring normal body function. The brain is the central part of the CNS and interacts with other components of the nervous system at various levels. The brain receives information from the nervous system through the spinal cord and cranial nerves. This information covers signals from sensory organs that transmit external stimuli (sounds, odours, touch, etc.) and internal stimuli that indicate the state of internal organs. The brain processes incoming information by analysing it and responding to it. It regulates functions such as movement, comprehension, speech, memory, emotions, coordination, and many other aspects of physiology and psychology. The brain interacts with the peripheral nervous system via the spinal cord and nerves. This interaction allows controlling movements, perception of sensory information, and regulation of internal organ functions. The brain is responsible for regulating the body's internal processes, such as body temperature, respiration, heartbeat, metabolism, hunger, thirst, etc. The brain coordinates the body's actions and responds to external stimuli and internal changes [62–64]. For example, it allows for responding to danger, performing movements to maintain balance, and performing many other functions. The brain interacts with the endocrine system, regulating the production of hormones that affect various organs and systems in the body. The brain needs a good blood supply for its normal functioning. The vascular system delivers blood and oxygen to the brain, which is crucial for functioning. The connections between the brain and the nervous system regulate a large number of physiological and psychological processes in the body, and understanding them helps to consider the importance of proper brain function for human health and well-being.

The brain affects the endocrine system by regulating the release and function of hormones in various situations and needs of the body. The hypothalamus is a part of the brain that affects the endocrine system through the production of various hormones and the regulation of the pituitary gland, which is very close. The hypothalamus produces releasing hormones that affect the production of hormones by the adenohypophysis, such as thyrotropin, ACTH, and a number of others that regulate the thyroid gland, adrenal glands, and other structures of the endocrine system. The pituitary gland, located at the base of the brain, produces and removes hormones that control the function of other endocrine glands. The hypothalamus affects the pituitary gland through release factors, and this allows the brain to regulate the release of hormones such as thyrotropin, ACTH, luteinising hormone, and follicle-stimulating hormone, among others.

The brain, and in particular, the hypothalamus, activates the body's stress response by releasing stress hormones such as epinephrine and norepinephrine. These hormones affect the functions of the adrenal glands that regulate the response to stressful situations [65]. The brain affects the regulation of sleep and the body's activity rhythms by controlling melatonin, a hormone that regulates the sleep and wake cycle. This hormone is secreted by the pineal gland, which is located in the brain. The hypothalamus also plays an important role in regulating hunger and thirst. It controls appetite and affects when and how much we eat, producing hormones that regulate feelings of satiety and hunger. The brain affects the endocrine system by regulating the release of hormones such as cortisol according to emotions and responses to stressful situations. In general, the brain plays an important role in regulating the functions of the endocrine system through nerve signals that affect the production and release of hormones in response to needs and changes in the body.

Brain development is a complex process that occurs throughout a person's life. The first stages of brain development begin in the womb. The processes of neurogenesis and migration of neurons occur during the fetal period. The brain undergoes intense growth and formation. After birth, brain development actively continues. New connections between neurons appear, and the brain develops through environmental research [66]. An important period for learning a language and learning is considered to be from three to seven years. During this period, the frontal lobes of the brain, which are responsible for planning, making decisions, and controlling behaviour, are actively developing. The learning process and the formation of social skills are key aspects of development. The brain continues to develop into adulthood (Fig. 7). Especially important is the impact of learning, new experiences, and maintaining mental activity. With age, there are natural changes in the brain, such as a decrease in brain mass and associated changes in brain function. However, the brain remains plastic and can adapt to new conditions even in old age. Brain development depends on genetic factors and heredity but is also strongly influenced by the environment, learning, emotions, and experience. Maintaining a healthy lifestyle, proper nutrition, physical activity, regular training, and psychological state can contribute to optimal brain development throughout life.

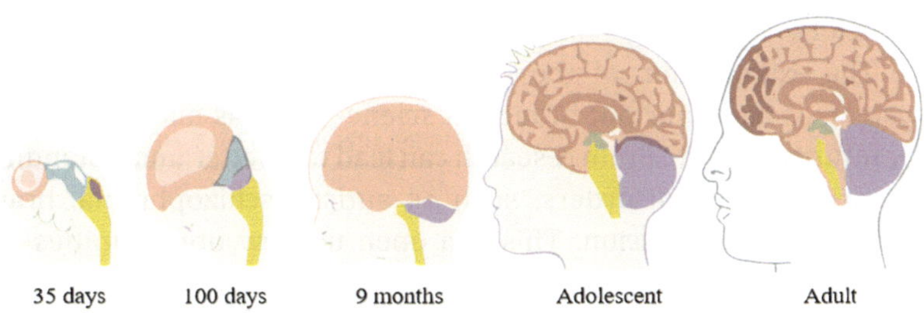

Source: University of Queensland [328].

Figure 7 Brain development in ontogenesis: Stages of brain development from the embryonic period to adulthood.

Brain development in early childhood is extremely important because it is the period when the basic structures and connections of the brain are formed. After birth, up to 6 months, the baby continues to develop its brain, especially those parts of it that are responsible for basic functions such as breathing, sucking, and motor skills. Under the influence of external stimuli, such as sounds, light, and touch, the brain forms neural connections [67]. From 6 months to 2 years, speech and language skills are developed. Speech promotes the development of the frontal lobes of the brain, which are responsible for speech and cognitive function. Children learn to communicate and express their thoughts clearly. Between the ages of 2 and 6, emotional and social skills are developed, and the brain develops systems that are responsible for empathy, understanding emotions, and perceiving social situations. Children learn to interact with other children and adults. In primary school, until the age of 12, cognitive skills such as logical thinking, math skills, and reading are developed. The brain is actively working on information processing and learning. During adolescence, between the ages of 12 and 18, children experience intense growth and development in the frontal areas of the brain that are responsible for decision-making, planning, and impulse control. This period can be important for the formation of identity and abilities. From the age of 18, brain development continues into adulthood, and this is an important period for choosing a profession, studying, and starting a family. The brain can adapt to new challenges and learn new information [68]. These periods of brain development in early childhood determine the foundations of a child's cognitive, emotional, and social skills. This development is important for further learning and adaptation in society.

The examination of the brain and its interaction with other systems is an important branch of science and has many prospects for future research. Brain research can lead to a better understanding of various brain disorders, such as autism, schizophrenia, brain injuries, and depression. This can open up new opportunities for treating and diagnosing such conditions. Brain research helps improve treatments for various brain diseases. For example, the development of new drugs and neurosurgery techniques may improve the treatment of neurodegenerative diseases such as Parkinson's disease or Alzheimer's disease. The development of brain–computer interface technologies opens up opportunities for communication and control of equipment using thoughts. This can be especially useful for people with physical disabilities.

Understanding how the brain works can help develop better teaching and learning methods. Innovative learning methods that take into account brain processes can improve student learning outcomes [69]. One of the key standpoints is to understand the nature of consciousness and intelligence. Brain research can lead to the discovery and understanding of the mechanisms underlying these phenomena. Brain research also includes analysis of the effects of the environment, stress, nutrition, and other factors on brain development and function. This is important for maintaining brain health and preventing negative effects on the brain. Examining the brain provides information that can be used in the development of artificial intelligence. Insights into how the brain works can help create smarter and more advanced artificial intelligence systems. These are just a few areas where brain research can have a substantial impact. Understanding how the brain works is key to many aspects of our lives, and scientists continue to investigate this complex system to make discoveries and advances.

Analysing brain anatomy is of great importance and has an impact on many aspects of our lives and science. Knowledge of brain anatomy helps doctors diagnose and treat various brain disorders, including neurodegenerative diseases, injuries, and psychiatric disorders [70]. Understanding how the brain processes information and controls our behaviour is essential for psychology and psychiatry. The examination of the brain leads to the development of new technologies in the field of neuroscience, such as brain–computer interfaces and methods for treating neurological disorders. Knowledge of brain anatomy helps improve learning methods and develop more effective approaches to education. The examination of the brain reveals and understands the nature of consciousness and intelligence, which is one of the most important mysteries of science [71]. Understanding the impact of factors such as nutrition, stress, and sleep on the brain helps people maintain brain health and prevent disorders. Brain research provides insights that can be used to develop new technologies, including artificial intelligence. Overall, the knowledge of brain anatomy is an important part of scientific research and has a profound impact on various aspects of our lives, from medicine to technology and psychology. Future areas of research in the field of brain anatomy include various aspects that are aimed at expanding our understanding of this complex organ [72–74]. Continuation of projects that aim to create detailed brain maps, including connections between different regions. This will help understand how the brain works in general. Exploring the nature of perception, consciousness, and intelligence, including exploring different levels of consciousness and their impact on our perception of the world.

1.3 Neuroplasticity and Neurogenesis

Neuroplasticity and neurogenesis are two important concepts in neurobiology that indicate the brain's ability to adapt and change over the course of life. Neuroplasticity occurs in different parts of the brain. Neuroplasticity is widespread in the cerebral cortex, where many higher cognitive functions such as perception, speech, memory, and decision-making occur. The cerebral cortex is made up of different regions that can change their structure and function according to learning and experience. The hippocampus is responsible for forming new memories and spatial navigation. Hippocampal neuroplasticity allows the creation of new connections between neurons and information storage [75]. The apical cortex is a key area for processing sensory information. Neuroplasticity in this part of the brain allows for adaptation to new stimuli and learning new skills. The basal ganglia are responsible for coordinating movements and performing habitual actions. Neuroplasticity in these structures may be important for restoring motor function after injury or illness. The limbic system includes structures that regulate emotions and motivation. Neuroplasticity in the limbic system may be key to regulating emotions and learning new strategies to deal with emotions. In general, neuroplasticity can occur in different parts of the brain and plays an important role in learning, adapting, and restoring function after an injury or illness.

The Phineas Gage case is a noteworthy example of neuroplasticity in action [76]. A railway foreman named Phineas Gage suffered a catastrophic brain injury in 1848 after an iron rod pierced his head, seriously harming his frontal lobes (Fig. 8). Phineas Gage showed significant personality changes even though he survived the trauma and maintained his fundamental cognitive and motor skills. His once responsible and industrious manner gave way to impatience and unpredictable conduct. This example offers preliminary proof of the important role the frontal lobes play in social behaviour and personality regulation. The brain's ability to heal itself after damage shows how flexible neuronal circuits are and how neuroplasticity makes it possible for compensatory processes to happen. This is how Phineas Gage was able to keep doing important things even though he was disabled. Further research on comparable situations has demonstrated how distinct brain regions can replace lost functions, especially with focused rehabilitation and cognitive training.

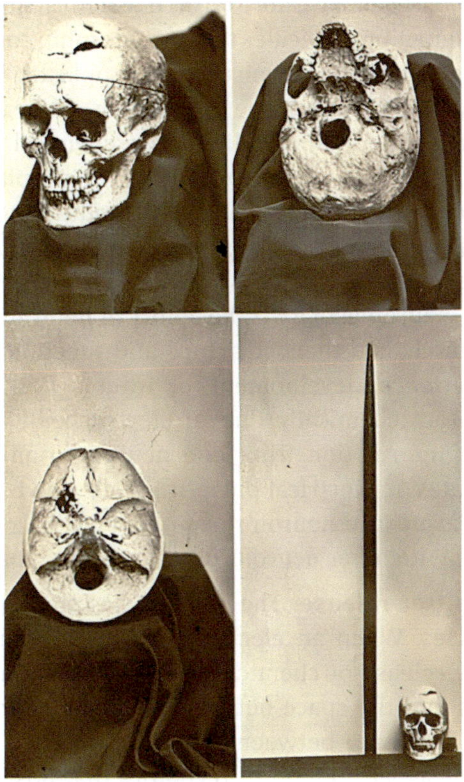

Source: R. V. Teles [76].

Figure 8 Skull of Phineas Gage with damage to the frontal lobes of the brain.

The chemistry of neuroplasticity involves a number of chemical processes and molecular mechanisms that underlie the ability of neurons and synapses to change their structure and function during learning, memory, recovery, and adaptation. Synapses are the main points of contact between neurons [77]. Changes in synapse strength and efficiency can be achieved through changes in the number and function of receptors, synaptic proteins, transmitters (chemical transmitters), and other molecules that affect the synapse. Various molecular signalling pathways and proteins, such as protein kinases and protein phosphatases, play a key role in signalling that regulates neuroplasticity. Neurotrophic factors, such as brain-derived neurotrophic factor (BDNF), affect the survival and development of neurons and the formation and strengthening of synapses. Synapses can exhibit long-term or short-term plasticity. This plasticity is regulated by signalling molecules that include inhibitors or activators of synaptic processes. The chemistry of neuroplasticity also involves genetic changes that affect the expression of genes associated with neuroplasticity. Understanding these chemical processes and mechanisms allows scientists to consider neuroplasticity at the molecular level and develop approaches to modulate these processes in various clinical contexts, such as neurorehabilitation or treatment of neurological diseases.

In general, neuroplasticity is the ability of neurons and synapses in the brain to change their structure and function as a result of learning, experience, development, or trauma [78]. Neurons and synapses interact chemically due to electrochemical signals that help transmit information from one neuron to another. When a neuron generates an electrical pulse, it travels down a neuron fibre known as an axon. The neuron's axon ends at the points where it makes contact with other neurons or cells, called synapses.

Neurotransmitter release: The synapse has small sacs known as synaptic vesicles. When an electrical impulse reaches a synapse, it triggers the release of chemicals called neurotransmitters from these vesicles into the space between neurons. Neurotransmitters pass through the space between synapses, known as the synaptic gap, and are designed to bind to receptors that host neurons. This causes a change in the electrical potential that the neurons receive. After neurotransmitters are attached to the receptors of the receiving neuron, its electrical potential changes, and this can affect the decision of the receiving neuron to transmit the signal further. After interacting with neurotransmitters, excess material from the synapse must be removed to prepare the synapse for the next signal. This process involves the return of neurotransmitters to synaptic vesicles and their recycling. This process of electrochemical signal transmission helps neurons communicate with each other and regulate brain and nervous system functions.

Various neurotransmitters are involved in the transmission of nerve impulses, which are chemicals that help transmit signals from one neuron to another [79]. Acetylcholine is the main neurotransmitter used in the nerves that control muscle contraction. It plays a role in controlling muscle movement and function.

Dopamine is responsible for regulating mood, reward, movement, and coordination. Dopamine deficiency is associated with various disorders, such as Parkinsonism and schizophrenia. Serotonin affects mood, sleep, and appetite and regulates many other body functions. It is a key neurotransmitter associated with depression and other disorders. GABA is the main inhibitory neurotransmitter in the CNS. It reduces the activity of neurons and affects the stress response. Glutamate is the main excitatory neurotransmitter in the brain and is responsible for learning, memory, sensory systems, and other functions. It plays a role in the stress response and sleep. These are just a few examples of neurotransmitters, and there are others that perform different functions in the brain and nervous system. The interaction between these neurotransmitters allows neurons to transmit information and regulate brain function.

Neurotransmitters are produced in neurons, which are the main cells of the nervous system. The process of neurotransmitter synthesis occurs in several stages. Neurons are able to capture the necessary chemical compounds, which are the starting materials for the synthesis of neurotransmitters. These chemical compounds are called precursors [80]. In the next step, the precursors are converted to neurotransmitters themselves in the process of chemical reactions that occur in a cellular structure called the endoplasmic reticulum. Synthesised neurotransmitters are stored inside the neuron in special structures called synaptic vesicles. When a neuron is ready to transmit a signal to another neuron, synaptic vesicles fuse with the cell membrane at the synapse site, and neurotransmitters are released into the synapse. The released neurotransmitters interact with special receptors on the membrane of this neuron, causing a response in this neuron. Once released, neurotransmitters must be reabsorbed (reloaded) again into vesicles or decomposed by enzymes so that the signal in the synapse does not remain too long.

Each type of neurotransmitter has its own precursor and unique biochemical synthesis pathway. This process is important for the normal function of the nervous system and regulates the transmission of nerve signals between neurons. Neurotransmitters in the nervous system propagate and transmit signals between neurons through a process that involves several key steps. When a neuron is ready to transmit a signal, it releases neurotransmitters from synaptic vesicles in the synapse. This happens through a process that involves the fusion of vesicles with the membrane of the neuron (there can be several large steps in this process). Once released, neurotransmitters diffuse into the synaptic cleft, which separates the releasing neuron from the receiving neuron. The released neurotransmitters interact with receptors on the membrane of the receiving neuron. Each type of neurotransmitter has its own receptors with which it interacts. This causes changes in the membrane potential of the receiving neuron [81]. The interaction between neurotransmitters and receptors leads to the generation of a nerve impulse or subsequent transmission of a signal through the receiving neuron. Once the signal is transmitted, neurotransmitters must be redistributed. This can occur due to reabsorption (the return of neurotransmitters to the secreting neuron) or decomposition by enzymes. This process allows transmitting signals between neurons in the nervous system and regulating the function of neural circuits. Notably, the exact dynamics of these processes may vary for different types of neurotransmitters and synapses.

This means that the brain can reprogram itself to adapt to new conditions. Neuroplasticity plays an important role in learning, memory, and restoring brain function after injuries or illnesses.

It allows for adapting to new situations and improving skills. There are several types of neuroplasticity, including synaptic neuroplasticity (changes in synapses between neurons), structural neuroplasticity (changes in the structure of neurons), and functional neuroplasticity (changes in brain function). The term "neuroplasticity" was first coined in the 1890s by Norwegian neurophysiologist Thorleif Sollesnes. Sollesnes is a Norwegian neurophysiologist, researcher, and physician known for his research in neurobiology and brain function. His research is aimed at understanding various aspects of the work of neurons and the brain in general. He developed and improved various methods of examining the brain. In particular, he applied interference microscopy and holography techniques to investigate the structure and activity of neurons. The author investigated the phenomenon of neuroplasticity, particularly how the brain can change and adapt as a result of learning, experience, and rehabilitation after head injuries. He also conducted research aimed at understanding the basic principles of brain function, including synaptic connections between neurons and information processing [82]. Overall, Thorald Sollesnes made important contributions to understanding neurophysiology and neurobiology, helping to expand knowledge about brain and neuron function.

However, the term gained popularity and wide distribution due to the research of the scientist and plastic surgeon P. Bahurist in the 1960s and 1970s. Bahurist investigated the use of vibrational skin stimulation to restore lost functions in people with visual impairments and brain damage. His work has given rise to further research on neuroplasticity and the development of methods for restoring brain function after injuries and neurological disorders. Neuroplasticity, or the brain's ability to change and adapt, is an important area of research in neurobiology. Many scientists made substantial contributions to the examination of this process. M. Merzenich, a well-known neuroscientist, worked on the analysis of neuroplasticity in the context of the auditory and sensory cortex of the brain. His research has shown that learning and experience can affect brain structure and function. The recognised American neurobiologist and neurophysiologist has made important contributions to the research of neuroplasticity and the development of neurorehabilitation methods. M. Merzenich has conducted extensive research on neuroplasticity, which means the brain's ability to change its structure and function as a result of experience, training, and rehabilitation. The scientist established that even in older adults, brain plasticity is possible, which has become the basis for developing new treatments and rehabilitation for neurological disorders. Merzenich was focused on developing and implementing innovative neurorehabilitation techniques for patients with various neurological disorders, such as stroke, speech, and hearing disorders [83]. The researcher examined how different parts of the brain are responsible for different functions and developed detailed brain maps. These studies have contributed to a better understanding of the functions of various brain regions. He investigated how neuroplasticity can be used to restore lost functions, especially in the field of speech and hearing; collaborated with patients who had problems with these functions; and developed rehabilitation methods based on the investigation of neuroplasticity. M. Merzenich examined how intensive rehabilitation can affect the recovery of functions after brain injuries, strokes, and other neurological diseases. He emphasised the importance of regular and targeted rehabilitation to promote neuroplasticity. M. Merzenich collaborated with a team of researchers to create computer programmes and technological tools for rehabilitation based on the scientific foundations of neuroplasticity. These drugs helped patients restore brain function. Neuroplasticity was also examined by the scientist in the context of child development. He emphasised the importance of early intervention for children with developmental disabilities and provided advice to parents and educators on how to maintain neuroplasticity in children. He has made a great contribution to understanding neuroplasticity and its use to improve the rehabilitation of patients with various neurological diseases and speech and hearing problems. His research and practical contributions to medicine and rehabilitation continue to influence modern neurorehabilitation practices. Merzenich actively introduced new technologies to examine the brain and develop rehabilitation and treatment methods using advanced neurotechnology. For his contributions to neuroplasticity and neurobiology, M. Merzenich has received numerous awards and recognition in the field of neuroscience.

E. Tolman, a psychologist and neuroscientist, analysed neuroplasticity in the context of learning and memory in animals [84]. He believed that the brain interacts with the environment and stores maps of the world that can change. E. Tolman was a well-known American psychologist and neurobiologist who conducted substantial research in the field of learning and memory and made important contributions to understanding neuroplasticity. Tolman first introduced the concept of a "cognitive map", which indicates that animals and humans create mental models of their environment. These maps are important for understanding learning and navigation. Tolman developed the concept of "cognitive map", which is important for understanding how the brain displays information about the environment. He examined how the brain encodes and stores information about the spatial landscape. Tolman investigated how changes in neural networks contribute to learning and memory. His work focused on trying to understand how neuroplasticity affects how the brain responds to learning and knowledge acquisition. Overall, E. Tolman has made important contributions to understanding learning, memory, and neuroplasticity through his research in psychology and neurobiology.

The well-known "Hebb theory", formulated recently, predicted that neuroplasticity occurs based on changes in synaptic connections between neurons. It states that synaptic connections change as a result of activity and learning. Hebb's theory, also known as plasticity theory, is a fundamental theory in the field of neurobiology that emerged from the contributions of renowned psychologist and neurobiologist D. O. Hebb in the mid-20th century. This theory defined the basic principles of studying how neurons are organised and function and became the basis for understanding neuroplasticity [85]. The founder of the theory, D. O. Hebb, formulated its main provisions in 1949. The basic idea of Hebb's theory is that neurons form new connections (synapses) based on experience and learning, and this leads to changes in the structure and function of the brain. The main provisions of Hebb's theory are as follows:

1. The principle of "joint activation–joint reinforcement" (Hebbian learning). This principle states that if two neurons are activated simultaneously and connected by a synapse, the connection between them is enhanced. In other words, if the activity of one neuron is often accompanied by the activity of another neuron, then the synaptic connection between them becomes stronger.
2. Synapse plasticity. Hebb considers the ability of synapses between neurons to change. Namely, they can become stronger or weaker depending on activity and learning. This allows the brain to adapt to new information needs.
3. Formation of neural circles. Hebb's theory indicates that neural circuits or pathways can form specific connections by strengthening synapses between activated neurons.
4. The role of theory in learning and memory. Hebb's theory is used to explain learning and memory processes in the brain. It argues that neurons form new connections through learning, and such connections can store information. Hebb's theory was an important step in understanding neuroplasticity and analysing how the brain adapts to its environment and learning. Its principles are relevant and applied in modern research in neurobiology and neurorehabilitation.

Scientists such as E. Kandel, a Nobel Prize–winning scientist, have also made substantial contributions to neuroplasticity in the context of learning and memory. His research points to the role of synaptic changes in the formation of long-term memory. N. Doidge, a Canadian psychiatrist and author of The Brain That Changes Itself, highlighted the importance of neuroplasticity and highlighted cases of people who changed their brain functions through exercise and rehabilitation. M. Diamond is a scientist who has conducted research on the effects of learning and stimulation on brain structure. Her work has shown that neuroplasticity can affect brain development, even in adults. Vilayanur Ramachandran, also known as V. S. Ramachandran, made major contributions to the investigation of neuroplasticity through his research in neuroscience and the visual cortex.

A. Pascual-Leone is an outstanding researcher in the field of transcranial magnetic stimulation (TMS) and its application to the examination of neuroplasticity. These researchers and many others contribute to understanding the nature of neuroplasticity and its importance for human learning, rehabilitation, and overall development. Neuroplasticity helps us understand that the brain can change as a result of learning and experience. This encourages the desire to learn and develop throughout life.

Neuroplasticity helps patients who have suffered a brain injury or have a neurological condition restore brain function through rehabilitation and training. Knowing about neuroplasticity can help people manage stress, improve their mental health, and develop positive habits. Understanding how the brain can change affects teaching and learning methods [86]. This helps to develop more effective pedagogical approaches. Some treatments for mental illnesses are based on stimulating neuroplasticity to correct brain function disorders. Knowing about brain neuroplasticity helps parents and teachers understand how children learn and develop and how to support their success. Understanding that neuroplasticity can continue even in old age helps maintain a sharp mind and a healthy brain.

Therefore, knowledge of neuroplasticity is essential for overall development, learning, health, mental health, progress of scientific treatments, and rehabilitation. One of the most well-known examples of neuroplasticity is the recovery of brain function after an injury or illness. For example, patients who have suffered traumatic brain injuries may regain some lost functions due to neuroplasticity.

Another example is learning and developing skills. When a person learns new things or develops a skill, it is accompanied by changes in the brain, in particular, the creation of new connections between neurons and the strengthening of existing ones. Neuroplasticity helps the brain adapt to new challenges and store information. Even in everyday life, neuroplasticity can be noticed. For example, when a person learns a new language, music, or instrument, the brain adapts to this process by creating new neural connections.

Modern techniques such as fMRI and EEG allow scientists to observe brain activity in real time and examine neuroplasticity in living organisms. Neuroplasticity research is important for understanding how the brain adapts to new learning, recovers from injuries, and what factors influence the development of neurological diseases. It has become a key concept in modern neuroscience and medicine and has opened up new opportunities for the development of rehabilitation and treatment methods for patients with various neurological conditions.

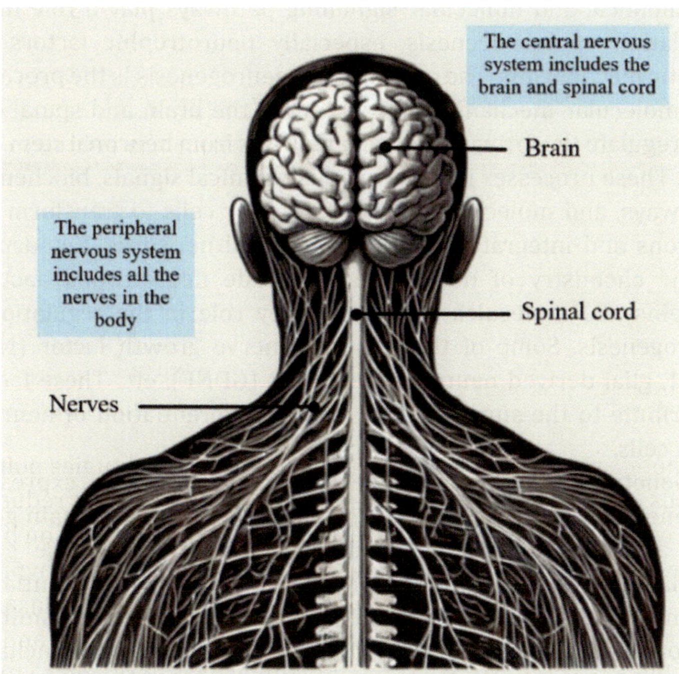

Figure 9 Spinal cord and its connections with the brain.

Neurogenesis is the process of forming new neurons from neuronal stem cells. This process occurs in certain areas of the brain and spinal cord throughout life, especially in early childhood and in some parts of adulthood (Fig. 9). The hippocampus is a structure located in the brain's notch known for its role in learning and memory. Active neurogenesis occurs in the hippocampus throughout life. Neuronal stem cells located in certain areas of the hippocampus can create new neurons. The subventricular zone is located along the wall of the lateral ventricles of the brain. It also contains neuronal stem cells that can be secreted into new neurons.

Some studies point to the possibility of neurogenesis involving glial cells. Glia are cells that are not neurons but can interact with neurons and participate in neurogenesis processes [87]. The process of neurogenesis involves the proliferation of neuronal stem cells, the differentiation of these cells into neurons, and their integration into an existing network of neurons. This process is important for maintaining and restoring brain function and adapting to new environments and learning.

Chemical and molecular signalling pathways play a role in the regulation of neurogenesis, especially neurotrophic factors and genetic mechanisms. The chemistry of neurogenesis is the processes and molecular mechanisms occurring in the brain and spinal cord that regulate the formation of new neurons from neuronal stem cells [88]. These processes involve various chemical signals, biochemical pathways, and molecular events that allow cells to transform into neurons and integrate into existing neural networks. Key aspects of the chemistry of neurogenesis include neurotrophic factors–proteins and molecules that play a key role in the regulation of neurogenesis. Some of them include nerve growth factor (NGF), BDNF, glial-derived neurotrophic factor (GDNF), etc. These factors contribute to the survival, growth, and differentiation of neuronal stem cells.

Genetic mechanisms include the regulation of the expression of genes that control neurogenesis processes [89]. The main genes that affect neurogenesis include Notch, Wnt, etc. These genes regulate various aspects of cellular development and differentiation. Chemical signals such as molecules, hormones, neurotransmitters, and other molecules affect various stages of neurogenesis, including migration, differentiation, and integration of new neurons into neural networks. Intercellular interactions play an important role in the regulation of neurogenesis [88]. This includes interactions between neuronal stem cells, adjacent cells, and matrix cells that provide a microenvironment for the growth and development of new neurons. Neurogenesis chemistry is a complex interdisciplinary science that helps understand the processes of creating and functioning neurons in the brain. Research in this area is important for understanding brain development and developing possible treatments for neurodegenerative diseases and brain injuries.

Thus, neurogenesis is the process of creating new neurons or generating neurons that occurs in certain areas of the brain, even during adulthood [90]. Major areas of neurogenesis include the hippocampus and lateral spur, although some evidence points to limited neurogenesis in other parts of the brain. Neurogenesis is important for maintaining cognitive function, learning, memory, and adaptation to new situations. This process can repair lost neurons and maintain brain flexibility.

Many scientists made a substantial contribution to the development of the theory of neurogenesis. J. Altman was one of the first to actively study neurogenesis in the adult brain in the 1960s and 1970s [91]. His work on the number of new neurons in the brain was an important contribution to this field. J. Altman was an important figure in the examination of neurogenesis—the process of forming new neurons in the brain after birth. He made substantial contributions to understanding this phenomenon, and his work helped substantially rethink old concepts of neurogenesis and neuroplasticity. Initially, Altman refuted existing ideas that neurogenesis is completed at an early age and showed that new neurons can form in the adult brain. This discovery was substantial because it opened up new possibilities for investigating the role of neurogenesis in brain health and disease. Further in his studies, Altman showed that neurogenesis can occur in different parts of the brain, including the hippocampus, an area important for memory and learning. He also investigated the effects of environmental factors on neurogenesis and established that physical activity and learning can promote the formation of new neurons. His work has made major contributions to modern neurobiology and neuroplasticity and continues to influence the understanding of processes occurring in the brain. Altman was instrumental in opening up new horizons for research on brain structure and function in the context of neurogenesis.

A. Brega, an Italian anatomist and neuroscientist, made important contributions to understanding neurogenesis and the anatomy of the nervous system [92]. He was one of the first scientists to offer ideas about the possibility of neurogenesis, that is, the formation of new neurons after birth and throughout life. In the mid-19th century, A. Brega argued that the nervous system can develop and change throughout life, even in adulthood. He conducted research on pigeons, where he examined the structure of their brains and also noted the presence of new neurons in certain areas of the brain. Although many of his ideas were highly speculative and not always supported by scientific evidence at the time, A. Brega opened up new horizons for research on neurogenesis and neuroplasticity. His views became the basis for further scientific research and made an important contribution to the development of modern neurobiology.

Interest in these concepts arose because of their importance to science and medicine. Understanding neuroplasticity and neurogenesis opens up opportunities for the development of new treatments for various neurological and mental illnesses and contributes to improving learning methods and maintaining cognitive functions throughout life [93]. The question of neurogenesis, the process of creating new neurons, is being examined by many scientists and researchers in the field of neurobiology. Elizabeth Gould was an American neurobiologist who made important contributions to the examination of neurogenesis, the process of forming new neurons in the brain. Gould investigated neurogenesis in adult monkeys. She conducted a series of studies in which she examined the hippocampus of the monkey brain and found evidence that neurogenesis occurs in this area of adult monkeys. Her work was the basis for accepting the idea that neurogenesis can also occur in the human brain, which substantially rethought traditional ideas about this process. Gould also investigated the effect of environmental factors on neurogenesis [94]. The researchers showed that physical activity, learning, and social interaction can contribute to the formation of new neurons in the hippocampus. These findings are important for understanding how neurogenesis can be maintained and improved in adults. In general, Elizabeth Gould's research made an important contribution to understanding the processes of neurogenesis in the brain and its regulation.

A well-known American neurobiologist who also devoted his research to neurogenesis and the creation of new neurons in the adult brain and showed the importance of this process for memory and learning, Fred G. Gage made a great contribution to the examination of neurogenesis, namely, to the investigation of the process of formation of new neurons in the adult brain. One of his outstanding contributions to this area is to examine neurogenesis in the hippocampus, a brain structure that plays an important role in memory and learning. Gage and his colleagues determined that new neurons continue to form in the hippocampus throughout a person's life, and this can affect memory and cognitive function. Fred Gage's studies also covered the effects of physical activity, social environment, and stress on neurogenesis. He established that various factors can stimulate or inhibit the formation of new neurons. Fred Gage continues to explore neurogenesis and its effects on brain function, and his work is important for our understanding of the brain and its ability to repair and plasticity. The author was also one of the pioneers of using stem cells to repair neurons in the brain, which has great potential for treating various neurological diseases.

A Mexican scientist who examined neurogenesis in the brain and the importance of this process for brain development and regeneration made substantial contributions to understanding the structure and function of brain neurogenesis. A. Alvarez-Buylla is a recognised scientist in the field of neurobiology, specialising in the study of neurogenesis, that is, the formation of new neurons in the brain, in particular, along the neuroventricular system channel. The researcher and his colleagues have conducted numerous studies to understand how and when new neurons are formed in the adult brain and how this process can be regulated. His work is important for understanding the mechanisms of neurogenesis and its effect on brain plasticity. Alvarez-Buylla examines neurogenesis in various parts of the brain, such as the hippocampus, olfactory tract, etc. His research also focuses on possible applications of neurogenesis in the treatment of various neurological diseases and brain injuries. All of these studies by A. Alvarez-Buylla contribute to a better understanding of neurogenesis and its role in brain health and disease.

Researchers examine neurogenesis using a variety of techniques, including cell labelling, experiments with mice and other animals, molecular biology, and neuron imaging. They also investigate the effects of factors such as exercise, exercise, stress, illness, and treatment on neurogenesis. These studies are of great practical importance, as understanding neurogenesis can lead to the development of new methods to improve memory, treat neurological diseases, and regenerate brain tissue. The examination of neurogenesis is an important aspect of the study of the nervous system and brain, and it has deep roots in the history of science. Neurogenesis, the process of creating new neurons (neuroblasts), occurs at various stages of the body's development, including the early ones and after birth. In the Middle Ages, scientists first discovered that newborn animals have fewer neurons than adults and made assumptions about the process of creating new neurons in early development.

In the mid-20th century, scientists began to investigate the process of neurogenesis in more detail, using modern techniques, including cell labelling and the use of microscopy. In the 1960s and 1970s, studies were conducted on adult rodents, which showed that neurogenesis can occur even in adult brains and not just in early childhood. This brought important insights into the process. Modern molecular research revealed more processes that control neurogenesis, including the role of genes and signalling pathways. The examination of neurogenesis is of great practical importance for medicine. Understanding this process may lead to the development of new treatments for neurological diseases and brain tissue regeneration [95]. Today, researching neurogenesis is an active branch of neurobiology, and it continues to develop thanks to modern research, technologies, and methods that help to better understand this important process in the nervous system.

The industry is constantly evolving, and new research continues to identify and understand the processes of creating new neurons in the brain. An example of neurogenesis is the process of creating new neurons in the brain's hippocampus. The hippocampus is a structure associated with learning and memory. Over the course of life, under certain circumstances (e.g., when learning new information or recovering from certain injuries), new neurons can be created in the hippocampus. This process is called neurogenesis. Other examples include the creation of new neurons in the limbic system under stress or the creation of new neurons in the body after learning a new language or skill. Neurogenesis plays an important role in brain plasticity and adaptation to changes in the environment. The hippocampus is an area of the brain responsible for memory and learning. New neurons in the hippocampus can form throughout a person's lifetime. This process is called hippocampal neurogenesis and plays an important role in the formation of new memories and the restoration of functions after brain damage. The olfactory bulb is the part of the brain responsible for olfactory sensations. It has a high level of neurogenesis, and new neurons are formed to replace old ones. Neurogenesis can also occur in the context of diseases or brain injuries.

The body may try to repair damaged neurons by forming new ones. Physical activity is known to stimulate neurogenesis in some parts of the brain. For example, regular exercise can promote the formation of new neurons in the hippocampus. Some studies suggest that stimulating an environment rich in new experiences and challenges can boost neurogenesis. These examples suggest that neurogenesis is an important function of the brain and can occur in different areas and under different conditions.

In conclusion, it can be stated that the examination of the anatomy and physiology of the brain plays an important role in understanding the functions of this complex organ. The brain is responsible for a number of important functions, including perception, processing and storing information, controlling movements, regulating internal body functions, and many other aspects. Studying anatomy helps understand exactly how different parts of the brain interact to perform these functions. Neuroplasticity and neurogenesis show that the brain can change over the course of a person's life. Neuroplasticity allows neurons to reconfigure their connections, which is important for learning and adapting to changes in the environment. Neurogenesis, in turn, opens up the possibility of forming new neurons, which can be useful in restoring function after injuries or diseases. Knowledge of these processes is important for developing treatments and rehabilitation for various neurological diseases and injuries.

Understanding the anatomy and physiology of the brain and the processes of neuroplasticity and neurogenesis contributes to the further development of neuroscience and improves people's quality of life. Research on neuroplasticity processes that allow the brain to adapt to new conditions and restore lost functions. Development and introduction of new technologies and methods for examining the brain, such as functional magnetic resonance imaging and optical stimulation. Research on the mechanisms of development and treatment of various brain disorders, including neurodegenerative diseases, epilepsy, and psychiatric disorders. New methods for integrating the brain with computers and artificial intelligence technologies are being developed to develop more efficient interfaces. Development of individual-oriented medicine, considering the genetic and neurological characteristics of the patient for optimal treatment and prevention. Research on the impact of factors such as environmental pollution and stress on brain health and the development of disorders. Future research on brain anatomy promises to uncover even more of the mysteries of this neurobiological miracle and find applications for this knowledge in medicine, technology, and psychology.

Chapter 2

Hypnosis and Its Mechanisms

Hypnosis is an intriguing and enigmatic phenomenon that has captured the attention of scientists, physicians, and psychologists for many centuries. Hypnosis is a state or process during which an individual exhibits an elevated susceptibility to influence and suggestions. It can be utilised for various purposes, including psychotherapy, pain management, enhancing concentration, and even for entertainment and amusement shows. The history of hypnosis spans millennia of observations and the use of various techniques to achieve this state. Ancient Egyptians, Greeks, Indians, and other ancient civilisations employed similar methods of hypnosis for healing and spiritual practices. Over time, as science progressed, notable figures such as F. Mesmer, J.-M. Charcot, and I. Pavlov endeavoured to understand the nature of hypnosis and its potentialities. Modern psychology and medicine utilise hypnosis for treating various mental and physical ailments, and this method continues to be the subject of scientific inquiry. This chapter delves into the history of hypnosis, the various methods and approaches used in modern hypnosis, and the role of prominent figures such as Sigmund Freud and Milton Erickson in the development of this field.

Mechanisms of Brain Function and Adaptive Potential
Liana Spytska

ISBN 978-981-5129-91-5 (Hardcover), 978-1-003-71697-6 (eBook)
www.jennystanford.com

DOI: 10.1201/9781003716976-3

2.1 Review of the History and Psychology of Hypnosis

Hypnosis, as a phenomenon, has always evoked interest and fascination. Its origins trace back to ancient times, and the development and exploration in the field of hypnosis have come a long way, becoming an integral part of the history of psychology and medicine. This realm of knowledge not only unveils incredible facts and possibilities of the human mind but also serves as an important tool for addressing various psychological and physical issues. The roots of hypnosis extend from ancient civilisations such as the ancient Egyptians, Greeks, and Indians. Many great researchers made substantial contributions to the field of hypnosis. The history of hypnosis spans millennia of development and encompasses numerous stages and discoveries, illustrating the unique path of this phenomenon and revealing its wide range of applications across different cultures and epochs. From ancient rituals to modern medicine and psychotherapy, hypnosis has proven to be an effective tool for altering consciousness and treating various conditions. Since ancient times, people have shown interest in trance states and deep subconscious influence. Shamans, priests, and physicians employed various rituals and techniques to alter consciousness, which facilitated healing and understanding of mental states.

The earliest records indicating the use of hypnosis date back to ancient civilisations such as the Egyptians, Greeks, Indians, etc. They used various methods of hypnosis for healing and spiritual practices. For instance, ancient Egyptians believed in the existence of two souls: the "ka" (individual soul) and the "ba" (personal intellect or will). They employed diverse methods of hypnosis and trance for communicating with spirits, obtaining prophetic dreams, and healing. Hypnotic techniques were integral to Egyptian spiritual and healing practices, including the use of hypnotic states during treatment. Various methods of hypnosis and trance existed in ancient Egypt for communicating with spirits, receiving prophetic dreams, and healing. Music and ritualistic sounds played a substantial role in ancient Egyptian ceremonies and religious practices. During musical rituals and prayers, individuals entered trance states, making their consciousness more susceptible to interaction with spirits and visitors. Ancient Egyptians conducted religious rituals during which prayers, invocations, and rituals were performed. During such rituals, people experienced deep trance states, believing they could communicate with spirits and gods. Ancient Egyptians consumed special herbal narcotics or utilised various meditation and trance techniques before sleep to induce prophetic dreams. These dreams were considered prophetic and used to predict future events and solve problems. Ancient Egyptians also used hypnotic states to treat various illnesses. These methods possibly included the use of subconscious influences through words and other forms of psychotherapy. Ancient Egyptians believed in the power of hypnosis and trance for spiritual communication, healing, and future prediction. They developed their own methods and practices, which became part of their religious and magical legacy.

Ancient Greeks utilised terms such as "hypnos" and "somnambulism" to describe deep trance states. This state was known as "ὕπνος" (hypnos), meaning sleep, and was used for healing purposes and entertainment. Hence, the term "hypnosis" originates, which was later introduced into the medical lexicon by the French physician and physiologist J.-M. Charcot in 1842. He used this term to denote a state similar to sleep that occurs during special procedures, now known as hypnotherapy. Ancient Greeks used methods of hypnosis and trance for various purposes, including healing, religious rituals, and entertainment. In ancient Greece, methods and practices such as temple sleep (Hippocratic Sleep) were widespread. Ancient Greeks examined various treatment methods, including hypnosis techniques, to alleviate physical and mental suffering. The word "sleep" in this context did not necessarily mean physical sleep. It indicated a state of concentration and relaxation that could be induced to alleviate suffering or perform healing rituals. Substantial attention was given to treating the sick in temples and sacred places. The sick could be influenced by specialised prayers, rituals, and special techniques to improve their health condition. In religious rites and rituals, ancient Greeks used hypnotic states to communicate with gods and engage in spiritual practices. There is evidence of using meditative methods to achieve states of deep tranquillity and concentration. Ancient Greeks staged theatrical performances where actors and choirs used rhythmic singing, music, and special movements to create hypnotic effects on the audience. Although modern conceptions of hypnosis differ from how it was used by ancient Greeks, it can be seen that even then, there was a conscious attempt to induce states of consciousness that promoted the improvement of physical and mental well-being.

In the Indian spiritual legacy, meditation and trance techniques that allow individuals to achieve states of deep tranquillity and concentration exist. Yoga and transcendental meditation (TM) are examples of Indian practices that can induce trance states akin to hypnosis. Ancient Indians utilised various methods of hypnosis and trance within their spiritual and healing practices. For example, Dharana and Dhyana are terms denoting concentration and meditation in the Indian context. Dharana involves the process of focusing on a single object or thought, which can lead to a deep state of concentration. Dhyana is the subsequent stage where the meditator transitions into a state of heightened perception or trance. Pranayama is a practice of breath control used to achieve states of concentration and relaxation. Some forms of pranayama encompass diverse breathing techniques that can facilitate deep relaxation. Indians used mantras (specific sounds or words) for meditation and concentration. Repetition of a mantra can help calm the mind and transition into a trance state. The practice of yoga encompasses various asanas (poses) and breathing exercises that promote physical and psychological concentration. Some forms of yoga, such as Kundalini Yoga, aim to achieve higher states of consciousness. TM is a popular contemporary meditation practice based on ancient Indian traditions. During TM, the meditator uses a mantra to attain a state of trance and concentration. The state of Samadhi is a state of ecstasy or higher communion with the spiritual aspects of life. In Indian philosophy, it is considered the highest state of spiritual development. These methods and practices were integral parts of the Indian spiritual heritage and can lead to trance states where individuals can explore their inner worlds and attain spiritual insights.

The history of hypnosis in ancient China and among Shaolin monks represents an intriguing phase in the development of this phenomenon. Notably, this period lacks objective sources, and information often relies on legends. In Chinese history, there has been a tradition for thousands of years of applying various forms of meditation and concentration to achieve inner peace and develop various skills. In Shaolin monasteries, renowned for their martial arts, specific methods of meditation and focused attention were practised. One of the legends about hypnosis in China involves the use of hypnotic methods by tea masters. According to legend, they were capable of deeply concentrating their students' attention, transferring their consciousness to another state. This served to enhance their abilities and attain inner enlightenment. Regarding Shaolin monks, their meditation practice, which involved deep concentration and inward-focused attention, can be perceived as a form of hypnosis. These monks honed their physical and mental abilities and cultivated inner control. It is important to consider that the history of hypnosis in ancient China and among Shaolin monks remains a subject of research and interpretation, and many details may be lost over time due to cultural nuances.

The history of hypnosis has its roots in ancient cultures and civilisations, including ancient northern tribes, both in its modern understanding and in its various forms. However, examining the specific history of hypnosis in ancient northern tribes can prove challenging due to the limited historical records and sources. In ancient northern tribes, such as the Vikings and Germans, various traditions and rituals related to religious and magical concepts existed. Shamans and seers used various methods of altered states of consciousness for communication with spirits, obtaining answers to questions, or healing. One possible method associated with hypnotic elements was the trance ritual, involving music, dance, and mantras [96].

The history of hypnosis among seers, magicians, and sorcerers encompasses some interesting aspects that can be traced in various cultures and periods. Seers, often associated with shamans and magicians, used various methods and rituals to perform their magical or healing practices. Elements resembling hypnotic states can be identified in their rituals. One aspect of this field is the use of music, dance, movements, and mantras to induce altered states of consciousness. This may resemble hypnotic techniques aimed at altering consciousness and facilitating interaction with the spirit world or other realities. Additionally, seers may have employed psychological and physical methods to induce trance states. This included attentional focus, the use of hypnotic rituals and symbols, and the ability to control imagination and psychic processes. An important aspect of the history of hypnosis among seers is its religious context. In many cultures, seers performed religious and ritual functions, and their practices may have included hypnotic elements for communication with spirits or other realms.

The history of hypnosis among shamans is part of the broader context of religious and magical practices found in various cultures worldwide. Shamans, acting as intermediaries between the physical and spiritual worlds, often utilised altered states of consciousness for intuitive insights, healing, and communication with spirits. One of the methods shamans used to induce trance states was ritual. This could involve the use of ritualistic music, dance, special objects, or plants with potential altered properties. Shamans may have used mantras and specific movements to achieve a certain level of concentration and induce a trance state. In the trance state, shamans could experience interaction with other realities, communicate with spirits, journey to other worlds, or exhibit extrasensory abilities. In altered states of consciousness, shamans possibly conducted journeys into the spirit world or performed healing ceremonies. Additionally, there may have been techniques inducing hypnotic states within the context of various rituals or engaging in magical practices. However, without specific written records or archaeological evidence, it is challenging to determine the exact nature and prevalence of hypnotic practices among ancient northern tribes. In general, altered states of consciousness had their place in the religious and magical traditions of ancient northern tribes and may have had elements resembling hypnosis. These actions encompass elements similar to hypnotic phenomena, such as heightened concentration, increased perception, and alterations in consciousness. Documented cases in history suggest that shamans in various cultures possibly employed hypnosis-like methods for healing, prophecy, and conflict resolution. These practices demonstrate a means of accessing the reserves of the psyche and interacting with the subconscious in religious and magical contexts.

The history of hypnosis in the context of ancient Slavs is not extensively documented, making it difficult to provide specific details due to the limitations of historical records and the absence of precise testimonies. However, it can be speculated that some forms of altered consciousness and magical practices existed within their traditions. Ancient Slavs had their own religious and magical system, associated with beliefs in various deities, natural forces, and spirits. Shamans and priests, known as "volkhvs", employed various rituals and ceremonies to communicate with the invisible world, seek answers to questions, and heal. Certain techniques aimed at altering consciousness may have been part of these rituals. The use of rhythmic music, dance, specific movements, and mantras may be linked to attempts to induce altered states of consciousness. Specifically, some ethnographic descriptions suggest that Slavic volkhvs possibly used techniques resembling hypnosis or trance states to achieve an ecstatic perception of the spirit world and interact with it. However, to provide accurate data on the history of hypnosis among ancient Slavs, additional research and archaeological findings are necessary to confirm these assumptions.

Hypnosis in the context of ancient healers and sorcerers often involved rituals, ceremonies, and specific techniques aimed at altering consciousness and interacting with the spiritual realm. These practices were associated with healing power, knowledge transmission, or interaction with other realities. Healers and sorcerers used ritual dances accompanied by music to induce trance states. Movements, rhythms, and sounds aided concentration and facilitated alterations in consciousness. The use of plant toxins or consciousness-altering plants was also a widespread method. Some plants had the ability to induce altered states of consciousness. Sorcerers and healers possessed specific verbal formulas used to influence the patient's consciousness. These formulas may have induced specific states of consciousness. Various healing rituals could include hypnotic elements that contributed to the restoration of physical or mental health. The use of specific symbols and amulets may be associated with creating an atmosphere or altering consciousness through focus. Some healers possibly used special rituals to induce patients into a state of hypnosis or trance, where they became more susceptible to their influences. Notably, the rituals and practices of ancient healers and sorcerers were rooted in their beliefs in connection with nature, spirits, and other worlds. The hypnotic elements of these practices were used to achieve specific goals, such as healing, knowledge transmission, or regulating events in their understanding of the world.

Molfarism, as an interpretation of the capacity for hypnosis within the context of traditional Slavic and Ukrainian beliefs, incorporates rituals, verbal formulas, and techniques aimed at altering consciousness and interacting with the spiritual realm. Molfars possess unique vocal techniques used to create a hypnotic effect. Special mantras and formulas may induce a trance state or hypnosis in listeners. Molfar rituals often involve the use of symbols, amulets, and objects with symbolic significance that can influence the psyche. This can create an atmosphere conducive to the emergence of a hypnotic state. Dynamic rituals, such as dances, may be used to stimulate psychic sensitivity and facilitate entry into a trance state. The movements of Molfars may have a rhythmic nature that aids in concentration. Molfars may use techniques of attention focusing and psychic influence to induce trance states. This may include deep breathing and verbal commands aimed at relaxation. Sacred places, natural objects, and other sacred elements may be used to enhance hypnotic effects. Molfars may embark on ritual journeys to alter states of consciousness. The use of symbols, allegories, and specific language can play an important role in hypnotic practices. Molfars may influence consciousness through carefully chosen words and imagery. Interaction with the spiritual realm and the ability to induce hypnotic states in the context of Molfarism were often associated with fulfilling the roles of healers, guides, or spiritual leaders within the community.

All these aspects paint a colourful mosaic image of the history of hypnosis, characterised by a variety of approaches and rituals and a specific context of religious and magical beliefs. Ancient practices of different civilisations share common elements such as a focus on spirituality, meditation, the use of symbols, rituals, and ceremonies. However, they may substantially differ in their specific methods, doctrines, philosophies, and culturally specific features. Different ancient civilisations had different belief systems and gods, influencing their spiritual practices. For example, ancient Egyptians believed in multiple gods, and their rituals and ceremonies were aimed at appeasing divine forces. Meanwhile, ancient Indians had their own belief system associated with Hinduism, which greatly influenced their practices. Various cultures had their own meditation and concentration techniques. Rituals and ceremonies varied depending on the culture. Rituals and ceremonies varied depending on the culture. For instance, ancient Egyptians had rituals associated with worshipping gods and the dead, while some other cultures developed rituals related to nature or sun worship. Different cultures used their own symbols and signs. For instance, Egyptians used hieroglyphs with specific meanings, while other cultures had their own symbols, such as the swastika or Om. Ancient practices also relied on different philosophical approaches. For example, the philosophies of Buddhism, Indian yoga, and Confucianism substantially differed from the philosophy of ancient Egypt or Greece. Although there are common features in ancient practices, each culture developed its own unique methods and spirituality, reflecting its worldview and history. All these ancient cultures used hypnotic states for various purposes, including healing, spirituality, and entertainment. Techniques and methods varied depending on the culture, but the common goal was an attempt to influence consciousness and the subconscious mind to achieve certain results.

The history of hypnosis in the Middle Ages has its roots in various cultures and practices, including mystical and religious beliefs that influenced the perceptions of consciousness at the time. A variety of factors determined the views and practices that contributed to the development of concepts of hypnosis in the Middle Ages. In various religious traditions of the Middle Ages, rituals containing mass ecstasy or trance states were observed. Religious leaders often used these states to establish contact with celestial forces. Practices of magic and sorcery, which encompassed the use of rituals, incantations, and magical objects, also contributed to the formation of beliefs about psychic states. Magicians often used hypnotic effects to impress their audience. Medical practices of the time often involved the use of various religious or magical elements. Physicians may have used hypnotic methods in medical procedures and healing rituals. In the Middle Ages, beliefs in the power of words and gestures already existed. Magical formulas and rituals gave words and gestures a special power that could influence consciousness and the mental state of a person. Occult and theurgical traditions also contributed to the development of the idea of influencing consciousness through rituals and specific practices. Although the term "hypnosis" was introduced much later, the beginnings of this idea can be traced back to religious, magical, and medical practices of the Middle Ages, where beliefs in supernatural influences on consciousness and the body were widespread.

The first scientific attempts to understand hypnosis began in the 18th and 19th centuries when the French physician and physicist Franz Anton Mesmer (1734–1815), known as a pioneer of the theory of "magnetism", developed the concept that vital energy and healing power were transmitted through "magnetic" influence. He studied medicine in Vienna and obtained a medical degree, becoming an important figure in the history of hypnosis. Mesmer developed a concept known as "magnetic current" or "mesmerism" and used it to treat patients through hypnotic sessions. F. A. Mesmer decided to examine a phenomenon he called "magnetic current". He believed that this magnetism permeated all things, including living organisms, and that an imbalance in magnetic current could lead to physical and psychological illnesses. Mesmer developed a treatment technique in which he used magnetic current to influence patients. This method involved the ritualistic use of magnetic rods and magnetic current. During sessions, patients sat or lay in group sessions, subjected to the influence of "mesmeric magnetism". During mesmerism sessions, Mesmer waved his hands around the patient and believed he could control magnetic currents to cure various ailments. Patients often entered a specific state of deep concentration and relaxation, similar to a trance. Mesmer's methods became very popular in the late 18th century, and he was even able to alleviate hysteria outbreaks in Paris during the French Revolution. However, his practices became the subject of debate and criticism, with scientists and physicians considering his methods fraudulent. Despite the controversy surrounding Mesmer's methods, his contribution lies in discovering the deep states of trance and hypnosis, which can induce profound concentration and relaxation. This state was later used in the development of modern hypnosis and psychotherapy. F. A. Mesmer passed away on March 5, 1815, in Saint-Georges-en-Laye, France. His name remains associated with the exploration of deep states of consciousness and hypnosis in the history of medicine and psychology [97].

In the 18th and 19th centuries, hypnosis was regarded as a remarkable and mysterious area, and its study and application underwent several stages of development. In France, physician and scientist Jean-Martin Charcot (1736–1813) was another important researcher of hypnosis. He studied the state of hypnosis and coined the term "somnambulism" to describe a deep trance state, making a substantial contribution to the development of hypnosis and neurology in the late 19th century. Charcot made an important contribution to the examination of somnambulism, particularly in the context of hypnosis and neurology. He did not "discover" somnambulism in a general sense but scientifically investigated and classified somnambulic states for the first time, including their relationship with hypnosis and other mental states. His research helped uncover the nature and phenomena of somnambulism in the context of psychological disorders and hypnosis. Charcot classified different forms of somnambulism and identified their common features. He distinguished between somnambulism accompanied by hypnosis and somnambulism associated with various mental states, including hysteria. Charcot conducted studies on patients in somnambulic states, during which they exhibited unusual behaviour and speech. His findings included analysing this phenomenon and attempting to understand the nature of these states. Charcot used hypnosis to induce somnambulic states in patients. He observed how patients transitioned into somnambulic states and what occurred during these states. Charcot's work in the field of hypnosis and somnambulism influenced the further development of psychoanalysis and the work of prominent psychoanalysts such as S. Freud. Charcot laid the foundation for further research into psychological disorders and the inner world of humans. Thus, J.-M. Charcot did not discover somnambulism, but he was one of the first to scientifically examine and classify this phenomenon and analyse it in the context of hypnosis and psychological disorders. His work in this area was a substantial step in the development of science and the understanding of psychology. His research and clinical experience in hypnosis garnered great recognition and influenced the further development of this science. Charcot was one of the first to investigate and classify hysteria, considering it a mental disorder. The scientists also applied hypnosis as a method of treating hysteria and other mental illnesses. His clinical experience with hypnotised patients was a substantial stage in the development of psychoanalysis and psychiatry. Charcot introduced scientific methodology into hypnosis research, demonstrating its real impact on psychological and physiological processes. He conducted a series of public lectures and demonstrations on hypnosis, sparking interest and viewing this phenomenon as scientific. Charcot examined the phenomenon of hypnosis in the context of neurology. He researched various states of consciousness and phenomena associated with somnambulism, ecstasy, and other states, including the influence of hypnosis on physiology. Charcot applied hypnosis as a method of treatment in his clinical practice. He analysed its effectiveness in influencing symptoms and the condition of patients, establishing the first clinical standards for the use of hypnosis in medicine. Charcot's work greatly influenced subsequent researchers of hypnosis and psychology, particularly the development of psychoanalysis. His studies formed the basis for the development and understanding of hypnosis as a scientific phenomenon. Due to his scientific approach and research, J.-M. Charcot became a substantial figure in the development of hypnosis and psychology, and his work continues to influence these fields to this day.

James Braid (1795–1860) was a Scottish physician and surgeon who made substantial contributions to the development of hypnosis and became an important figure in the history of hypnotherapy. J. Braid investigated and practised hypnosis as a method of treating both mental and physical illnesses. He believed in the potential of hypnosis to improve patients' health and reduce pain. Braid used hypnosis as a method of treating various disorders, including functional disorders and pain syndromes. He refined hypnotic techniques to influence psychosomatic disorders. J. Braid debunked popular myths about hypnosis and developed structured methods of hypnotherapy for treating patients. His approaches became the basis for further developments in hypnotherapy. Braid's work was crucial for the development of hypnotherapy and contributed to the interest in hypnosis among other physicians and scientists. He influenced subsequent researchers such as J.-M. Charcot, H. Bernheim, and S. Freud. Braid was the first person to examine hypnosis as a systematic method of treatment. His work is important for understanding the history and development of hypnotherapy as a science and medical practice. He renamed "magnetic current" to "hypnosis" and attempted to use hypnosis to treat various ailments after his discovery of what he previously called "magnetic current". His experiments and discoveries helped formulate a new understanding of this phenomenon and its use in medicine. Initially, J. Braid observed how patients responded to "magnetic current" procedures involving magnets. He believed that these magnets created a flow of effects that influenced and relaxed patients. Braid concluded that the magnetic flow actually depended on psychological factors such as attention and trust of the patient. He abandoned the concept of magnets and renamed this phenomenon to "hypnosis", derived from the Greek word "hypnos" (sleep). Braid renamed this phenomenon to reflect a state of increased attention and concentration that he observed in his patients during procedures. After renaming it, J. Braid began to use hypnosis as a method of treating various ailments. He believed that the hypnotic state could help alleviate pain, relax patients, and influence psychosomatic disorders. He developed structured methods of hypnotherapy to achieve these goals. J. Braid renamed "magnetic current" to "hypnosis" because he believed that this state depended on psychological factors rather than magnets. His work and attempts to use hypnosis to treat patients influenced the development of hypnotherapy as a science and medical practice.

Hypnosis in the 18th and 19th centuries garnered interest among both physicians and the general public, contributing to the development of its theories and practices. In the latter half of the 19th century, hypnosis became more popular in medical and psychotherapeutic applications. Physicians began using hypnosis for pain management, stress reduction, and the treatment of various mental disorders. By the end of the 19th century, various schools of hypnosis emerged, offering different methods and techniques. This was a time of experimentation and the development of various approaches to hypnotherapy.

Émile Coué (1857–1926) was a French physician and psychologist who made substantial contributions to the development of hypnosis and psychotherapy. He developed the theory of autosuggestion, which placed central importance on self-belief and the influence of thought on a person's physical state. É. Coué believed that thinking had a substantial impact on physical health and mental well-being. He developed a concept whereby individuals could use their internal beliefs and positive thinking to self-regulate their physical state and inner feelings. Coué believed that people could influence their physical health and mental state through conscious mental autosuggestion. This meant that through positive thoughts and beliefs, individuals could improve their physical condition and treat various ailments. Coué developed the method of autosuggestion, in which patients were required to repeat certain positive phrases and mantras aimed at improving their health. These phrases encompassed positive beliefs about rapid recovery and restoration. Coué used his theory and autosuggestion methodology to treat various physical and mental disorders. He argued that through the right positive thoughts and beliefs, patients could activate internal reserves for healing. Coué typically commenced treatment with a specialised consultation with the patient. During this consultation, he would establish a diagnosis and understanding of the patient's condition. A crucial goal of this stage was positive communication and fostering the patient's belief in the possibility of recovery. Coué explained the technique of autosuggestion to the patient, detailing how internal positive beliefs and phrases could influence the patient's physical state. Patients were explained that their own trust in recovery and positive thinking are substantial factors in the treatment process. Patients were taught the method of using autosuggestion. They were instructed to repeat certain positive phrases and mantras aimed at improving their condition. These phrases included thoughts of recovery, health, and a positive outlook. Coué emphasised the importance of psychological self-improvement. He convinced patients of the importance of positive changes in their way of thinking and contributed to their psychological self-improvement. He also provided advice on positive habits and mental hygiene. Coué emphasised an individualised approach to each patient. He adapted the autosuggestion methodology and psychotherapy to the specific needs of each patient and the nature of their condition. Coué provided support to patients and monitored their progress. He convinced them of the possibility of recovery and provided additional motivation to support a positive attitude towards treatment. E. Coué's treatment was based on positive psychotherapeutic influence and emphasised the importance of the patient's psychological state in the treatment process. His methodology allowed patients to influence their physical and mental health through their internal beliefs and self-confidence. É. Coué's work has had an impact on modern psychotherapy and psychosomatic medicine. His approach highlights the importance of positive thinking, self-confidence, and internal motivation in the treatment and maintenance of physical and mental health. The theory of autosuggestion was developed by É. Coué was a substantial step in the development of psychotherapy and highlighted the role of thought and psyche in overall human well-being [98]. Another important figure in the development of autosuggestion practices was Emile Gustave Gréber (1856–1936), a Swiss psychiatrist and psychologist who also investigated and implemented autosuggestion methods in the treatment of mental and physical disorders. Autosuggestion also influenced the development of various forms of psychotherapy, such as autogenic training, which was developed by German psychiatrist Johannes Heinrich Schultz (1884–1970) based on Coué's ideas. All these figures and their contributions to the development of autosuggestion practices helped create the foundations for understanding and using these methods in psychotherapy and self-improvement.

In the 20th century, hypnosis became part of psychoanalytic research. Scientists such as Sigmund Freud (1856–1939) used hypnosis to uncover subconscious processes and work with patients, making a substantial contribution to understanding psychological disorders. S. Freud was an Austrian neurologist and psychiatrist known as the father of psychoanalysis. He was born on May 6, 1856, in the city of Freiberg (modern-day Czech Republic) and died on September 23, 1939, in London. Freud made a huge contribution to the development of psychology and the understanding of the human psyche, and his ideas remain influential today. Indeed, S. Freud developed the theory of personality structure known as the "id, ego, and superego" structure. He believed that these three components interact and determine a person's behaviour and mental state. The "It" (id) is the primal and most basic structure of personality. The id is present from birth and represents the desire to satisfy immediate needs, such as hunger, thirst, relaxation, and other primal instincts. The id acts at the unconscious level. The "I" (Ego) emerges in early childhood and acts as a mediator between the id and the real world. The Ego considers social norms, realities, and the consequences of behaviour. It seeks to find a balance between the desires of the id and the constraints of the real world, working on the principle of realism. The "Above-I" (Superego) develops from awareness and internalisation of moral norms and values. The Superego represents an internal moral judgment and ideals. It reflects internal standards imposed by social institutions and authorities. This structure seeks to restrain the id and control the behaviour of the ego, relying on the concept of "moral duty". Freud believed that conflicts and decisions between these structures determine a person's behaviour and mental state. A deep analysis of these internal conflicts was a key aspect of Freud's psychoanalytic approach and his method of treatment, including dream analysis and the use of hypnosis. Freud developed the method of psychoanalysis, which involves uncovering unconscious aspects of consciousness and understanding psychological conflicts. Psychoanalytic sessions typically involve a conversation between the patient and the analyst to explore dreams, associations, and past experiences. Freud developed a theory of child sexuality and introduced the concept of developmental phases, such as oral, anal, phallic, latent, and genital phases. He believed that conflicts related to development in these phases influence subsequent behaviour and personality. Freud researched dreams and believed that they were a window into the unconscious aspects of the psyche. He viewed dreams as a way of expressing unfulfilled desires and conflicts. The theory of symbolism in dreams is also an important part of psychoanalysis. Freud first described the Oedipus complex, the conflict between the child and the parents, as a fundamental component of personality development. This concept became a substantial element of his theory. Freud examined the influence of religion and culture on the psyche and published works analysing the concepts of God and religious beliefs. Freud was an active physician and dedicated part of his career to researching and treating psychological disorders. S. Freud left a great legacy in modern psychology and society in general. He helped improve understanding of the human psyche and developed methods of psychotherapy and psychoanalysis, and his ideas continue to influence modern psychology and culture. S. Freud also investigated hypnosis and used it in the early stages of his career. Freud was interested in examining the nervous system and psychological disorders, and in this context, he became acquainted with hypnosis methods that were popular in his time. Freud worked with hypnotherapy methods, exploring their influence on patients' psyches. He even underwent training with J.-M. Charcot, a famous French physician and hypnotherapist. Freud used hypnosis as a method of treating psychological disorders and nervous conditions. He believed that hypnosis could help uncover unknown aspects of patients and resolve psychological conflicts. Over time, Freud abandoned hypnosis as the main method and focused on developing his own theory of psychoanalysis. He transitioned from hypnosis to the method of associative thinking and dream analysis. Essentially, Freud's psychoanalysis became an alternative to hypnosis, where the analysis of conversations and associations between different psychological phenomena played a critical role. Although Freud abandoned hypnosis as the main method of treatment, some ideas and methods he discovered while working with hypnosis influenced the development of his psychoanalytic theory. For example, attention to the role of the subconscious and the unknown in the human psyche was another direction of Freud's research, and it remained crucial in his later work. Thus, hypnosis played a certain role in the development of psychoanalysis, but Freud moved on to other methods that became the basis of his own psychoanalytic theory [96].

In the second half of the 20th century, hypnosis became a popular method for treating various mental and physical disorders. Doctors used it for pain relief, treatment of phobias and stress, and for enhancing psychotherapeutic sessions. Milton Erickson (1901–1980) was a prominent American psychiatrist and psychotherapist known for his work in the field of hypnosis and psychotherapy. His history and contribution to psychology and psychotherapy are substantial and interesting. He was the sixth of nine children in a large family. Having experienced difficulties in childhood, including polio, he developed a multifaceted view of life and could relate to different people. Erickson received his education at medical school, earning a medical degree. He specialised in psychiatry and psychoanalysis. Milton worked in psychiatric hospitals and treatment facilities, where he gained rich experience in communicating with patients of different age groups and with various mental disorders. Erickson investigated hypnosis and became an expert in this field. He first emphasised the importance of language and expression during hypnosis and its influence on the subconscious level. M. Erickson developed a unique therapy methodology based on the use of hypnosis and hypnotic communication. He was known for his ability to use metaphors and stories for therapeutic purposes. Erickson developed a new approach to psychotherapy known as "Ericksonian psychotherapy" or "Ericksonianism". His methodology considered the individuality of each patient and the use of hypnotic techniques to achieve positive changes in behaviour and attitude. The essence of Ericksonian hypnosis lies in the use of hypnosis and hypnotic communication to achieve therapeutic goals and positive changes in patients. Erickson placed great emphasis on the individual characteristics of each patient. He believed that each person has a unique world of awareness that should be understood and considered in treatment. Erickson used language and metaphors to influence the subconscious level of the patient. He often told stories and created images that helped patients change their perceptions and beliefs in the possibility of positive changes. Erickson believed that patients have negative beliefs and perceptions that can be changed through hypnosis. He helped patients rethink their fears, phobias, and limiting beliefs. Erickson analysed and used the internal resources of patients to help them find inner strength for change. He emphasised internal capabilities and self-care methods. Erickson considered hypnosis only as a tool for achieving therapeutic goals. He rejected the idea of hypnosis as a state of deep trance and developed methods that allowed the use of hypnosis at different levels of consciousness. Erickson encouraged active patient participation in therapy and the process of self-discovery. He encouraged patients to participate in their own healing and change process. The essence of Ericksonian hypnosis lies in an individualised and creative approach to influencing the subconscious level, beliefs, and perceptions of the patient to achieve therapeutic goals and positive changes. Erickson published numerous books and papers about his work and approach to psychotherapy. M. Erickson became a pioneer in the field of hypnotherapy and psychotherapy. He influenced many other psychotherapists and psychiatrists who use his methods and approaches in their work. M. Erickson left a substantial mark on the development of psychotherapy and hypnotherapy due to his innovative methodology and contribution to understanding human psychology. Additionally, M. Erickson is considered a pioneer of clinical hypnotherapy, and he used electroencephalography to examine changes in brain activity during hypnotic states. Research by Erickson indicates that hypnosis can influence brain electrical activity, and an EEG allows examining these changes in real-time. These studies stimulated interest in understanding the physiological changes that occur in the brain during hypnosis. Other researchers also use electroencephalography to examine hypnosis and its effects on brain activity, but hypnosis is a complex process, and understanding its mechanisms involves not only electrophysiological aspects but also psychological and sociocultural factors. M. Erickson used electroencephalography to examine brain activity in real time during hypnosis. He conducted experiments in which patients were in a hypnotic state, and their brain waves were recorded using an EEG. Various changes in brain electrical activity can occur during hypnosis, and electroencephalography allows these changes to be recorded in the form of an electroencephalogram. Specifically, changes in the frequency and amplitude of waves can be observed. Notably, Erickson investigated not only the physiological aspects of hypnosis but also developed his own methods of clinical hypnotherapy. He considered hypnosis a powerful tool for treating various mental and physical problems, and his research was an important contribution to understanding the influence of hypnosis on brain activity.

Researchers are beginning to use electroencephalography to examine the effects of hypnosis on brain activity. For example, American psychologist Ernst Rosenberg, who investigated hypnosis and its impact on consciousness, used EEGs to investigate changes in brain activity during hypnotic states. Rosenberg's study, aimed at understanding hypnosis and its influence on consciousness and uncovering the mechanisms of hypnotic states, included the use of electroencephalography to examine brain activity during hypnosis. One of his substantial contributions to the study of hypnosis is the theory of "splitting consciousness". One of his substantial contributions to the study of hypnosis is the theory of "splitting consciousness". He believed that the hypnotic state arises from the splitting or division of consciousness into two parts: the "hypnotically inclined" and the "not inclined to hypnosis". According to his theory, during hypnosis, one part of consciousness can focus on performing tasks while another part remains unconscious. In his research, E. Rosenberg used electroencephalography to record brain activity during hypnotic states. He investigated the effects of hypnosis on various aspects of consciousness, such as perception, memory, and bodily control. His research helped to uncover the phenomenon of hypnosis and became important for the development of theories about the nature of hypnosis and its influence on brain activity.

American psychologist and hypnotherapist Andre M. Vanderschaft also conducted research using electroencephalography to examine hypnosis. His contribution to the examination of hypnosis has had a substantial impact on the development of modern understanding of hypnotic phenomena. Vanderschaft examined the physiological and psychological aspects of hypnosis, including its influence on consciousness and brain activity. He used electroencephalography to objectively measure changes in brain electrical activity during hypnosis. His research included examining hypnosis as a tool for altering consciousness and improving mental health. Vanderschaft is also the author of books and papers in which he has extensively examined hypnosis as a psychophysiological phenomenon. His works contributed to expanding knowledge about hypnosis and influenced further developments in hypnotherapy and the investigation of consciousness. A. M. Vanderschaft made important conclusions by examining changes in EEGs during hypnosis.

One of the substantial studies in the field of altered consciousness during hypnosis is the work of French psychologist and physician Pierre Janet, who was one of the pioneers in the investigation of hypnosis and psychological disorders at the end of the 19th and early 20th centuries. Janet conducted numerous studies investigating the phenomenon of amnesia during hypnosis and revealing the mechanisms of the emergence and disappearance of states of alternative consciousness. He concluded that under the influence of hypnosis, a person can experience various states of consciousness that can affect the perception of time, events, and one's own identity. One of the important conclusions made by the researcher was that during hypnosis, consciousness can be divided into subconscious and conscious parts, leading to changes in the perception and interpretation of events. Janet also investigated psychological disorders, such as dissociative states, and explained them as a result of the influence of subconscious processes. Janet's research made a substantial contribution to understanding the phenomenon of hypnosis and the states of alternative consciousness and amnesia, which remain important for contemporary psychology and psychotherapy. P. Janet examined the change in consciousness, including during hypnosis, by carefully observing patients and analysing their states. He examined various aspects of consciousness. Janet investigated the phenomenon of amnesia, when a person, while in a hypnotic state, forgets events that occurred during this state. He reviewed the mechanisms of this amnesia and tried to understand why some informational events do not remain in consciousness after coming out of hypnosis. Janet noted that hypnosis can induce dissociative states, where consciousness is divided into different parts, each with limited access to information. He researched these states and their impact on the perception and functioning of the individual. Janet analysed how, under the influence of hypnosis, a person can experience different states of consciousness, including states where they feel younger or older, change their identity, or relive events from the past. He observed changes in the language, behaviour, and perception of these individuals. Through these observations and analyses, Janet sought to uncover the mechanisms that lead to changes in consciousness during hypnosis. His research yielded important conclusions about how consciousness can be altered in hypnotic states and became important for further research in the field of psychology and psychotherapy.

David Spiegel is a recognised American psychiatrist and researcher who examined hypnosis and its application in medicine. He has substantial experience in investigating the psychophysiological aspects of hypnosis and uses various methods, including electroencephalography, to examine the effects of hypnosis on brain activity. His research covers various aspects of hypnotherapy, including the use of hypnosis in treating mental and physical illnesses. D. Spiegel researched the effects of hypnosis on pain, stress, anxiety, and other mental states. Regarding the use of electroencephalography in his studies, this method allows for the objective measurement of changes in brain activity during hypnotic states. This may include analysing frequency spectrum, wave synchronisation, and other parameters that help better understand the physiology of hypnosis and its impact on the brain. David Spiegel's research contributes to the development of scientific understanding of hypnosis and the determination of its potential clinical applications. These researchers and other scientists have made a substantial contribution to understanding how hypnosis affects brain activity and how this may be related to changes in consciousness and mental state [96].

Hypnosis has been the subject of examination and research for several centuries, undergoing several stages of development and using different approaches. Throughout the 20th century, psychotherapy shifted from predominantly using hypnosis to developing other psychotherapeutic methods, such as psychoanalysis, psychointerpretation, and cognitive-behavioural therapy. Hypnosis remained one of the alternative therapeutic methods. In the second half of the 20th century and the beginning of the 21st century, hypnosis again became the subject of interest for researchers and psychotherapists. It is used as one of the methods of psychotherapy and in modern research in the field of neuropsychology and hypnotherapy. Thus, hypnosis has undergone evolution throughout history, changing its place in medicine and psychotherapy but remaining an important method in the examination of consciousness and the human psyche. In the 21st century, hypnosis remains the subject of intensive research. Modern scientific methods, such as functional magnetic resonance imaging (fMRI), allow for identifying its effects on brain activity and psychological processes [97].

The development of the history of hypnosis demonstrates its evolution from magical rituals to recognition as a scientific and medical instrument. Its role in modern medicine, psychotherapy, and science testifies to the importance of this phenomenon in the examination and treatment of the human mind and body. Modern hypnosis is used as a method of treatment and psychotherapy for entertainment and improving psychophysical well-being. Hypnotherapy is used to treat various phobias, habits, and sleep problems. Hypnosis is also the subject of research aimed at revealing its mechanisms and applications. The history of hypnosis testifies to the gradual development and understanding of this phenomenon, which is of great importance in medicine, psychology, and psychotherapy. Modern hypnosis includes various methods and techniques used to achieve a hypnotic state and achieve specific goals, such as treatment, psychotherapy, improving physical well-being, and many others [98]. Hypnotherapists use the hypnotic state to help clients overcome fears, phobias, depression, and anxiety, and even reduce pain and accelerate recovery during medical procedures. People can learn to enter into self-hypnosis using special techniques and exercises. Self-hypnosis is often used to increase self-confidence, improve self-control, and solve personal problems. One such method of hypnosis, Ericksonian hypnosis, was developed by the eminent American psychotherapist Milton Erickson. He used implicit hypnosis and hypnotic communication to influence the patient's subconscious level and promote improvement in their condition. It is used to treat various conditions and problems, including depression, post-traumatic stress disorder (PTSD), addictions, etc. [99].

A substantial number of professionals in the fields of psychotherapy and medicine investigate and utilise hypnosis methods, expanding the understanding and application of this tool in treatment and psychology. Many other psychotherapists, physicians, and researchers continue to research and use hypnosis for various psychotherapeutic and medical purposes [100–102]. Their contribution to the development of hypnosis can be highly substantial. The modern method of neurohypnosis integrates hypnosis with brain research and neuropsychology. The use of neurohypnosis allows for a better understanding of how hypnosis influences brain activity and neurophysiological processes. Physicians and psychotherapists use hypnosis to treat physical and psychological conditions. This may include the treatment of chronic pain, stress disorders, addictions, and other medical conditions. Suggestion and positive affirmation techniques are used to shape positive beliefs and alter subconscious thinking through the hypnotic state. Contemporary hypnosis is based on various approaches and methodologies and is used to treat various conditions, achieve personal development, and enhance the quality of life. Each hypnotherapist may have their own approaches and methodologies that cater to the needs of their patients.

2.2 Mechanisms of Hypnosis

The mechanisms of hypnosis constitute a complex and intriguing field of science that examines the phenomenon of enabling influence on human consciousness and psyche. Hypnosis is known for its ability to create a trance state in which the subject becomes more susceptible to influence and information processing [103]. It is utilised as a tool for psychological treatment, investigation, and exploration of various aspects of consciousness. Scientists have been researching for many years how exactly hypnosis affects individuals and the processes occurring in the brain and psyche during a hypnotic state. The mechanisms of hypnosis are explored from various perspectives, including physiological, psychological, chemical, and neurophysiological approaches [104].

For instance, concerning the physiological aspects of hypnosis, characteristic changes in brain electrical activity are observed during hypnosis. This may include a reduction in wave frequency, particularly an increase in theta waves and a decrease in beta waves. Hypnosis can influence muscular activity and pain responses, promoting muscle relaxation and reducing muscle tension. It can affect the autonomic nervous system by decreasing sympathetic system activity (responsible for stress reactions) and increasing parasympathetic system activity (responsible for relaxation) [105]. The psychological aspects of hypnosis can involve changes in awareness, reduced critical thinking, and increased susceptibility to suggestions and instructions. Hypnosis may allow access to the subconscious, activating resources and addressing psychological issues such as phobias or negative beliefs. In hypnosis, there may be an ability for automatic task execution that typically requires volitional control, such as analgesia (reduction in pain sensation). Research indicates that specific receptors in the brain interact with certain molecules during hypnosis, providing evidence for the neurophysiological aspects of hypnosis. Specific brain regions, such as the anterior hypothalamus, play a key role in hypnotic phenomena [106–110]. The chemical aspects of hypnosis represent a complex interplay of various neurochemical processes in the brain occurring during hypnotic states. These processes encompass the work of neuromodulators, chemical receptors, and changes in electrochemical signals between neurons. Hypnosis can affect the level of serotonin, a neuromodulator that regulates mood, sleep, and other mental states. Changes in serotonin levels can induce calmness and relaxation, which are typical signs of a hypnotic state. The influence of hypnosis can alter dopamine levels, potentially causing positive emotions and feelings. Gamma-aminobutyric acid (GABA) is the main inhibitory neurotransmitter in the central nervous system. An increase in GABA levels can lead to relaxation and pain relief. Hypnosis may enhance the activity of GABA-A receptors, resulting in increased inhibitory action of GABA, reduced neuronal activity, and relaxation. Hypnosis can also affect NMDA receptors, which play a role in memory formation and learning [111]. Changes in their function can influence cognitive functions during hypnosis. Hypnosis can alter brain electrical activity, particularly by reducing alpha and beta wave activity, which is characteristic of a deep relaxation state (Fig. 10). During hypnosis, the production of endogenous opioids, such as endorphins, which affect pain sensation and euphoria, may occur. Hypnosis can cause changes in synapses, increasing or decreasing the efficiency of signal transmission between neurons. Although the chemical aspects of hypnosis are still far from fully understood, research in this area helps to reveal the underlying mechanisms of this intriguing phenomenon. In summary, hypnosis is a complex phenomenon that integrates physiological, psychological, chemical, and neurophysiological aspects. Investigating these aspects allows for a better understanding of the mechanisms of hypnosis and its use in various medical and psychotherapeutic contexts.

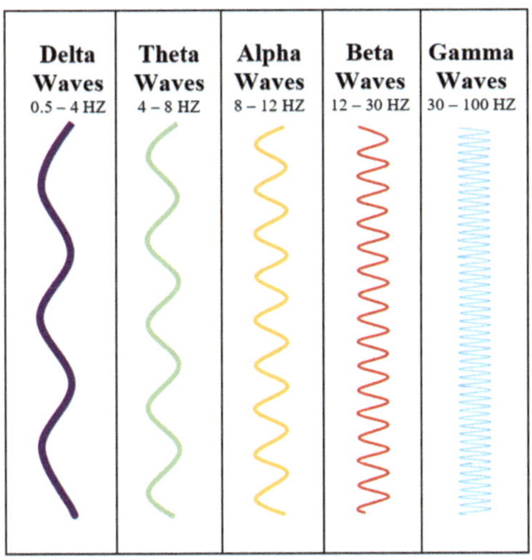

Figure 10 Types of brain waves (delta, theta, alpha, beta, and gamma).

Hypnosis can be classified according to various criteria, including methods of use, purposes, and the depth of the hypnotic state. Hypnosis can be applied through different methods and for various purposes. For instance, hypnotherapy uses hypnosis as a treatment method, aiding psychotherapeutic sessions. Recreational hypnosis is used for entertainment or personal development, offering opportunities for relaxation and enjoyment. Experimental hypnosis is employed in research to examine consciousness, brain activity, and other aspects [112].

Hypnosis can be classified according to its intended purpose as clinical, psychological, or recreational. Clinical hypnosis is applied to treat mental and physical illnesses; psychological hypnosis is used to address personal problems, self-awareness, and self-change; recreational hypnosis provides entertainment and amusement for the audience. This diverse range of applications demonstrates the versatility and potential effectiveness of hypnosis in various contexts [113].

The classification of hypnotic states by depth has significant practical importance and can be applied for different purposes. Light hypnosis is characterised by a relaxed state, where the individual can perceive external stimuli and respond to them. This level can be useful for stress relief and improving general well-being. Medium hypnosis leads to a deeper hypnotic state, where the consciousness becomes more focused on the hypnotist, and the sense of reality is partially lost. This level can be used to focus on specific questions or tasks. Deep hypnosis is marked by profound immersion in the hypnotic state, loss of awareness of the external world, and the ability to influence the subconscious. This level can be utilised for exploring subconscious processes or in therapeutic sessions aimed at psychotherapy and self-development. There is also direct hypnosis, where the hypnotist gives instructions and commands directly to the patient, asking and receiving responses. Indirect hypnosis involves the hypnotist using more general instructions, riddles, or metaphors to induce the hypnotic state [100]. This classification helps understand the variety of hypnosis applications in different life and research areas. It is important to note that hypnosis can have different effects on different individuals and should be used according to specific needs and circumstances.

Hypnosis can lead to changes in a person's consciousness. During the hypnotic state, an individual may experience deep relaxation and an altered perception of the world around them. This may include altered states of consciousness that allow for different ways of perceiving information. Hypnosis can induce changes in consciousness through various factors and mechanisms. Notably, hypnosis is a state of deep relaxation and altered perception occurring under the influence of the hypnotic process. Under the influence of hypnosis, an individual can enter a state of heightened concentration and awareness. This helps alter the way they perceive the world around them. A person may become more sensitive to internal processes, notice more details, or perceive information differently. Hypnosis can create the impression of a change in the speed of time. For someone in a hypnotic state, it may seem as though time is slowing down or speeding up. Hypnosis can intensify internal experiences and emotions. This helps individuals better understand their thoughts, experiences, and states. The hypnotic state can lead to deep physical and mental relaxation. A person feels deep inner tranquillity and comfort. During hypnosis, more vivid visualisations, fantasies, and sensations may arise. This can lead to immersion in specific images and scenarios. Hypnosis can affect memory, making it easier for individuals to recall information or change their perception of it. Changes in consciousness during hypnosis are explored from various perspectives, and the precise mechanisms are not yet fully understood. However, hypnosis can create a unique state of consciousness that can be beneficial in psychotherapy, self-discovery, and the development of certain skills [114].

Hypnosis is the subject of extensive research in psychology, psychiatry, and neuroscience. Researchers aim to understand how hypnosis can lead to changes in human consciousness. During hypnosis, individuals become less critical of the hypnotist's instructions. This means that they may accept suggestions and information that would typically be rejected or ignored in standard circumstances. Hypnosis can enhance the ability to focus attention on specific ideas, images, or memories, leading to deep awareness or altered states of consciousness. It can alter the way individuals perceive the world, their emotions, and their experiences. It can induce various states, from relaxation to euphoria or even regression. Research in this area continues, helping to better understand the qualitative and quantitative aspects of how hypnosis affects human consciousness. Understanding these mechanisms can be beneficial for the development of modern psychotherapy and for a general understanding of human consciousness functioning [102].

Inducing an altered state of consciousness through hypnosis requires specific skills and understanding of hypnotherapy techniques. Firstly, it is necessary to create a conducive environment. This may be a therapist's office or any other calm and comfortable setting. Ensuring silence and the patient's comfort is crucial. Building rapport and trust with the patient is essential. Conducting a preliminary conversation, explaining what will happen during hypnosis, and answering all questions are important steps. Directing the patient's attention to a specific focus, such as an object, sound, or their own breathing, can help divert their attention from external stimuli. Detailed instructions about what is expected from the patient during hypnosis can be provided. For example, soft, soothing language can be used to guide attention. Gradually, the patient should be guided into a deep relaxation state, for instance, through requests for relaxation of different muscle groups or visualisation of peaceful places. After achieving deep relaxation, the patient can be guided into a hypnotic state. This may appear as deep tranquillity, a state resembling sleep, or focused attention on specific ideas or images. During hypnosis, work can be conducted to achieve specific goals chosen by the patient. This may include resolving psychological issues, changing beliefs, or reducing stress. It is always important to bring the patient out of the hypnotic state gradually and carefully, allowing them to return to a standard conscious state. It is worth remembering that hypnosis requires training and experience, and it is best used under the guidance of a qualified hypnotherapist. Not all individuals perceive hypnosis equally, and it is not always the best treatment method [103].

Hypnosis is an important tool in psychotherapy, but like any medical or psychological technique, it can have certain risks and complications. In rare cases, hypnosis may lead to deterioration of the patient's mental state, especially if the hypnotherapist is unqualified or unprofessional. In some instances, hypnosis may lead to the emergence of memories that are actually falsified or untrue. This is particularly crucial in the context of using hypnosis to recover memories of events related to crimes or traumas. Some patients may experience amnesia regarding what happened during hypnosis, both in terms of specific instructions given during hypnosis and the content of the session itself. Hypnosis is not always effective for all individuals or all types of problems. Some patients may prove to be resistant to hypnosis, or hypnosis may be ineffective in addressing certain psychological or physical issues [104, 115]. In rare cases, hypnosis may evoke emotional reactions such as anxiety or feelings of loss of control. These reactions may arise due to individual patient characteristics or flaws in the hypnosis procedure. Hypnosis can influence memory and perception of events, potentially leading to distortion of memories, which is crucial in understanding potential complications. Extensive study and professional training of the hypnotherapist can help avoid most complications. Notably, hypnosis is an effective treatment method for many psychological and physical problems, but it should be used cautiously and, if necessary, in the context of other treatment methods.

Hypnosis can work by affecting subconscious processes. During a hypnotic state, the subconscious may become more susceptible to influence and positive beliefs. Hypnosis works by influencing subconscious processes, relaxing the conscious filter, and directly addressing the individual's subconscious. Under the influence of hypnosis, there is usually a decrease in the criticality of consciousness. A person becomes less likely to doubt or reject instructions during a hypnotic state. This allows ideas and instructions to enter the subconscious more easily. Hypnosis can trigger the activation of subconscious knowledge and experiences that may be hidden under normal conditions. This can be useful, for example, in the case of identifying memories of events or even skills that the person did not have a conscious awareness of. Hypnosis can affect beliefs and stereotypes that are usually held back by the mind. This can be useful for reformatting negative perceptions or helping overcome phobias. Hypnosis can have a positive effect on one's emotional state and mood. This can be achieved by introducing positive instructions or creating positive images in the subconscious mind. Hypnosis can help activate internal resources, such as self-healing powers, creativity or energy reserves. This can improve the patient's overall condition and ability to overcome difficulties. However, the mechanisms of hypnosis are not fully understood, and their effects may vary [105, 116]. Hypnosis remains the subject of active research to identify its exact mechanism and action.

Suggestion plays a key role in hypnosis, which is the main mechanism of influence on the human psyche during the hypnotic state. Suggestion involves providing instructions, hints, or ideas that are perceived by the patient as valid or real. Hypnosis can make a person more vulnerable to suggestions, reducing criticality and resistance to consciousness. In this state, a person can more easily accept the ideas and instructions that are given to them. Suggestion is aimed at influencing the subconscious mind, bypassing the conscious filter. This allows ideas and instructions to penetrate directly into the depths of the human psyche, which can lead to changes in perception and behaviour. Suggestion can be used to create positive images, idealise or promote specific mental states. For instance, the doctor might guide the patient towards reducing pain or enhancing well-being. Suggestion helps guide the patient's attention by directing it to specific ideas or impressions. This can be used to create a deeper recess in the hypnotic state or to change the perception of objects. Suggestion can help shape automatic responses to certain stimuli. This can be important for changing skills, cognitive biases viruses, or habits. Suggestion can be used to support medical treatment, improve medication effectiveness, or promote patient recovery. Suggestion in hypnosis is an essential element that helps create favourable conditions for changing a person's awareness, behaviour, and physiological responses [106].

People can also learn to enter a hypnotic state on their own by autosuggestion, using methods of mental self-control. Autosuggestion techniques are not related to a specific inventor or scientist, as they have developed over a long time and have been implemented in various traditions and techniques. However, the ideas of self-suggestion and internal influence on the human psyche have been investigated and developed in different cultures and traditions.

There are several ways in which people can independently enter a hypnotic state through autosuggestion. Calm and deep breathing can help reduce stress levels and create a physiological environment for hypnosis. The person can slowly inhale and exhale, focusing on the rhythm of breathing. Lateral relaxation of various muscle groups of the body can also contribute to entering a hypnotic state. A person can consciously focus on relaxing their muscles by doing this in stages. Using repetitive positive thoughts or phrases (autosuggestion) can help create a positive attitude and facilitate hypnosis. For example, a person may repeat phrases like "I feel relaxed" or "I am calm". Imagining different scenarios can help create a sense of virtual reality. A person can imagine themselves in a quiet place or perform actions that are perceived as relaxing. Focusing on a single thought or object can lead a person out of a certain state and create favourable conditions for hypnosis. Using audio recordings with positive autosuggestion or hypnotic inductions can help a person enter a hypnotic state without the direct involvement of a therapist. Notably, the depth and effectiveness of the hypnotic state by autosuggestion can vary. Skills and experience in using these methods can improve results [107, 117].

The level of openness to hypnosis can depend on various factors, and each person can respond to hypnosis differently. The belief that hypnosis is an effective method can have a positive effect on a person's readiness for hypnotic influences. Biases or negative beliefs about hypnosis can make it difficult to enter a deep state of hypnosis. People who are open to new experiences and experiments may be more susceptible to hypnosis [108]. Trust in the person performing hypnosis is crucial. If a person feels confident and safe, they may be more inclined to discover hypnotic influences. Individual characteristics, such as age, degree of concentration, and ability to visualise, can affect the depth of the hypnotic state. People who have a positive experience with hypnosis may be more open to reusing this method. Fear of losing control or awareness of hypnosis can affect openness to this method. Hypnosis is an individual process, and its effectiveness can vary depending on the unique characteristics of each person.

There are several interesting cases related to the mechanism of hypnosis that have attracted researchers. In some cases, hypnosis can cause catalepsy, a condition in which the muscles become clumsy. It was also proven that hypnosis can affect the sensation of pain, even causing analgesia (pain relief). A notable example of using hypnosis for pain relief is the historical practice of performing certain surgical procedures without general anaesthesia, where patients relied on hypnosis to manage pain and stress. A prominent instance of this application of hypnosis is the hypnosis-assisted surgeries performed by Dr. David Spiegel in the late 20th century. Dr. Spiegel used hypnosis for patients who underwent a mastectomy (removal of the breast) without general anaesthesia. This was particularly significant for women who were unable to undergo anaesthesia due to medical or personal reasons. Patients who participated in these procedures were immersed in a state of hypnosis, which allowed them not to experience pain and stress during the operation. This approach not only helped relieve pain but also reduced the number of complications and contributed to a faster recovery of patients. Notably, such practices require appropriate training and experience from the doctor and are not always suitable for all types of surgical procedures [109].

In certain instances, hypnosis is employed to retrieve memories or events that might be suppressed or forgotten. However, it is important to consider that hypnotic memory can be subject to distortion and influence. Notably, the use of hypnosis to restore memories is becoming the object of criticism and close consideration in professional circles and human rights organisations. Hypnotherapy to restore memories can create problems with the reliability and accuracy of the resulting memories. For instance, in the 1990s, a woman sought help from a hypnotherapist, believing she had traumatic childhood memories that were buried in her mind. She claimed that she had witnessed a serious crime and that she wanted to restore those memories to assist in the investigation. The hypnotherapist conducted a detailed consultation with the woman, identifying her reasoning and making sure that she understood the limitations of hypnosis and the possible impact on memories. During hypnosis, the hypnotherapist guided the patient to recover childhood memories, specifically concentrating on the period when she claimed to have witnessed the crime. Under the influence of hypnosis, the woman began to restore memories that, according to her, were previously lost. She tried to name the place, the people, and the circumstances. The hypnotherapist had to cautiously consider the potential for creating unverified or false memories under the influence of hypnosis. The memories obtained were provided to law enforcement agencies for further investigation. In this particular case, refutations of the accuracy and reliability of the memories occurred at the forensic stage. Doubts were raised about the credibility of memories retrieved during hypnosis, and they were not recognised by law enforcement in the legal proceedings. This case underscores the complexities and risks associated with using hypnosis to recover memories in legal contexts.

Hypnosis can cause impressive control over physiological functions such as heart rate, respiration, and blood pressure. This can be used to treat certain diseases. Hypnosis is often used to treat phobias, anxiety, and stress. This can help people overcome the limitations associated with certain fears. Hypnosis can sometimes induce a state of amnesia where individuals may not recall specific events or information. There are instances of hypnotic somnambulism, where individuals can follow a hypnotist's instructions while in a deep hypnotic state. These instances illustrate that hypnosis can influence diverse aspects of an individual's psychophysiology. However, it is crucial to note that responses to hypnosis are individualistic, and the reasons behind certain effects are not always evident [110].

The effectiveness of hypnosis substantially depends on the experience and personality of the hypnotist. An experienced hypnotist who knows a wide range of hypnosis techniques and strategies can achieve better results. The hypnotist's experience also includes the ability to understand and adapt to the client's individual reactions and needs. The effectiveness of hypnosis increases if there is trust and mutual understanding between the hypnotist and the client. A hypnotist who shows empathy and provides support can create a favourable environment for hypnosis. The hypnotist must be an excellent communicator and effectively use the language and tone of voice to achieve certain effects. The ability to maintain one's attention and focus is important for successful hypnosis. A hypnotist can use a variety of creative techniques to achieve certain results. An experienced hypnotist can adapt their approach to the needs and characteristics of a particular client. A hypnotist may have various hypnosis techniques in their arsenal, such as induction, suggestion, and relaxation. Overall, the hypnotist's experience and personal qualities play an important role in determining the effectiveness of hypnosis. Favourable interaction and compliance with professional standards can contribute to the achievement of positive results in clients seeking hypnotherapy [111].

Hypnosis can affect people of all ages, including children, but the effects and responses may vary depending on individual characteristics and situations. It is important to keep in mind that hypnosis is an individual experience, and its effectiveness can depend on many factors, including openness to hypnosis, health status, relationship with the hypnotist, etc. Many adults note that hypnosis helps them achieve a deep state of relaxation and calm. It can be useful for relieving stress, improving sleep quality, and relieving pain. Hypnosis can also be used to change skills and attitudes and break bad habits. Hypnotherapy can be useful for children with various problems, such as anxiety, phobias, and sleep problems. Hypnosis in children can enhance creativity and imagination by fostering deep concentration. Some studies show that children may be more susceptible to hypnosis because their imagination and openness may be higher. However, this does not happen in all cases. Hypnosis can be used in pediatric hypnotherapy to treat physical and mental problems in children. It is important to emphasise that hypnosis is an individual experience, and it is not always effective. It is also important that hypnosis is performed by an experienced professional who adheres to ethical standards, especially regarding children [112].

The mechanisms of hypnosis are being examined quite intensively, and so far, no unambiguous theory has emerged that would answer all questions about its action. The theory of suggestion and influence on subconscious processes remains fundamental, but the exact mechanisms of hypnosis are still the subject of further research. The mechanism of hypnosis covers the interaction of physiological, psychological, and neurophysiological factors. Suggestion plays a key role in creating a hypnotic state, where influential restructuring of ideas and physiological responses is possible. Thus, although the mechanisms of hypnosis remain the subject of active research, its impact on various aspects of human functioning has already been recognised and investigated.

2.3 Examining How the Brain Responds to Hypnotic States

Hypnosis is defined as a state of mental concentration and concentration that causes profound changes in consciousness and perception. Observing how the brain responds to hypnotic states is becoming an object of interest for scientists and researchers in the field of neuroscience. Examining this aspect of hypnosis opens up unique opportunities for understanding the fundamental processes that occur in the brain during changes in consciousness.

The investigation of hypnosis covers the use of various methods and research approaches. Standardised psychometric tests are usually performed to measure elevated or altered states of consciousness. In addition, they survey patients or people who are experiencing hypnosis to get their impressions and experiences. Physiological techniques such as electroencephalography, electrocardiography (ECG) and fMRI are used to examine changes in brain activity and other physiological parameters during hypnosis. Researchers conduct experimental studies and controlled experiments to investigate specific aspects of hypnosis, such as the effects on memory, attention, or other cognitive functions, and the effects of hypnosis on physical and mental processes [118]. Clinical cases and the use of hypnosis in treatment scenarios are being investigated to understand its effectiveness in treating various medical and mental conditions. Neuroimaging techniques are used to analyse the activity of different brain regions during hypnosis and changes in interactions between different brain regions. They assess the socio-psychological aspects of hypnosis, such as the influence of suggestion, belief in hypnosis, and individual differences in responses to hypnotic techniques. These methods are often used in combination to comprehensively understand hypnosis and its impact on the human psyche and physiology.

Various testing and survey methods are used to review hypnosis and its effects on the individual. Scales such as the Spielberger hypnotic propensity scale (Stanford Hypnotic Susceptibility Scale) or other modifications are used. The level of a person's tendency to hypnosis is assessed based on their reactions to standard hypnotic procedures. Standard psychometric tests are used to measure cognitive and emotional parameters that can change under the influence of hypnosis, and tests aimed at examining changes in memory, attention, creativity, etc. Researchers create questionnaires to obtain the impressions and opinions of people experiencing hypnosis, in which questions can relate to both physical sensations and emotional aspects of the hypnotic experience [119]. During hypnosis, scales can be used for self-assessment of various aspects of hypnosis, such as the depth scale of hypnosis, the sense of realisation scale, etc. Analysing patients' reactions during hypnotherapy or experimental sessions, considering changes in behaviour, speech, and physiological parameters, is also important for understanding the effect of hypnosis on the human brain. These methods allow researchers and therapists to gain a comprehensive understanding of the hypnotic experience and its effectiveness in various contexts [120].

Hypnosis can affect various physiological functions, including the heart, blood vessels, and other internal organs. Hypnosis can lead to changes in heart activity, such as a decrease in heart rate. A deep hypnotic state can cause deep relaxation, which affects the autonomic nervous system and heart activity. Some studies point to the possibility of hypnosis affecting blood pressure. Stress reduction and relaxation caused by hypnosis can lead to some effects on blood pressure. Hypnosis can affect one's breathing rate. A deep hypnotic state can lead to slow and deep breathing, which promotes relaxation. Hypnosis can affect stomach function, reducing tension and the possibility of some stress-related functional disorders. Some studies show that hypnosis can affect the immune system, in particular, by altering the production of certain cytokines and immune responses [121].

It is important to keep in mind that the response to hypnosis can be individual, and the impact on organs can vary depending on the methods of hypnosis, the type of session, and the characteristics of a particular person. A more detailed understanding of these influences may arise from the results of specific studies.

The hormonal state may undergo changes during hypnosis, although the details of these changes may vary depending on the specific study and individual characteristics. Several hormones may be involved in responses to hypnosis. During hypnosis, there may be a decrease in the level of stress hormones, such as cortisol. Hypnotic states are often associated with a decrease in the activity of the stress system. Hypnosis can promote the production of endorphins, which are natural analgesics and can cause feelings of euphoria and relaxation. Oxytocin, a hormone known as the "love hormone" or "uniting hormone", can increase under the influence of hypnosis. It is associated with strengthening social connections and a sense of trust. Another hormone that can undergo change during hypnosis is somatotropin, which is responsible for growth and interaction with metabolism. Some studies suggest that hypnosis can affect levels of serotonin and dopamine, neurotransmitters associated with mood and emotions. It is important to remember that the effect of hypnosis on the hormonal state may depend on the individual characteristics of the person, the context of the hypnotherapy session, and the methods used. More detailed and specific data may arise from the results of specific studies in this area [122].

Neuromedical research techniques such as fMRI and structural MRI (sMRI), diffusion tensor imaging (DTI), and conectomy analysis allow for in-depth analysis of interactions between different brain regions during hypnosis. These studies help uncover the internal mechanisms that determine changes in memory, attention, perception, and other cognitive functions in the hypnotic state.

The history of examining the effects of hypnosis on the brain and its response to hypnotic states has a long way to go. In the 20th century, scientists began to investigate the effects of hypnosis on the brain more systematically using various methods, such as neurophysiological measurements and imaginative brain mapping.

Neurophysiological measurement is a scientific approach aimed at measuring the functions and activity of the nervous system. This method allows researchers to obtain objective data on the functioning of the brain and other parts of the nervous system. The main methods of neurophysiological measurement include electroencephalography, in which the electrical activity of the brain is recorded using electrodes placed on the surface of the scalp. An EEG is used to examine electrical fluctuations associated with different phases of brain activity. Electroencephalography is used to examine brain activity, and in the context of hypnosis, researchers have conducted a number of studies using this method [123]. The technique of using electroencephalography during hypnosis involves several key stages. In the beginning, the patient is prepared, during which electrodes are attached to their head, possibly using conductive gels to ensure effective skin contact. The electrodes are then placed on specific areas of the head in accordance with an international system that determines their location in the anatomical areas of the head. After that, signal recording begins, and the patient is informed about the need to remain calm and motionless during the measurements. During hypnosis, especially if the examination is aimed at investigating hypnotic states, psychophysiological parameters can be monitored. This may include measuring changes in wave frequency, amplitude, and other characteristics of the electroencephalography signal. The collected data is analysed to determine changes in brain activity that may be associated with a hypnotic state. This process provides detailed insights into brain interaction during hypnosis and helps uncover its impact on various aspects of cognitive function. This technique allows examining the electrophysiological aspects of hypnosis and helps in understanding how this condition affects brain activity.

The main findings relate to changes in brain activity that can be observed during hypnotic states. During hypnosis, there may be changes in the frequency spectrum of brain activity. For example, there may be an increase in the frequency of alpha waves and a decrease in beta waves, which is characteristic of a deep relaxation state. Hypnosis can affect the synchronisation of brain waves. Some studies show increased coordination between different parts of the brain during hypnotic states. There may be changes in the amplitude of waves that indicate the degree of brain activity. In uncontrolled hypnosis, there may be a decrease in the amplitude of beta waves and an increase in the amplitude of alpha waves. These results suggest that hypnosis can affect brain physiology, and this approach allows obtaining objective data on hypnotic states using techniques such as electroencephalography [124].

Neurophysiological measurement also includes MRI and fMRI, which measure changes in the brain's magnetic field to examine its structure and functional activity during various tasks. In the context of investigating the effects of hypnosis on the brain, researchers use MRI and fMRI to examine how hypnosis affects brain activity, structure, and interaction between different brain regions during hypnotic states. The findings may help identify the mechanisms of hypnosis and its impact on brain physiology. Resonance imaging techniques such as MRI and fMRI provide a detailed review of hypnotic states at different levels of the brain. MRI scans provide detailed images of brain structure, including various areas and their anatomical features, which can indicate possible changes during hypnotic states. fMRI scans measure changes in blood supply to different parts of the brain, indicating their activity. During hypnosis, there are changes in the response of various brain regions. fMRI scans can also indicate interactions between different brain regions during hypnosis, helping to understand which parts of the brain interact during this condition. MRI studies can discover changes in interaction networks between different brain regions that may be associated with hypnotic states [125]. Thus, these techniques provide an opportunity to examine in detail not only the structure of the brain but also its functional activity and interaction between different parts during hypnosis. Performing MRI and fMRI during hypnosis requires a specific technique to ensure the accuracy and reliability of the results. A general technique consisting of preparing the patient, providing peace of mind and comfort to the patient, and performing hypnosis itself is provided below. Hypnosis can be triggered using standard methods, such as language commands and attention fixation [126]. Once the patient is in a hypnotic state, an MRI or MRI scan is performed, and data on brain activity and structure is collected. The data obtained are analysed to examine the effects of hypnosis on brain activity and structure. The use of MRI allows identifying areas of brain activity during hypnotic states, while MRI helps to examine the structural aspects of the brain. These techniques allow researchers to determine correlations between hypnosis and physiological changes in the brain.

There is also a single-cell recording method in which electrodes are inserted directly into neurons to record their activity. This method allows examining the activity of individual neurons. The single-cell recording method is a highly accurate method used in neurophysiology to examine the activity of individual neurons in a living organism. The electrodes are inserted directly into brain tissue or a separate neuron. The electrode is fixed at the insertion point to stabilise and ensure a reliable connection to the cell membrane. After the electrode is inserted, electrical signals generated by the neuron are recorded [127]. The resulting signals are analysed to examine the characteristics of neuron activity, such as emission frequency, amplitude, time intervals between emissions, etc. The single-cell recording method is a method of examining brain activity at the level of individual neurons, using electrodes to record electrical signals from individual cells. This method allows examining how specific neurons respond to different stimuli and conditions. Regarding the use of the single-cell recording method in hypnosis, it serves to establish how the activity of individual neurons can change under the influence of hypnotic states. Neurophysiological studies of this kind can help in identifying the mechanisms by which hypnosis affects brain activity at the cellular level. This method allows examining the activity of specific neurons and their interaction with other cells. The single-cell recording method can be used to examine a neuron's response to various stimuli, elucidate its functional properties, and respond to various external influences.

Measurements are also taken of the breakdown of radioactive substances introduced into the blood to assess flow and metabolism in different parts of the brain and examine the effects on the brain. Positron emission tomography (PET) is a highly accurate brain imaging technique that uses radioactive markers to examine blood flow and metabolism in different parts of the brain. The basic idea is that positron-emitting substances (radioactive markers) introduced into the body emit positrons that interact with electrons in the body, resulting in the emission of two photons that can be measured. From this information, an image of the internal structure of the brain can be recreated. PET can detect even very small changes in brain activity and provides functional data that allows examining activity and metabolism in real time. PET imaging can be used to individualise treatment in patients with various brain diseases [128]. This method allows assessing various aspects of brain function, including blood flow, metabolism, and the activity of individual brain regions. Using PET to examine the effects of hypnosis on the brain provides important information about the functional activity of the brain during hypnotic states. PET detects the distribution of radioactive markers that indicate the activity of different parts of the brain. PET allows measuring blood flow to specific areas of the brain, which indicates their activity. During hypnosis, there may be changes in the blood supply to certain brain regions, which may indicate changes in their functional activity and identify which specific brain structures become active during hypnotic states. This can be useful for understanding how hypnosis affects different parts of the brain and their interactions. PET allows examining the metabolism in the brain, which can also change under the influence of hypnosis. This is important for understanding the energy processes that occur in the brain during hypnotic states. PET can detect changes in the distribution of chemicals in the brain, such as neurotransmitters. This allows investigating the chemical aspects of hypnosis and their impact on mental processes. The procedure for performing PET during hypnosis involves preparing the patient and placing it on the tomograph table. Before starting hypnotic exposure, the baseline level of brain activity is assessed to compare with the data obtained during hypnosis [129]. During hypnosis, the patient can perform tasks that trigger activity in certain areas of the brain. During tasks performed by the patient, data on the distribution of radioactive material in the brain is collected. The data obtained are analysed and processed to determine the active areas of the brain. The results are compared with baseline activity levels to determine changes caused by the hypnotic state. This method may vary depending on the specific research task and the equipment used. An important part of the procedure is to try to create conditions that best match the conditions of hypnosis and provide reliable data on brain activity. All of these aspects provide a deeper understanding of how hypnosis affects the brain at various levels, from structural and functional changes to chemical processes.

These methods provide researchers with the opportunity to obtain detailed data on the functioning of the brain and nervous system at the physiological level. Neurophysiological measurement plays a key role in examining brain function and understanding the mechanisms of various mental and neurological conditions, including hypnosis.

Imaginative brain mapping (or neural mapping) is a concept that analyses how different parts of the brain relate to functions or activities in the body. This means that certain areas of the brain are responsible for performing certain tasks or functions. Brain imaging techniques, such as fMRI and PET, allow researchers to identify active brain regions when a person is performing a specific task or acting. For example, if a researcher wants to examine how the brain responds to visual stimuli, they can use MRI scans to visualise the activity of different parts of the brain when a person looks at an image. Such studies help to determine how different brain regions interact and perform their functions, which contributes to a better understanding of the interaction of brain structures and their impact on physiological and mental processes [130].

A new field of study has emerged and is rapidly developing—the neuroscience of hypnosis, which combines approaches from neuroscience and psychology to identify the mechanisms of hypnotic phenomena. The neuroscience of hypnosis is an interdisciplinary field that combines approaches from neuroscience and psychology to examine the mechanisms of hypnotic phenomena. The main goal of this field is to understand how hypnosis affects brain activity and mental processes. Key aspects of the neuroscience of hypnosis include investigating the effects of hypnosis on brain structure and function using brain imaging techniques such as MRI, PET, fMRI, and electroencephalography. Using these methods, it is possible to understand the influence of hypnotic states on mental processes such as awareness, memory, imagination, and expression of will and examine the genetic aspects and individual diversity that can influence the tendency to hypnotic states. It is also possible to determine ways that indicate brain synchronisation and interaction between different brain regions during hypnotic phenomena. Using methods such as MRI, PET, fMRI, and electroencephalography, the effect of hypnosis on biological processes such as biorhythms, hormonal status, immunity, etc., can be analysed [117].

The neuroscience of hypnosis attempts to identify the complex interactions between brain processes and hypnotic phenomena to provide a more scientifically based understanding of this state of consciousness. Current research focuses on changes in brain activity, structure, and function that occur during hypnosis. This may include considering the effects of hypnosis on pain perception, attention, memory, and other cognitive processes. These studies contribute to understanding the nature of hypnosis and its possible applications in medicine and psychotherapy. Research of the effects of hypnosis on the brain, conducted using various methods such as electroencephalography, MRI, fMRI, and PET, provides valuable information about the impact of hypnotic states on the neurophysiological and structural aspects of the brain. Studies using electroencephalography indicate changes in brain activity during hypnosis. A decrease in alpha-rhythm and an increase in beta-rhythm may be associated with a deep hypnotic state. Brain imaging using MRI can show structural changes, such as increased activity in certain areas of the brain associated with switching on and off various functional networks. fMRI allows examining functional connectivity between different brain regions. Hypnosis can affect the interaction between brain regions and functional networks. PET allows examining blood flow and metabolism in different parts of the brain during hypnotic states, indicating possible changes in biological processes. Individual variations in the brain's response to hypnosis can be determined by genetic and individual factors, the analysis of which is important for understanding response variability [113].

The general assessment of the effects of hypnosis on the brain using various methods provides a multidimensional understanding of this condition, considering neurophysiological and structural aspects and functional relationships between brain regions. A general analysis of the effects of hypnosis on the brain using various techniques such as electroencephalography, MRI, fMRI, and PET provides a multidimensional understanding of this condition.

An EEG measures the electrical activity of the brain, enabling the determination of changes in the level of consciousness and the nature of interactions between different brain regions during hypnosis. MRI scans provide detailed images of the brain's structure, identifying changes in the location and volume of different regions. fMRI allows visualising brain activity during specific tasks or states, which indicates functional changes in real time. PET is used to examine blood flow and metabolism in the brain, identifying areas that are activated during hypnosis.

The analysis of the effects of hypnosis on awareness, memory, and experience provides insight into the psychological aspects of this condition. The examination of the effects of hypnosis on neuromodulators and chemical processes in the brain reveals the chemical mechanisms of its action. MRI techniques can detect possible changes in the structure of the brain that may occur under the influence of hypnosis. The investigation of group hypnosis and individual responses identifies the variety of responses and possible variations in the effects of hypnosis on the brain. This allows understanding of the interaction between the conscious and subconscious levels of the brain, determining the role of subconscious processes in the formation of hypnotic phenomena. All these aspects interact to create a comprehensive understanding of the effects of hypnosis on the brain, which is constantly expanding thanks to new discoveries and research in the field of neuroscience [131].

The neuroscience of hypnosis uses advanced techniques and technologies to examine how hypnosis affects brain activity and mental processes. An EEG allows measuring the electrical activity of different parts of the brain, determining how hypnosis affects the work of individual areas. MRI allows visualizing the structural aspects of the brain, while fMRI examines the activity of certain areas of the brain during hypnosis. PET is used to examine metabolism and blood flow in the brain, which can establish additional details about physiological processes during hypnosis.

Studies of the effects of hypnosis on memory, attention, perception, and other cognitive functions can be performed using psychometric tests and tasks. Tests that evaluate the memorisation and reproduction of verbal information during and after the hypnotic state can be used to examine the effect of hypnosis on verbal memory.

Research may include experiments that require memorising images, faces, or other nonverbal stimuli. Attention separation testing may involve the use of tasks that examine a person's ability to focus on multiple tasks at the same time. Research may include assessing the ability to focus on a specific task for a long period. Studies can assess how hypnosis affects the perception of sensory stimuli, such as sound, light, or touch [114]. Experiments can explore how the hypnotic state affects the interpretation of information and the perception of shapes, colours, etc. Research may cover the effects of hypnosis on cognitive functions such as planning, decision-making, and problem-solving. Some studies may include analysing the coordination and accuracy of performing movements during a hypnotic state. When conducting such studies, it is important to consider methodological aspects such as control groups, randomisation, identification, and variable management. In addition, the individual characteristics of the response to hypnosis, which may depend on the characteristics of the individual and previous experience with hypnosis, are notable.

The neuroscience of hypnosis evolves by focusing efforts on understanding how hypnosis affects neuromodulators, such as serotonin, dopamine, and GABA, which play a vital role in regulating neuronal activity and mental processes. Notably, research in this area is challenging, and many questions remain open. However, there are certain observations and theories that review the interaction of hypnosis and neuromodulators. Hypnosis can affect the level of serotonin, which is a key neuromodulator that regulates mood. Changes in serotonin levels can affect the emotional state and perception of events during the hypnotic state. Research shows that hypnosis can affect dopamine, which is responsible for reward, motivation, and pleasure. Changes in dopamine levels can affect the feeling of pleasure from a hypnotic experience. Hypnosis can also affect GABA, the main inhibitory neuromodulator in the central nervous system. Changes in GABA levels can affect the level of excitability and control of neural activity during hypnosis. Hypnosis can affect the levels of stress hormones such as cortisol. Reducing stress levels during a hypnotic state can lead to various physiological and mental effects. Research in this area often uses visual techniques such as fMRI and electroencephalography to examine brain activity during hypnosis. However, it is important to consider that the individual response to hypnosis may vary, and research results may be context- and personality-dependent [99]. A scientific understanding of the interaction between hypnosis and neuromodulators opens up the possibility of developing new treatments and helps in understanding the mechanisms underlying hypnotherapy and hypnotic phenomena.

Investigating brain activity indicators plays an important role in understanding how hypnosis affects different levels of consciousness and brain activity. For this purpose, a variety of methods are used, including fMRI, electroencephalography, and other techniques for imaginative examination of the brain. fMRI examinations can help determine how hypnosis affects the activity of the frontal cortex, which is responsible for behaviour planning and decision-making. The activity of the frontal cortex associated with performing executive functions, such as controlling other brain functions, can also be assessed. Examining structural changes in the corpus callosum (a large bundle of white fibres connecting the two hemispheres of the brain) may reveal how hypnosis affects cooperation between the left and right hemispheres. Functional connections between different brain regions can be examined using MRI techniques to understand how hypnosis modifies communication networks in the brain. This may include investigating how hypnosis affects the activity of sensory areas of the brain in response to various stimuli, such as sound, light, or touch. Examining the electrical activity of the brain using an EEG can help determine changes in the frequency of waves (e.g., alpha, beta, delta), which may indicate changes in the level of excitability and state of consciousness during hypnosis. Examining the effects of hypnosis on cognitive functions such as attention and memory can determine the activity of the brain regions responsible for these processes. Investigating these aspects allows for a better understanding of how hypnosis affects brain function at various levels, from structural changes to functional activity. These data can serve as a basis for the development of new methods of hypnotherapy and an in-depth understanding of the psychophysiology of hypnosis [108].

Conectomy analysis is a method of examining structural and functional connections between different brain regions, especially interactions between regions through networks of nerve fibres. Various analysis methods are used to understand the effect of hypnosis on brain connectivity, including fMRI and sMRI. The analysis of the resting state of the brain in patients under hypnosis using fMRI allows determining baseline activity levels and contributes to the examination of network interactions in the absence of external stimuli. Examining brain activity when performing specific tasks in a state of hypnosis allows the establishment of how interaction occurs between different areas under the influence of a hypnotic state. Analysing the structure of brain fibres using DTI in a hypnotic state allows identifying changes in the connectivity and degree of structural organisation of neural networks. SMRT analysis may include a review of how hypnosis affects the structure and integration of white matter that provides communication between brain regions. The use of network analysis techniques allows identifying the degree of connection and ordering of the brain network during hypnosis and key "nodes" or areas that interact most intensively. Quantitative metrics, such as clustering coefficient and average pathway length, are used to quantify the effectiveness of brain connectivity during hypnosis. Evaluating dynamic changes in the brain's network reflects how hypnosis can affect connectivity over time, providing information on how quickly and effectively changes occur in interactions between different brain regions. Conectomy analysis provides a comprehensive view of the interaction between different brain regions during hypnosis, considering both structural and functional aspects. This helps to better understand the mechanisms of the hypnotic state and its effect on brain function [131].

Neuroscience provides insight into the deeper mechanisms and physiological changes that occur in the brain during hypnosis, opening up new perspectives for using the condition in medicine, psychotherapy, and other fields. The investigation of how the brain responds to hypnotic states opens up a fascinating maze of the inner world of consciousness and its interaction with hypnosis (Fig. 11). The results of the studies conducted using a variety of methods, including fMRI and sMRI, psychometric tests, and conectomy analysis, delve deeply into the mysteries of how the brain responds to hypnotic influences. Hypnosis was identified to cause specific changes in brain activity, interactions between different brain regions, and structural adaptations, indicating its effects on cognitive functions such as memory, attention, and perception. Functional studies at different stages of hypnosis describe the dynamics of changes in brain activity, indicating how hypnotic states can modulate conscious and subconscious processes. Conectomy analysis expands the understanding of hypnosis by identifying how different brain regions interact in the hypnotic state. Research on neuromodulators suggests that hypnosis can affect the functioning of chemical transmitters in the brain, altering their levels and interactions. All of this contributes to creating a comprehensive image of the interaction between the brain and hypnosis, determining important aspects such as individual response variations, response mechanisms, and potential applications of hypnosis in mental health and treatment. Due to these studies, researchers today have an understanding of how hypnosis affects the work of our most complex organ, which opens up new opportunities for the development of methods of influencing the consciousness and subconscious of a person.

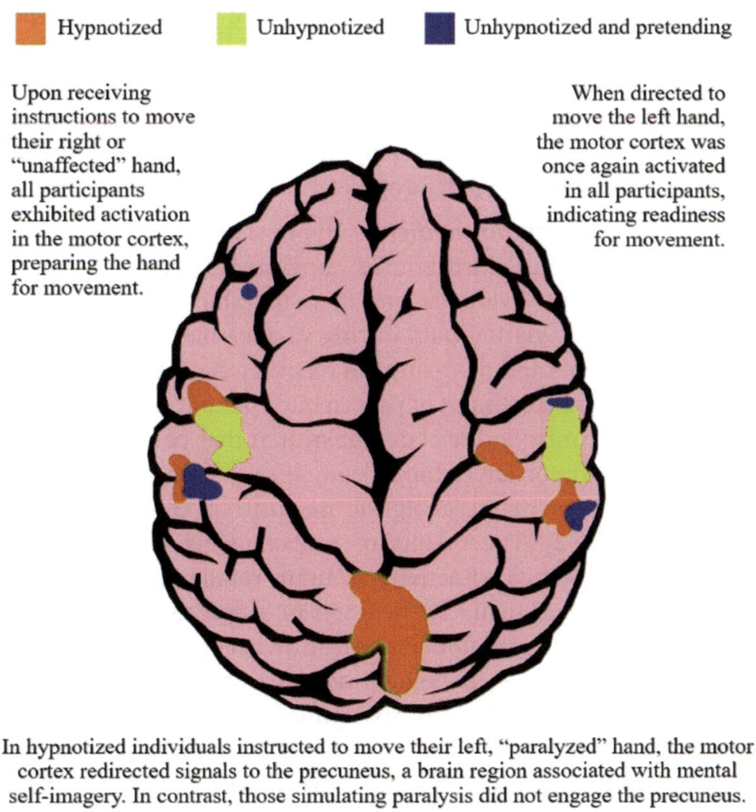

Figure 11 Brain response to hypnosis.

2.4 Application of Hypnosis in Medicine and Psychology

The use of hypnosis in medicine and psychology is an exciting field of research that opens up new perspectives in the treatment and understanding of the human brain and psyche. Hypnosis, as an ancient method of influencing consciousness, is experiencing a renaissance in recognition of its potential clinical applications and psychotherapeutic benefits. Hypnosis can be used to treat various mental and physical illnesses. Hypnotherapy can be used to reduce pain, stress, and other symptoms, and it is also integrated into modern medical and psychological practice. The use of hypnosis is becoming an increasingly innovative and dynamic direction in modern medicine and psychology, so examining this subject is important for the development of effective treatment methods and the improvement of patients' quality of life. Hypnosis is used in various fields of medicine and psychology as an additional or alternative method of treatment and psychotherapy. Notably, the effectiveness of hypnosis may vary depending on individual characteristics and the specific application.

The use of hypnosis in medicine. Hypnosis can help in *reducing pain* during medical procedures, operations, and for chronic diseases and pain. Evidence has demonstrated the effectiveness of hypnosis in alleviating pain during various medical procedures, chronic conditions and post-surgical recovery. A notable application of this is in the management of pain in burn patients. The findings of a randomised controlled trial demonstrated that patients receiving hypnotherapy reported reduced pain levels, with a concomitant reduced requirement for analgesic medication, in comparison to the controls [132]. The results of neuroimaging studies indicate that hypnosis modulates activity in brain regions associated with pain perception, particularly the anterior cingulate cortex and the thalamus. These findings underscore the neurological underpinnings of hypnosis-induced pain relief, reinforcing its clinical efficacy in pain management. Beyond its applications in burn treatment, hypnosis has been successfully employed as an adjunct to surgical anaesthesia. Patients undergoing neurosurgical procedures under hypnosedation have reported effective pain control and reduced anxiety, thereby minimising the risks associated with general anaesthesia while promoting faster postoperative recovery [133]. This approach reduces reliance on pharmacological anaesthetics and aligns with broader trends in multimodal pain management strategies.

The mechanisms by which hypnosis can affect the sensation of pain are complex and depend in part on the individual responses of the patient and the specific features of the medical intervention. However, some possible mechanisms include reducing the sensation of pain through the central nervous system, meaning hypnosis can affect pain perception by modulating pain signals in the brain. It can alter changes in activity in neurons that perceive and process biologically important information signals, including those that cause pain [134]. Hypnosis can promote deep muscle relaxation and reduce tension in the body, which can lead to less pain. Hypnosis can help change a patient's focus by shifting their attention from pain to other experiences or thoughts. This can change the way people perceive pain. In some cases, hypnosis can trigger a placebo effect, which means that the patient perceives less pain due to a psychological factor. As for specific examples, hypnosis can be used in areas such as surgery [135]. Patients can use hypnosis before surgery to reduce the duration of recovery and pain after surgery.

Using hypnosis during labour can help reduce stress and pain levels in pregnant women. Hypnotherapy can be used to treat chronic pain, such as migraines, arthritis, or back pain. Hypnosis can be used to reduce the duration and hassle of dental procedures. Notably, the effectiveness of hypnosis varies depending on individual responses, and not all patients experience the same degree of relief [136].

The application of hypnotherapy for the *treatment of functional disorders* can be useful for treating functional disorders such as intestinal irrigation syndrome, fibromyalgia, asthma, etc. The effectiveness of hypnosis for the treatment of functional disorders may be related to a number of mechanisms and psychophysiological changes that occur as a result of hypnotherapy. For example, when using hypnotherapy for irritable bowel syndrome (IBS), a decrease in stress and anxiety levels is noted, which may be factors contributing to an increase in IBS symptoms [137, 138]. Hypnosis can change the perception of pain and sensitivity of the intestines, reducing unpleasant sensations and discomfort. Hypnosis can affect the perception of pain and promote muscle relaxation, which can ease the symptoms of fibromyalgia. Hypnotherapy can help improve sleep quality, which is an important aspect in the treatment of fibromyalgia. Hypnosis can teach patients to control their breathing, reduce the degree of bronchial spasms and relieve asthma symptoms. Asthma is often associated with anxiety, and hypnosis can affect this aspect by helping to reduce anxiety responses. Examples include hypnotherapy sessions, during which patients learn techniques for relaxing, concentrating, and changing consciousness to influence physiological processes in the body. However, it is important to remember that the effectiveness of hypnotherapy may vary depending on the individual characteristics and response of the patient. In addition, hypnosis is not a substitute for traditional methods of treatment, and its use should be conducted under the supervision and guidance of a qualified specialist [139].

Recovery in the postoperative period can be expedited due to the use of hypnosis. The use of hypnosis to improve the rate of recovery and healing after surgery and other medical interventions is based on the influence of the hypnotic state on the physiological processes of the body. Hypnosis can be used to relax the patient and reduce the feeling of pain, which promotes healing and reduces the effects of stress on the body (Fig. 12) [140]. Hypnotherapy can be used to reduce the need for analgesics or the intensity of pain. Hypnosis can affect the immune system by increasing levels of certain cytokines and other factors that can improve tissue healing. This aspect is important for rapid recovery after surgery. Hypnotherapy can affect the severity of inflammatory processes, providing a lower level of inflammation in these areas of the body [141]. This can help improve healing and measure recovery. Hypnotherapy can help activate the body's internal resources, including self-healing and tissue regeneration processes. A patient's mental state can affect their physical health. Hypnotherapy can help relieve anxiety and stress, which promotes healing and improves the patient's psychoemotional state. For example, hypnosis can be used to relieve pain and anxiety before surgery. During surgery, hypnosis can help the patient stay calm and control pain and stress, which can have a positive effect on the healing process after surgery. Notably, the effectiveness of hypnosis and its use may vary depending on the individual characteristics of the patient and the specific medical situation. Before using hypnosis for medical purposes, it is recommended to consult a doctor or licensed hypnotherapist.

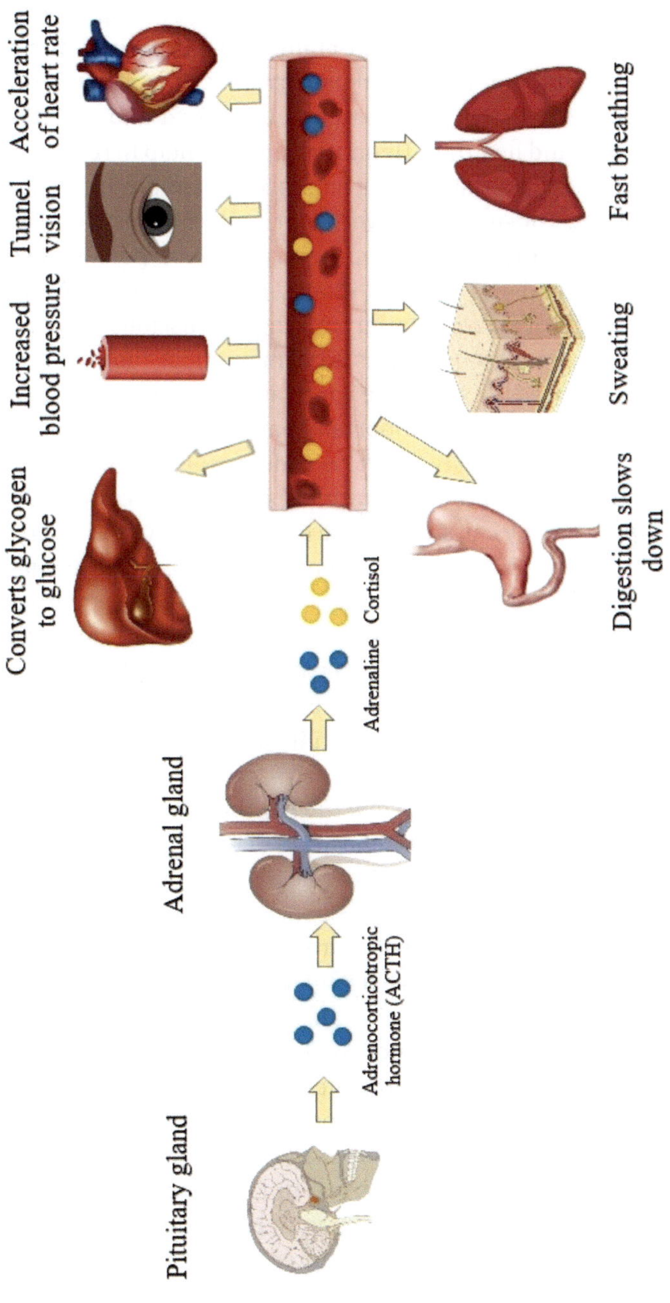

Source: D. Guidolin et al. [45].

Figure 12 The impact of stress on the body.

Hypnotherapy can be used as part of a comprehensive *treatment to control weight and maintain a healthy diet.* Hypnotherapy is used in the context of weight control and maintaining a healthy diet to influence the mental and emotional aspects that may affect eating habits and weight [142]. It helps patients treat food and nutrition more consciously and healthily. Hypnotherapy can help in retraining the brain to develop healthier eating habits. For example, it can help avoid harmful foods or switch to lower-calorie alternatives. Hypnosis can influence the patient's consciousness to reduce portion sizes and set healthy standards for food intake. Hypnotherapy can help manage emotional overeating by promoting a conscious response to stress or other emotional states that may affect food choices. Hypnosis can be used to stimulate motivation and pleasure from physical activity, which is an important part of a healthy lifestyle [143]. Some hypnotherapists work to reduce appetite or hunger by affecting the neuromotor apparatus responsible for regulating appetite. Hypnotherapy can support the development of self-control and the will to eat. Examples of hypnotherapy sessions include visualisation, positive programming, working with emotional blocks, affirmations, and other techniques that aim to positively change attitudes to food and maintain healthy eating habits. It is important to consider that the effectiveness of hypnotherapy may vary depending on the patient's personality and the professionalism of the hypnotherapist [144].

Hypnosis can help in *relaxing and the improvement of one's sleep quality.* Hypnosis promotes deep relaxation of the physical body and brain. It can help reduce stress and stress levels, which helps create a favourable environment for falling asleep. Hypnosis can change the state of consciousness, the transition from the highest wakefulness to a calmer state [145]. This can be useful for those who have trouble falling asleep due to overexertion or anxiety. A hypnotherapist can use autosuggestion or the introduction of certain ideas that promote relaxation and fall asleep. This may include positive statements or images that promote restful sleep. Hypnosis can include controlled breathing techniques that help reduce stress levels and activate the relaxation response, helping the person to fall asleep. Examples of hypnotic input scenarios that can help with relaxation and improve sleep quality include a hypnotherapist entering an image of a natural landscape, such as a calm beach or forest, to promote relaxation [146].

Using positive instructions to create positive associations with sleep. Instructions for consistent relaxation of different muscle groups can help bring the body into a state of relaxation. Recommendations for focusing on calm and regular breathing to reduce physical and mental stress. Notably, the effectiveness of hypnosis can vary from person to person, and for best results, it is recommended to work with an experienced hypnotherapist or use audio recordings of hypnosis designed to improve sleep.

Hypnotherapy can be used in auxiliary treatment for a number of mental disorders as an additional method to the main psychological approach. Notably, hypnosis is not the only form of treatment for these disorders and is often used in combination with other methods. In ***psychology,*** hypnotherapy is used to treat various psychiatric disorders, such as depression, anxiety, phobias, and PTSD [147]. Hypnotherapy is used to treat various mental disorders, offering patients an alternative method of influencing the consciousness and subconscious. Hypnotherapy can be used to work with the subconscious mind and change negative thinking patterns associated with *depression*. Hypnosis techniques can help reduce symptoms of depression and support emotional well-being. Hypnotherapy can help reduce the level of *anxiety* by working with subconscious stress processing mechanisms and learning relaxation techniques. People suffering from *phobias* can use hypnotherapy to change the associations and stereotypes associated with fears. This can help the patient to gradually tame the situations that previously caused anxiety. Hypnotherapy can be part of a comprehensive treatment approach to the treatment of PTSD. It can help patients relax, reduce nervous system hyperactivity, and work to process traumatic events. Hypnotherapy is used to *improve motivation and help overcome addictions* such as tobacco, alcohol, or food [148]. Hypnotherapy can be an effective adjunct to traditional treatments for mental disorders. However, it is important to contact a qualified therapist who has experience in hypnotherapy and understands the specific needs of each patient.

Hypnosis is also used in mental health treatment, specifically for PTSD: a clinical study of war veterans showed that hypnotherapy decreased nightmare, anxiety, and flashback symptoms by 40% over an eight-week period, and MRI scans showed increased connectivity in emotion-regulating areas, specifically between the prefrontal cortex and amygdala [149]. These results indicate that hypnosis not only alleviates PTSD symptoms but also induces neuroplastic changes that are beneficial for emotional regulation. In addition to PTSD, hypnosis is used to treat a variety of psychiatric conditions, such as anxiety disorders, phobias, and depression. Hypnosis helps patients manage anxiety by using subconscious stress-processing mechanisms and relaxation techniques. For phobias, it can alter associative fear responses, allowing patients to approach previously distressing situations with less anxiety.

Hypnosis can help with relaxation and reduce anxiety. During hypnosis, relaxation responses in the body can be activated. The hypnotherapist can use special instructions and imaginative techniques to encourage the patient to relax the physical body deeply. Hypnosis can affect the activity of stress responses in the brain, reducing the production of stress hormones and reducing the activity of the nervous system. Hypnotherapy can help change negative cognitive patterns and shift attention to positive, light images and thoughts [150]. Hypnosis can help the patient develop greater awareness of their condition and teach them to adopt more effective strategies to control anxiety. Hypnosis can be used to create a positive emotional state that promotes a sense of calm and reduces anxiety. These mechanisms may vary depending on the individual characteristics and reactions of a particular person to hypnosis. It is important that hypnosis be performed by a qualified professional who can consider the needs and circumstances of each patient.

Hypnosis is used to help people quit smoking, alcohol, and other bad habits through various strategies and approaches. Hypnotherapy has proven effective in the modification of behaviour, particularly with regard to smoking cessation. A study comparing hypnosis to behavioural therapy found that hypnosis was twice as effective in helping smokers to quit. Neuroimaging data revealed that hypnosis-induced changes in the insular cortex were associated with reduced cravings [151]. These findings underscore the role of hypnosis in altering deeply ingrained behavioural patterns, making it a valuable tool in addiction therapy. Furthermore, hypnosis has been employed in the context of weight management by influencing eating behaviours and promoting mindful eating. The application of techniques such as positive reinforcement, visualisation, and subconscious reprogramming has been instrumental in fostering healthier dietary habits and portion control.

A hypnotherapist can use hypnosis to increase a person's internal motivation to give up bad habits. This may include highlighting the positive aspects of health, improving the quality of life, and achieving personal goals. Hypnotherapy can help change the psychological associations connected with smoking or alcohol consumption. This may include reprogramming thinking and changing the perception of these habits. Hypnosis can help develop the ability to manage stress and anxiety without using bad habits. This is important because many people resort to smoking or drinking alcohol as a means of relaxation [152]. Hypnosis can help introduce positive changes to the subconscious mind, which contributes to the persistent rejection of bad habits. This can include positive affirmations and images that support a new, healthy lifestyle. Each person is unique, and the hypnotherapist can customise the approach according to the individual needs and characteristics of each client. It is important to emphasise that hypnotherapy is commonly used as part of a wide range of therapeutic strategies. This method may be effective for some people, but it does not always work for everyone. Treatment of alcohol or drug addiction should consider the individual characteristics and needs of the patient. Notably, the effectiveness of hypnosis can vary depending on the individual characteristics and feedback of a particular person. Hypnotherapy is used as part of a comprehensive approach to problem-solving and can be successful, especially with the appropriate attitude and desire of the patient to give up bad habits.

Hypnosis can help in *developing a positive self-perception and improving self-esteem.* Hypnosis can influence the development of positive self-perception and improve self-esteem through various mechanisms that include psychological and subconscious aspects [152, 153]. Hypnotherapy can be used to work with the subconscious mind and reprogram negative beliefs or self-stereotypes that can hinder positive self-perception. Hypnosis can promote a sense of inner strength and increase self-esteem by influencing one's mental state and attitude toward oneself. Hypnosis can be used to activate positive emotions and inner feelings of happiness, contentment, and peace, which can help improve one's overall emotional state and self-perception. Hypnotherapy can help a person understand their values, goals, and unique qualities, which contribute to the formation of positive self-determination. Hypnosis can be used to create positive psychological images that help improve interaction with the world around the person and achieve success. Hypnosis can teach relaxation and self-regulation techniques that can be useful for managing stress and improving overall well-being. Achieving a positive effect from hypnosis depends on the individual characteristics of the person, the depth of the hypnotic state and the professionalism of the hypnotherapist. Some people may notice substantial improvements in their positive self-perception after hypnotherapy, while others may experience less impact.

Developing creativity and concentration with the help of hypnosis is also possible. Hypnosis can be used to stimulate and increase creativity, concentration, and memory. The use of hypnosis to stimulate creativity, concentration, and memory is the subject of research, and the results may vary depending on the individual. However, some studies and clinical cases point to the possibility of a positive effect of hypnosis on these cognitive functions. Hypnosis can stimulate creative processes by reducing conventional barriers and opening up opportunities for new ideas and approaches [148]. During hypnosis, a person can be encouraged to experience greater freedom of thought and experiment with their thoughts and ideas. Hypnosis can be used to increase one's level of attention and concentration on a specific task or information. Especially in hypnotherapy sessions, the focus may be on specific aspects of tasks or training materials.

Hypnotherapy can be used for *memory stimulation* by activating and improving access to certain memory traces [152]. Individual hypnosis sessions can help bring hidden or forgotten memories to the surface. While there are indications that hypnosis can have some effect on cognitive function, it is important to understand that effectiveness can vary depending on personal characteristics and the nature of the tasks. In addition, it largely depends on the quality of hypnotherapy sessions and individual responses. Examples of specific cases and the effectiveness of hypnosis in this context may vary, as it is individual and depends on many factors. Applying hypnosis requires the knowledge and experience of a specialist, such as a doctor or licensed hypnotherapist. In each case, it is important to consider the individual characteristics of the patient and determine the most effective way to use hypnosis in a particular context.

Hypnosis-related techniques have a substantial impact on a person's mental and physical health, opening up new opportunities for treating and supporting patients. In a medical context, hypnosis has been shown to be effective in treating pain, stress, and psychosomatic illnesses. It finds its use in psychotherapy to address a variety of mental health problems, including anxiety disorders, phobias, and depression. Research on the effects of hypnosis on cognitive functions such as memory and attention points to potential benefits in these areas. In addition, the use of hypnosis in psychology allows a better understanding of the work of the human mind and the interaction of different levels of consciousness. Brain activity analysis and conectomy open up new horizons for scientists, identifying the mechanisms that determine hypnotic states. Despite the fact that this method causes certain discussions and controversy, its application in medicine and psychology continues to develop, opening up new opportunities for complementing and optimising the therapeutic process.

Chapter 3

Neural Networks and Their Role in Perception

This chapter is aimed at exploring complex interactions in neurophysiology that underlie the phenomenon of hypnosis. Modern research in the field of neurobiology provides a unique opportunity to better understand how the functioning of neural networks in the brain affects the perception of hypnotic states. This chapter uncovers key aspects of the structure and functions of neural networks and examines how these processes are related to the human brain's ability to become deeply immersed in hypnotic states. By examining the relationship between neural networks and hypnosis, people's understanding expands not only to the physiological aspects of these processes but also to their psychological and clinical dimensions.

3.1 Basic Concepts of Neural Networks

The basics of neurophysiology cover the structure and functions of the nervous system, particularly focusing on neurons as the basic biological nodes that transmit information in the form of nerve impulses. A neuron is the basic unit of the nervous system. It has a nucleus, a cell body, and dendrites and interacts with other neurons through axons. Information transmission between neurons occurs at

Mechanisms of Brain Function and Adaptive Potential
Liana Spytska

ISBN 978-981-5129-91-5 (Hardcover), 978-1-003-71697-6 (eBook)
www.jennystanford.com
DOI: 10.1201/9781003716976-4

synapses, where electric charges move at a speed of 0.001 seconds and can repeat 500 times per second [154]. Understanding these basics helps to explore the fundamental principles of brain function and its interaction with other systems of the body. Neurons, which are the basic building blocks of the nervous system, interact through axons and dendrites, responsible for receiving, transmitting, and processing information. Synaptic connections play a key role in signal transmission between neurons using neurotransmitters. Nerve impulses, which are electrical signals, convey information along nerve fibres. The central nervous system (CNS), which includes the brain and spinal cord, interacts with the peripheral nervous system (PNS), serving other parts of the body. Neurophysiology analyses reflexes, automatic reactions to stimuli that occur without the involvement of conscious thought. These basic concepts allow for a better understanding of the interaction of the nervous system with the environment and the performance of various functions, such as perception, movement, and higher mental functions. The neural networks of the human brain are complex systems that use electrochemical signals to process information and ensure various aspects of the organism's functioning.

The nucleus of a neuron is its central part, located in the cell body, and is responsible for the key process of generating the energy needed for the nerve cell to function. The nucleus is where the synthesis of biological molecules occurs, and the transmission of genetic information is regulated, which is critical for ensuring the cell's vital activity. This structurally functional nucleus is the site where the coordination and control of the most important biological processes that ensure the neuron's life activity occur.

A neuron, which is the fundamental functional unit of the nervous system, has a complex structure that includes a cell body and different types of processes (Fig. 13). Among these are dendrites, which are short and densely branched processes playing a crucial role in receiving nerve impulses to the cell body. It is important to note that the number of dendrites in a neuron can vary from one to many. Each nerve cell also has a single long process known as an axon, through which nerve impulses are directed away from the cell body. Axons can be impressively long, reaching several tens of centimetres. They group together into bundles, forming nerves. A particular feature of some processes, especially axons, is that they are

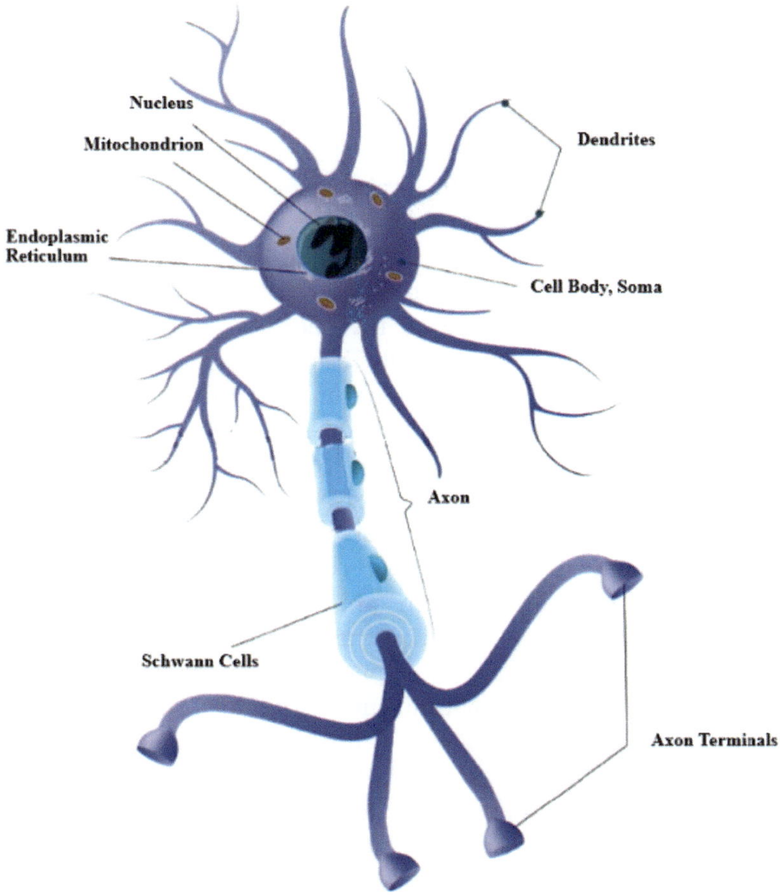

Figure 13 Neuron structure.

covered with a myelin sheath. Myelin, which is composed of proteins and fats, performs a vital function in protecting and insulating axons, thereby enhancing the efficiency of nerve signal transmission. In the CNS, clusters of myelinated axons form the white matter of the brain and spinal cord. The axon terminal or synaptic bouton is located at the end of the neuron's axon and is divided into terminals, which function to connect with other neurons and form a synapse. In these terminal boutons, neurotransmitters are concentrated in small storage structures called vesicles. A synapse is the transfer of these vesicles from the terminal boutons of one neuron to the dendrites of another. This process is a critical moment in the connection between neurons and their interaction in the signal transmission system of the nervous system.

Conversely, short processes such as dendrites and the neuron bodies themselves do not have a myelin sheath, giving them a grey colour. Their clusters form the brain's grey matter, highlighting the diversity of structures necessary for the nervous system's optimal functioning [155]. Dendrites are structures that function as the "hands of the neuron". These processes form small branches that emerge from different parts of the neuron's soma or cell body. Typically, there are many dendritic branches whose sizes depend on the neuron's function and location. The primary function of dendrites is to receive stimuli from other neurons. Dendrites have a complex structure, and their surface is equipped with a large number of specialised receptor sites known as dendritic spines. These structures receive chemical signals from other neurons through synaptic connections. The received signals are transformed into electrical impulses, which are then transmitted along the dendrites and to the cell body of the neuron. One of the features of dendrites is their ability to be plastic, meaning they can change according to external environmental conditions and introduced stimuli. This makes them a key component in adapting neuronal networks to new conditions and learning. Dendrites also play an important role in forming synapses, where they connect with the axons of other neurons. This contact allows for the transmission of information between neurons and determines the effectiveness of communication in neuronal networks.

The cell body is a fundamental part of a neuron and is defined by its important physiological significance, as it is where the synthesis and generation of most neuronal molecules occur. The cell body serves as a central platform for various biochemical processes that support the life and functions of nerve cells. This space is where proteins, lipids, and other molecules necessary for the normal functioning of the neuron are synthesised. The main biochemical processes occurring in the cell body include DNA replication and expansion, transcription and translation of genetic information, and the synthesis of enzymes and other important molecules necessary for neuronal functions. This area is essential for providing energy and resources to other neuronal structures, such as axons and dendrites, and for performing key functions related to the transmission and processing of information in the nervous system. Thus, the cell body can be considered the most important part of the neuron, where processes that define its life and functionality occur.

Mitochondria in neurons, as in other cell types, play an important role in energy production and other cellular processes. However, the importance of mitochondria for neurons is particularly great due to their activity and functional characteristics. Neurons, especially those in active parts of the brain, consume a significant amount of energy for information transmission and maintaining electrical activity. Mitochondria in neurons generate adenosine triphosphate (ATP), which is the main source of energy for the cell. Neurons interact with calcium ions for signal transmission. Mitochondria regulate calcium levels, which is important for proper neuronal function. They may also participate in inducing apoptosis if the cell is damaged or programmed cell death is needed [156]. Mitochondria in neurons play a role in various metabolic processes, such as the oxidation of fats and glucose, and can also affect the structure and shape of neurons, providing them with the necessary energy to maintain their functions and form. The ability of mitochondria to regulate their number and placement in the cell according to its needs at a particular time is important. Such dynamics of mitochondria are crucial for ensuring the efficient function of neurons under conditions of variable activity and energy needs.

Neuroglia, along with neurons, play an important role in the functionality of the brain and spinal cord (Fig. 14). Neurons, being extremely specialised cells, cannot independently provide all the necessary functions for their own survival. Thus, a neuron surrounds itself with other cells that perform these vital functions. This type of cell includes astrocytes, oligodendrocytes, and microglia, each of which performs unique tasks to support and protect neurons. Astrocytes are cells that play an important role in the functioning of the nervous system. Their main function is to support and provide the necessary resources for neurons, enabling them to work efficiently. Astrocytes provide the supply of oxygen and glucose needed for the energy needs of neurons. In addition, astrocytes actively participate in the removal of excess chemicals and toxins that may arise from the activity of neurons. This process, known as neuroglial metabolism, is important for maintaining the health and functions of nervous

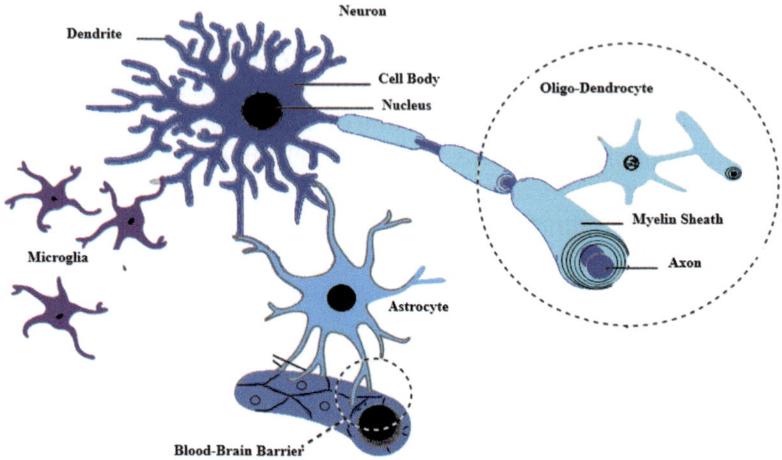

Source: Z. Bayram-Weston et al. [143].

Figure 14 The structure of a neuron and its environment.

tissue. Overall, astrocytes are an important component of neuroglia, providing comprehensive support and protection for neurons in the CNS. Oligodendrocytes are cells that perform an important function in the nervous system, namely the formation of myelin sheaths. Myelin is a fatty insulation that surrounds the axons of neurons, forming myelin sheaths. This myelination process is key to optimising the transmission of nerve impulses. Oligodendrocytes interact with axons, where they create and maintain myelin sheaths. Myelin helps retain and direct electrical signals along axons, increasing the speed of nerve impulse transmission. This process is critical for effective communication between neurons and the rapid response of the nervous system to stimuli. Thus, oligodendrocytes play a key role in supporting and optimising the functioning of the nervous system, ensuring efficient signal transmission through myelinated fibres. Microglia are specialised cells that perform an important function in the immune system of the nervous system. One of their key roles is to protect the brain and spinal cord from infectious agents and other harmful factors. Microglia interact with the surrounding environment, detecting and neutralising potential threats to nervous tissues. In addition, microglial cells play an important role in the tissue repair process [157]. When nerve structures are damaged or injured, they are activated to restore and regenerate the affected areas. This repair mechanism supports the integrity of the nervous system and ensures its functionality under external influences or injuries. Thus, microglia perform complex tasks in the immune and regeneration systems, helping to preserve and restore the optimal functioning of nervous tissues.

Schwann cells are key components of the PNS and also play an important role in myelinating axons. The process of myelin formation is facilitated by oligodendrocytes in the CNS and Schwann cells in the PNS. This complex biomaterial performs an essential function in supporting the efficiency of neural signal transmission and protection from external influences. The function of Schwann cells is crucial for the effective operation of the PNS. Each Schwann cell specialises in covering a specific segment of an axon. This creates separated segments known as myelinated compartments and interspersed segments that lack myelin or have only partial coverage. Such a structure is referred to as a Node of Ranvier. Myelinated compartments and Nodes of Ranvier play an important role in accelerating the transmission of neural signals, ensuring efficient and rapid information processing. Schwann cells also perform support and protective functions for axons, which are necessary for their integrated function in the neural signal transmission system.

Neuroglia is essential for maintaining homeostasis, as well as for protecting and repairing neural tissues. It interacts with neurons, providing optimal conditions for their function and recovery while also performing protective and regulatory functions that help maintain the normal physiological state of the nervous system. Neuronal regeneration is the process of restoring or replacing damaged or lost neurons in the nervous system. Neurons are responsible for transmitting information within the body, and their recovery can be crucial for restoring the functions of the nervous system after injuries, diseases, or other damage. Some regenerative processes in the nervous system are associated with the activation of stem cells. These are special cells that can differentiate into various cell types, including neurons. In some cases, neurons may try to repair damaged axons or grow new ones. Other neurons may attempt to compensate for lost functions by altering their structure and function, and this aspect of regeneration may be related to changes in synapses, where signal transmission between neurons occurs. Despite these potential mechanisms, neuronal regeneration in the CNS is significantly limited compared to other tissues. Some neurogenic methods and research focus on developing strategies to support or enhance neuronal regeneration in conditions of damage.

Neurons, as the primary functional units of the nervous system, exhibit various types depending on their shape and functions [158]. Based on their structure, neurons are divided into several main types:

1. *Multipolar neurons.* This type of neuron has many dendrites and one axon. They are the most common type and are typically found in the CNS.
2. *Bipolar neurons.* Such neurons have two processes—one dendrite and one axon. They are often located in specialised structures, such as sensory organs, where they are responsible for transmitting signals to the CNS.
3. *Pseudounipolar neurons.* This type has one process that divides into two branches, one functioning as dendrites and the other as an axon. These neurons are usually part of the PNS, transmitting signals from receptors to the CNS.
4. *Anaxonic neurons.* These neurons lack visible axons and specialise in transmitting information between dendrites and other cells in their immediate vicinity.

This classification by structure reflects the various forms that neuronal architecture can take, responding to diverse functions in the nervous system (Fig. 15). There is also systematisation of neurons depending on the role they perform:

1. *Sensory neurons.* This type of neuron is responsible for transmitting signals from organs (receptors) to the CNS, represented by the spinal cord and brain. The bodies of such neurons are located outside the CNS, usually in nerve nodes or ganglia. Nerve nodes are clusters of nerve cell bodies outside the CNS.
2. *Motor neurons.* Motor neurons are responsible for transmitting impulses from the spinal cord and brain to muscles and internal organs. They determine the transmission of signals from the CNS to the organs of movement. By executing the commands of the brain, motor neurons play a key role in regulating movements and organ functions.

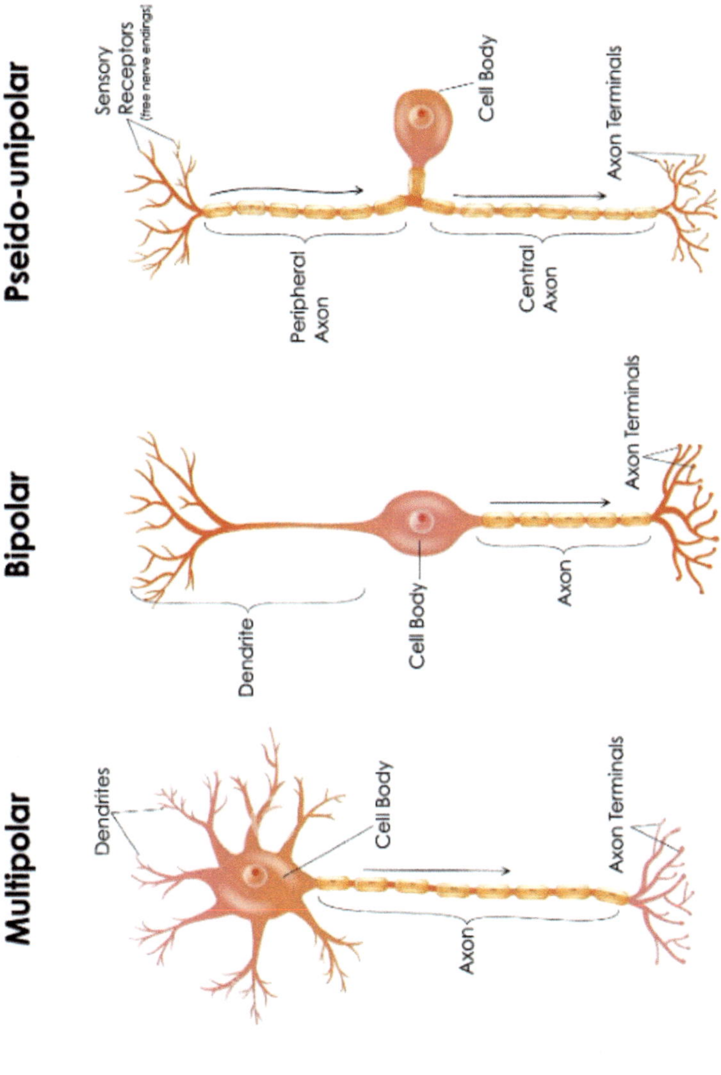

Figure 15 Structural classification of neurons.

3. *Interneurons*. Interneurons are responsible for connecting sensory and motor neurons, facilitating coordination and interaction between these two types (Fig. 16). This relationship is implemented through synaptic contacts in the spinal and brain. Interneurons are located within the CNS, meaning both their bodies and processes remain within the brain. This component is an important part of the nervous system, coordinating information between different types of neurons for effective system functioning [159].

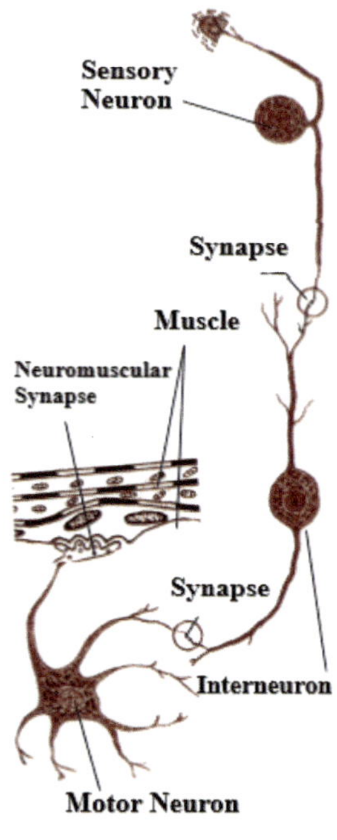

Figure 16 Interaction of different neuron types.

The neural network is a high-level structure that combines neurons of the CNS and ganglia of the PNS. These neurons interact through synapses, forming complex chains, ensembles, and systems to perform a variety of physiological functions. Each neural network consists of a group of neural chains, which are morpho-functionally related neurons. The interaction of neurons occurs through neural processes, such as long axons and short, branched dendrites. Due to the enormous number of neurons and synapses in the network, the structure becomes extremely complex. For example, the human brain can contain up to 100 billion neurons, each connected to a large number of other neurons through 1000 trillion synapses. Synapses, the points of contact between neurons, can be axo-dendritic chemical synapses, where the transmission of impulses occurs chemically or electrically through neurotransmitters and ions.

Neural networks define the functions of the brain and spinal cord, and understanding them has a significant impact on the development of artificial neural networks (ANNs) in artificial intelligence. These systems are widely used in mathematics and computer science, providing effective models of the nervous system's functioning. The functionality estimates of the brain are compared to a computer operating at a speed of 1 trillion bits per second, and the memory capacity can vary from 1 to 1000 terabytes.

The human brain is capable of complex cognitive functions such as perception, movement and higher mental abilities, and is comprised of interconnected neurons that facilitate these functions. The capacity for the brain to adapt to neurological damage through reorganisation and alternative pathway development is a demonstration of the complexity of these networks. An illustrative example is that of a 48-year-old woman who was diagnosed with cortical blindness as a result of extensive damage to the occipital lobe, but who exhibited the ability to perceive moving objects, a phenomenon referred to as Riddoch syndrome [160]. Functional imaging studies revealed activation in motion-sensitive regions, such as the middle temporal visual area (MT/V5), indicating that these areas can function independently of the primary visual cortex. This case underscores the brain's remarkable capacity for neural adaptation, supporting the idea that alternative neural pathways can compensate for damaged regions.

Scientists who first considered the role of neural networks in mental activity made significant contributions to the development of this field. Francis Bain (1873) and William James (1890) identified the interaction between brain neurons as the basis of cognitive processes. Bain believed that any activity activates a specific set of neurons, and the repetition of this activity strengthens the connections between neurons, forming memory. His concept has gained recognition, and modern research confirms its significance, particularly the brain's ability to process multiple tasks simultaneously. James's theory, similar to Bain's, included the element of electrical current between neurons as a factor in forming memory. James believed that electrical current could cause the process of memorisation without the need to establish connections between neurons for each action or event [161]. Although the theories of Bain and James were initially met with scepticism, modern scientific discoveries confirm the importance of their ideas in understanding brain function and memory mechanisms.

British physiologist Sherrington conducted important experiments in 1898 to test James's theory. During his research, he directed an electric current along the spinal cords of rats. Deviating from the expected strengthening of the current predicted by James's theory, Sherrington found that the electric current, on the contrary, gradually weakened over time. Sherrington's results played a crucial role in the further development of the theory of habituation. These experiments made a significant contribution to understanding how neural networks can adapt to constant influences and other factors affecting brain electrical activity.

In 1949, Donald Hebb proposed the idea known as "Hebb's rule", which significantly changed the understanding of neural network dynamics. According to this hypothesis, neural connections are not unpredictable but can be improved every time they are activated. Hebb's rule suggests that the learning process is not static and is not the result of fixed neuron characteristics; instead, it is determined by changes in their interaction over time. The basic concept of Hebb's rule is that groups of neurons tend to activate together in response to external stimuli. This synchronised activity can continue even after the stimulus is removed. Thus, an event that triggers the simultaneous oscillation of a group of neurons is fixed in memory as directed neurons. Hebb's discovery laid the foundation for further research in neurobiology and learning, and his ideas became key to understanding the mechanisms of neuroplasticity and memory formation.

A neural network is the basic element of the human nervous system. The mentioned basic concepts, such as neurons, synapses, nerve impulses, and components of the central and PNSs, reveal important aspects of the functioning of the brain and spinal cord. Neurons are the basic building blocks of the nervous system, and their interaction through axons and dendrites determines the transmission and processing of information. Synaptic connections have proven to be a key element where signal transmission occurs using neurotransmitters. Neural networks perform various functions, such as perception, information processing, movement control, regulation of internal organs, and higher mental functions, including reflexes, such as automatic reactions to stimuli that occur without the involvement of conscious thought. Scientists are currently examining phenomena in neurophysiology, such as ependymocytes and neuroglia, which play an important role in covering the brain ventricles and ensuring the functioning of the nervous system. Special attention is paid to myelin, which provides protection and insulation of axons and enhances the efficiency of signal transmission. Neural networks play an important role in the functioning of the human body and in exploring the principles of brain function.

3.2 Interaction of Neurons and Synapses

The interaction between neurons and synapses is a critical aspect of the functioning of the nervous system. Synapses act as sites for transmitting signals from one neuron to another, allowing information to pass from one point to another in the nervous system. The structure of the synapse, which includes the presynaptic terminal, synaptic cleft, and postsynaptic terminal, is detailed and important for understanding the mechanisms of signal transmission in the nervous system. The synapse is a structural and functional unit in the nervous system responsible for transmitting signals between neurons. Synapses play a key role in ensuring communication in the nervous system; their function is based on complex processes of chemical and electrical nature. The synapse consists of three main components.

The presynaptic terminal is the terminal part of a neuron's axon, which is the output process and contains specialised structures such as vesicles. These vesicles store neurotransmitters, chemical substances such as acetylcholine, serotonin, and noradrenaline. When a neuron is excited, the nerve impulse reaches the presynaptic terminal, causing the rapid release of these neurotransmitters into the synaptic cleft [162]. This process occurs through the fusion of vesicles with the membrane of the presynaptic terminal. Thus, the presynaptic terminal plays a key role in signal transmission between neurons by releasing neurotransmitters that initiate signal transmission across the synaptic cleft.

The synaptic cleft is an important part of the nervous system, representing the space between the presynaptic terminal (usually the terminal part of an axon) and the postsynaptic terminal (usually the dendrite of another neuron). This space, with a very small width of a few nanometres, plays a crucial role in signal transmission from one neuron to another. The synaptic cleft is where the key process of signal transmission occurs, where neurotransmitters are released from vesicles in the presynaptic terminal. These neurotransmitters cross the synaptic cleft and interact with receptors on the postsynaptic terminal, initiating further signal transmission in the neural network (Fig. 17). Despite its small size, the synaptic cleft is a key element that helps ensure efficient and accurate information transmission in the nervous system.

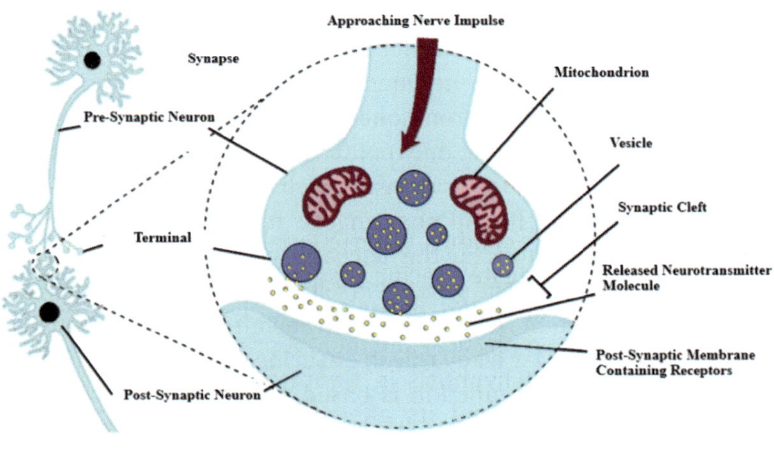

Source: Z. Bayram-Weston et al. [143].

Figure 17 Synapse.

The postsynaptic terminal is part of another neuron equipped with receptors for neurotransmitters. These receptors, which are specific to particular neurotransmitters, respond to their release by the presynaptic terminal during the transmission of the nerve impulse. In the case of successful interaction, this causes electrical changes in the postsynaptic terminal. These changes can trigger signal transmission along the neural network, which defines a key moment in neurocommunication and the performance of functions in the nervous system.

Given this detailed structure, one can understand how precisely the signal transmission system between neurons is organised and how it is regulated by chemical transmissions in synapses. Such detail is important for understanding the functioning of the nervous system and the interaction between its various components. Synapses play a critical role in the nervous system, allowing signals to be transmitted from one neuron to another. This process is the basis of neural communication and learning, and understanding the structure of the synapse helps explain the principles of the nervous system's functioning.

The process by which a nerve impulse reaches the presynaptic terminal is the result of membrane depolarisation of the neuron and subsequent signal transmission along the axon. The nerve impulse is generated in the neuron's cell body, after which membrane depolarisation occurs. This means that the distribution of ions on both sides of the membrane changes, including sodium and potassium ions. Depolarisation results in the generation of an electrical signal. Under the influence of depolarisation, the electrical charge moves along the neuron's axon. This process, known as an action potential (AP), allows the nerve impulse to move from the neuron's cell body to its terminal end or axon. When the AP reaches the presynaptic terminal (terminal end), specialised proteins and ion channels in the terminal's membrane are activated [163]. Under the influence of depolarisation in the presynaptic terminal, exocytosis occurs—the volumetric release of synaptic vesicles containing neurotransmitters. These neurotransmitters (such as acetylcholine) are released into the synaptic cleft. Neurotransmitters cross the synaptic cleft and interact with receptors on the membrane of the postsynaptic terminal, causing electrical changes in this terminal. This occurs through the opening of ion channels or changes in membrane permeability. Thus, the nerve impulse reaches the presynaptic terminal due to depolarisation, activation, and release of neurotransmitters into the synaptic cleft and further signal transmission through interaction with the postsynaptic terminal. If the interaction is successful, this leads to electrical changes in the postsynaptic terminal, initiating signal transmission.

Synapses are divided into several types, and each type can use different neurotransmitters. Chemical synapses are the most common type of synapses in the nervous system. Their peculiarity lies in the fact that signal transmission occurs through chemical neurotransmitters released by the presynaptic terminal and interacting with receptors on the postsynaptic terminal. Acetylcholine, serotonin, glutamate, gamma-aminobutyric acid (GABA), dopamine, and others are the most common neurotransmitters in chemical synapses. Chemical synapses are located in the synaptic cleft between the presynaptic and postsynaptic terminals. The neurotransmitter is released by the presynaptic terminal and interacts with receptors on the postsynaptic terminal.

Neurotransmitters are chemical substances released by neurons to transmit signals to other cells through synapses. Different neurotransmitters are produced in different parts of the nervous system and are used for different functions. For example, acetylcholine is localised in the central and PNSs and is released upon excitation of motor neurons and at motor synapses. Acetylcholine transmits signals between nerves and muscles and is also used in the CNS for various functions, including attention and memory. Serotonin is localised in the brain and spinal cord and is released upon activation of corresponding pathways in the brain to regulate mood, sleep, appetite, and other physiological processes. Dopamine is released in the brain upon experiencing reward or pleasure and is responsible for regulating mood, reward, and movement coordination. GABA is released when neuronal activity decreases in the CNS and inhibits nerve activity, helping to calm the nervous system. When neurons in the CNS are excited, glutamate is released, which is the main excitatory neurotransmitter responsible for signal transmission in the CNS. Norepinephrine, which is released during stress or high levels of activity, controls the sympathetic part of the autonomic nervous system and affects the regulation of sleep, mood, stress response, and other physiological processes. Adrenaline also controls the sympathetic part of the autonomic nervous system. It is released during stress and extreme situations, increases heart rate, raises blood pressure, and mobilises the body to respond to stress. Endorphins in the CNS are released during stress and pain and act as natural analgesics, reducing pain perception and improving mood. Adrenocorticotropic hormone (ACTH) is released during stress in the pituitary gland of the brain and regulates the function of the adrenal cortex, thereby stimulating the secretion of corticosteroids, which affect metabolism and the immune system. Dopamine, which is released during novelty and pleasant events, governs the CNS, especially in the areas of learning and motor control. It is responsible for motor control, learning, desire, and mood regulation. These neurotransmitters interact with each other and ensure the normal functioning of the nervous system, regulating various processes in the body.

In electrical synapses, there is no release of neurotransmitters, but signals are transmitted directly through connections known as gap junctions. These structures facilitate the direct exchange of molecules and other signalling substances between adjacent cells. This special type of cellular connection allows for the transmission of both electrical and chemical signals along nerve and muscle tissues. The main structure of a gap junction includes two components: gap junction channels, which are channels or openings between adjacent cells that create a direct path for the transfer of ions, molecules, and other signalling substances from one cell to another, and gap junction proteins, which are special protein structures that form openings in cell membranes and allow them to unite, creating a channel between cells. Gap junctions are important for the coordination of cell actions in various tissues, such as the heart and nervous system. They enable fast signal transmission and coordination of activity between adjacent cells, which is crucial for the overall functioning of the organism. In electrical synapses, neurotransmitters do not play a role, as the impulse is transmitted faster through direct electrical connections between neuron membranes.

Ionotropic synapses are a type of synapse in neuronal systems where signal transmission occurs through the movement of ions across ion channels. This type of synapse is characterised by rapid and direct signal transmission, as ions can quickly pass through ion channels, causing changes in the electrical potential of the postsynaptic neuron. The presynaptic terminal releases neurotransmitters into the synaptic cleft, but the actual release of neurotransmitters and their interaction with the postsynaptic terminal occur via ions, typically calcium (Ca^{2+}). Under the influence of a nerve impulse, the presynaptic terminal allows calcium ions (Ca^{2+}) to pass through open calcium channels. As a result of the calcium influence, vesicles containing neurotransmitters (such as glutamate or GABA) merge with the cell membrane of the presynaptic terminal, leading to the release of neurotransmitters into the synaptic cleft and their interaction with receptors on the postsynaptic terminal membrane. The electrical activity of the postsynaptic terminal can change depending on the nature of the neurotransmitter and the type of receptor [164]. For example, glutamate can induce excitatory effects, while GABA can induce inhibitory effects. The electrical signal spreads along the postsynaptic neuron, transmitting information within the neural network. Therefore, calcium plays a key role in the release of neurotransmitters and the initiation of signal transmission in the synaptic cleft. Located on the membrane of the postsynaptic terminal, ion channels open or close upon interaction with neurotransmitters. This leads to changes in ion permeability through the membrane. This type of synapse is fast and efficient, and its function is determined by the direct impact of ions on the electrical activity of the neuron.

Metabotropic synapses are a special type of chemical synapse in neural systems. Their main difference from other types of synapses is that they use metabotropic receptors for signal transmission, which involves a more complex process than ionotropic synapses. Metabotropic receptors are located on the surface of the postsynaptic terminal, and their activation triggers a complex sequence of intracellular events. The metabotropic synapse also consists of three main components that interact to transmit signals between neurons—the presynaptic terminal, which contains vesicles with chemical substances; the synaptic cleft, where the transmission of signals takes place using chemical intermediaries or neurotransmitters; and metabotropic receptors. Metabotropic receptors are one type of receptor found on cell membranes, including neurons in the nervous system. They are named for their signalling mechanisms involving a more complex cascade of biochemical events compared to ionotropic receptors. Metabotropic receptors interact with proteins known as G-proteins located on the inner side of the cell membrane. Activation of a metabotropic receptor leads to the activation of G-proteins, which, in turn, affect the operation of enzymes or open other signalling pathways. These events cause changes in cellular function. The same metabotropic receptor can interact with different effectors, leading to an impact on cellular processes over an extended period of time compared to the rapid effect of ionotropic receptors. Many neurotransmitters, such as serotonin, acetylcholine, and glutamate, interact with metabotropic receptors, regulating the function of neurons and signal transmission in the nervous system. Metabotropic receptors play a key role in many physiological and pathological processes, including neurotransmission, learning, and memory, and they regulate numerous cellular functions. The signalling process in a metabotropic synapse begins with the release of a neurotransmitter into the synaptic cleft under the influence of a nerve impulse. Neurotransmitters interact with metabotropic receptors on the postsynaptic terminal. Activation of metabotropic receptors triggers a series of biochemical reactions inside the cell. This signalling pathway may involve the work of G-protein-linked proteins and secondary messengers. Metabotropic receptor activation results in a change in cell activity, including changes in electrical activity, energy expenditure, and other biochemical responses. Thus, metabotropic synapses use a more complex signalling mechanism compared to ionotropic synapses, which expands the spectrum of possible effects on the postsynaptic neuron.

Neuromuscular (motor) synapses play a key role in transmitting signals from the nervous system to muscles, causing their contraction. Neuromuscular synapses are located at the junction between nerve tissue (the axon of a motor neuron) and muscle tissue. The ends of motor neuron axons form synaptic ends that are located near muscle fibres. When the neuron receives a stimulus from the CNS, an impulse is radiated along the axon of the motor neuron. At the ends of the synaptic ends of motor neurons, there are vesicles that contain the neurotransmitter acetylcholine. When the impulse reaches the end of the axon, the vesicles release acetylcholine into the synaptic cleft. Acetylcholine crosses the synaptic cleft and binds to receptors on the surface of the muscle fibres. The binding of acetylcholine to receptors causes changes in the membrane potential of muscle fibres, leading to the generation of an electrical potential in the muscle, known as an AP. The AP causes the opening of calcium channels in muscle fibres. Calcium causes muscle contraction, which ultimately results in a specific movement or action. This process is called neuromuscular transmission and is a key mechanism that allows the nervous system to control body movements. For example, the neurobiology and neurochemistry of muscle spasms occur at the level of neuronal and chemical interactions in the muscle system. It is important to pay attention to neurotransmitters, receptors, and synaptic mechanisms that regulate muscle activity. Neurotransmitters, such as acetylcholine, GABA, and others, play a key role in transmitting signals from neurons to muscles through synapses. Violations in the production, release, or interaction with the receptors of these neurotransmitters can lead to uncontrolled muscle contractions. The neurochemical aspects of muscle spasms are also associated with various molecules and ions, such as calcium, potassium, and sodium, which affect the electrophysiological properties of muscle cells. In this context, considering the neurochemistry of muscle spasms encompasses understanding the role of these substances in maintaining normal muscle function and their impact on muscle tone.

Each of these types of synapses plays a crucial role in signal transmission in the nervous system, and various neurotransmitters influence different aspects of neuronal interaction. Synapses play a key role in determining the strength and area of signals in the nervous system. They are also the site of signal modulation, adaptation to environmental changes, and dynamic regulation of neuronal activity. Researchers examine the relationship between neural networks and the sensation of pain to understand the mechanisms of pain and develop effective treatment methods [165]. However, the connection between these two aspects is complex and involves various aspects of neurophysiology and the psychology of pain. Pain arises when specialised neurons, called nociceptors, respond to stimuli such as injury or inflammation. These neurons transmit pain signals through neural networks in the spinal cord and brain. Various parts of the brain, particularly those responsible for perception, processing, and response to pain, include the somatosensory cortex, thalamus, and amygdala. These areas cooperate to create a sense of pain and determine its intensity and significance. Neuroplasticity, or the ability of neurons to change their activity and connections, affects the development of pain and can be key in the formation of chronic pain. The impact of neuroplasticity is analysed in the context of understanding the mechanisms of chronic pain and developing therapeutic strategies. Emotional state, psychological factors, and individual characteristics can influence the perception and sensation of pain. For example, stress, anxiety, or depression can enhance the feeling of pain. Pain transmission can occur through different types of synapses in neural networks. The two main types of synapses that play a key role in the transmission of pain impulses are chemical synapses and neuromuscular (motor) synapses. Chemical synapses transmit signals using chemical substances, and nociceptors (specialised neurons that respond to pain) interact with other neurons through chemical synapses to transmit pain [166]. Neuromuscular (motor) synapses are responsible for transmitting signals from nerves to muscles. Signal transmission in muscles can also be associated with the sensation of pain, especially during movement or mechanical impact on tissues. Other types of synapses, such as electrical synapses, can also participate in pain transmission, but chemical synapses are the main mechanism for signal transmission in neurons specialised in the sensation of pain. It is important to consider that pain transmission is a complex interactive process, and the role of synapses can vary depending on specific conditions, types of pain, and the context of stimulation. Understanding and exploring the connection between neural networks and pain is an active area of research that can lead to new methods of treatment and approaches to pain management, especially chronic pain. The phenomenon of phantom limb pain following limb loss can be viewed in terms of the operation of neural networks, representing a complex process of reorientation of neural pathways responsible for body sensation and pain. With the loss of a limb, there is a redistribution of brain maps that represent specific body areas, which can lead to changes in pain perception and processing. Changes in neuromodulation and the operation of the CNS caused by limb loss can affect pain intensity and its perception in the absent area. Phantom pain may be related to the preservation of a "phantom map" in the brain, where neurons that previously responded to sensations in the lost area continue to generate pain signals. Brain plasticity, which can adapt its structure and function under the influence of changes in input signals due to limb loss, plays a crucial role in this process. In addition, changes in peripheral nerves can influence the occurrence of phantom pain, as damage or changes in their operation can affect pain transmission and act as a factor in the formation of phantom pain. Understanding the neurophysiological aspects of phantom pain helps develop more effective treatment methods and support for people who encounter this type of pain after limb loss.

The process of impulse transmission in synapses and its connection to memory is a complex and intriguing aspect of neurobiology. In memory formation, neurons receive stimuli in the form of nerve impulses or electrical signals. The primary mechanism of interaction occurs when synapses are activated, and neurotransmitters such as acetylcholine, glutamate, serotonin, and dopamine interact with receptors on the postsynaptic neuron's membrane. Neurotransmitters are released into the synaptic cleft under the influence of stimuli (electrical signals) transmitted through synaptic connections. This leads to the activation of receptors on the postsynaptic terminal membrane. Neurotransmitters bind to receptors, causing electrical changes in the membrane of the postsynaptic neuron, which can generate an effective potential. The activation of these receptors causes electrical and chemical changes in the postsynaptic neuron, which can lead to changes in synaptic plasticity and memory formation. Therefore, stimuli in memory formation involve the transmission of nerve impulses and the release of neurotransmitters in synapses. These trigger a response in the postsynaptic neuron and potentially lead to changes in synaptic connections associated with information retention. Repetition of activity in certain synapses can lead to the strengthening of connections between neurons. This process, known as long-term potentiation (LTP), can cause stable changes in synaptic connections, shaping memory. At the molecular level, LTP may involve changes in gene expression, protein synthesis, and activation of various signalling pathways. Thus, impulse transmission in synapses plays a key role in the formation of memory through the strengthening of synaptic connections. LTP and other molecular mechanisms become the basis for storing information in neural networks that support memory and learning. An important mechanism for learning and memory formation in the nervous system is synaptic plasticity, i.e., the ability of synapses to change according to neuron activity [167]. Understanding the interaction between neurons and synapses is key to unravelling the principles of information processing in the nervous system and understanding its functions.

The interaction of neurons during stress is a complex phenomenon that involves physiological, psychological, and neurobiological aspects of the organism. Stress is a complex response of the organism to various factors perceived as threatening or causing tension. These factors can be external (e.g., work pressure, social situations) and internal (e.g., emotional state, diseases). During stress, chemical processes occur that are defined by the release of hormones such as cortisol, adrenaline, and norepinephrine. These hormones are activated by the sympathetic nervous system and affect the levels of neurotransmitters such as serotonin and dopamine, which can lead to changes in mood and emotional state. In parallel, the neurology of stress examines the impact of stress factors on the nervous system and the brain. Stress activates the sympathetic nervous system, leading to the release of adrenaline and norepinephrine, preparing the body for the "fight or flight" stress response. This also triggers the release of cortisol, which regulates physiological processes and interacts with the immune system. The overall response of the body to stress is determined by traces of chemical changes in the interaction of various systems, such as the endocrine, nervous, and immune systems. These processes modulate emotional state, physiological functions, and behaviour, creating a complex picture of the response to stressors. The body responds to stress by releasing hormones such as cortisol and adrenaline, which prepare it for action, increasing energy levels and providing a rapid reaction. Stress can activate the sympathetic nervous system, leading to increased arousal and readiness for action. Stress hormones can affect neural networks, modifying their activity and plasticity. Neurotransmitters released in synaptic clefts can modulate the stress response by affecting receptors in the brain and other parts of the nervous system. Chronic stress can affect synaptic plasticity, determining their ability to change, which is crucial for brain function and mental health. Overall, stress interacts with physiological and psychological aspects, as well as with neural networks and neurotransmitters, determining the body's overall response to challenges [168]. Chemical processes involve the release of various substances and changes in hormone and neurotransmitter levels, determining the physiological and psychological response to stress. The neurology of stress examines the impact of stress factors on the nervous system and brain, revealing the neurobiological aspects of the stress response, which include the activation of the sympathetic nervous system and the release of hormones that regulate various aspects of physiology and psychology.

Neuronal synaptic interactions in pain, memory, and stress are similar in that they all involve the transmission of signals between neurons in the form of electrical impulses and chemical signals at synapses, where the connection between neurons occurs. However, they differ in the specific processes that take place at these synapses and the molecular mechanisms that determine their function. For instance, in pain, neuronal synaptic interactions are often associated with specialised neurons such as pain neurons, nociceptors. They use chemical mediators like pain-sympathetic peptides and glutamate to transmit pain signals along pain pathways to the CNS. In pain, neurotransmitters and pain receptors, such as nicotinamide receptors or TRP channels, play a crucial role, responding to various stimuli and causing the sensation of pain. Regarding memory formation and retention, synaptic interactions include changes in the strength and efficacy of synapses, induced by LTP and other processes. The difference is that memory is associated with the modification of synaptic connections through processes such as LTP and long-term depression (LTD), which allow neurons to retain information for extended periods. Neurotransmitters like glutamate and dopamine play key roles in these processes. Stress affects the activity of the sympathetic nervous system and the release of stress hormones such as adrenaline and cortisol, which can also modify synaptic connections and influence neurotransmitters. However, stress involves additional aspects, such as the impact on the amygdala (a part of the brain responsible for processing emotions) and the regulation of stress hormones, which can affect cardiovascular function and immunity. All these reactions are closely interconnected, and exploring their mechanisms can help better understand the relationship between neurons and synapses in various processes [169].

The interaction between neurons and synapses is a key aspect of the functioning of the nervous system, determining the transmission of signals within the body. Synapses, which are the sites of connection between neurons, play the role of transmitting information through electrical and chemical signals. In the process of signal transmission through synapses, complex chemical and electrical events occur. Electrical impulses, rapidly propagating along the axons of neurons, reach the synaptic cleft. Here, they initiate the release of chemical mediators, such as neurotransmitters, from the presynaptic membranes. These neurotransmitters cross the synaptic cleft and activate receptors on the postsynaptic membrane, thereby transmitting the signal to the next neuron. This interaction between neurons and synapses determines various physiological and psychological functions, including perception, movement, memory, emotion regulation, pain sensation, and stress responses. However, this process is dynamic and depends on various factors such as the concentration of neurotransmitters, the structure of synaptic connections, and neuronal activity. Understanding these mechanisms of neuron and synapse interaction is crucial for a comprehensive understanding of the nervous system's operation and can have practical significance for developing treatments and approaches to various neurological and mental disorders.

3.3 Explaining How Neural Networks Contribute to Information Perception and Processing

A biological neural network (BNN) is a large system of neurons located in the brain and spinal cord of the CNS and in the ganglia of the PNS. These neurons interact and functionally integrate through synapses, forming complex neural circuits that collectively play a role in performing specific physiological functions. A neural network includes groups of neural circuits, which constitute morpho-functionally connected groups of neurons. Each neuron can interact with numerous other neurons using its neural processes: long axons and short, branched dendrites. The number of neurons and connections in the network can be astonishingly large—for example, the human brain consists of approximately 100 billion neurons, each of which can be connected to tens of thousands of others and interact through billions of synapses. The primary site of interaction between neurons is the synapses, which are the contact points between neurons. A synapse can be chemical in nature, being axo-dendritic, where impulse transmission occurs via neurotransmitters, or it can be electrical, with ions transmitting signals from one neuron to another. This complex signal transmission mechanism allows neurons to communicate effectively with each other. In addition, the number of possible synapses and interactions enables the network to perform a variety of functions and provides a high level of adaptability and plasticity. Mapping the brain's neural networks using diffusion MRI provides valuable insights into the structure and functioning of these complex systems, which is key for further research in neurobiology and artificial intelligence (Fig. 18).

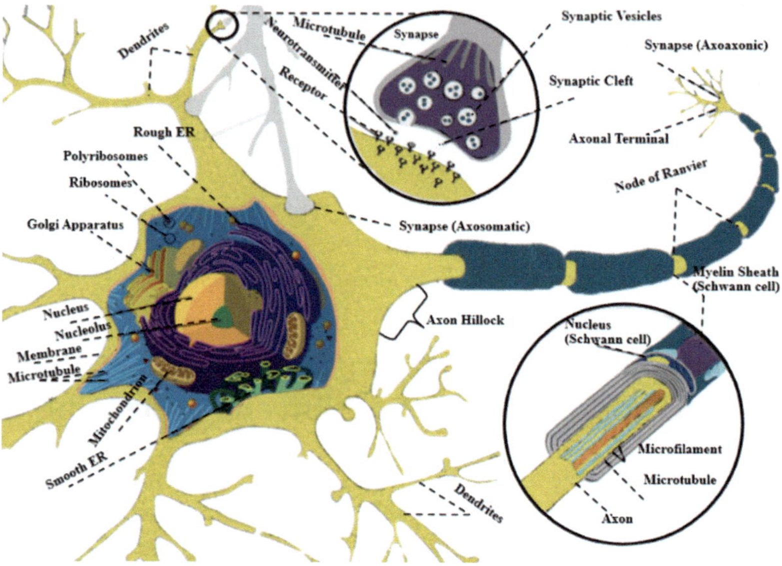

Figure 18 Neural system of the human body.

BNNs in the brain play a crucial role in perception and information processing through a complex set of electrochemical signals. Neurons interact through synapses, which are the points of contact between them. Starting with the reception of electrical signals on a neuron's dendrites, if the signal reaches the threshold, the neuron generates an impulse that propagates along the axon. Synapses play a key role in transmitting information between neurons. When the impulse reaches the end of the axon, neurotransmitters are released into the synaptic cleft. This activates receptors on the dendrites of the neighbouring neuron, causing a new electrical signal to appear in it. Thus, information is transmitted from one neuron to another, forming the basis of neural information processing in the brain. Neural networks operate in parallel, integrating a large number of signals. This allows the system to respond to various stimuli and perform complex cognitive functions such as perception, thinking, and decision-making. These networks, with a large number of neurons and synapses, create complex systems for signal transmission and processing.

Neural networks' parallel processing allows the brain to integrate many signals at once, facilitating complex cognitive processes like perception, reasoning, and decision-making as well as effective responses to stimuli. Research has indicated that deep learning algorithms and ANNs are based on these biological concepts. Convolutional neural networks (CNNs), for example, are employed in visual identification tasks to replicate the hierarchical processing of the visual cortex of the brain. According to MRI studies, CNNs' feature recognition layers match the brain's early visual regions (V1, V2), while their higher layers match the inferotemporal cortex, which is in charge of object recognition [170]. The way that brain structures interpret visual stimuli is better understood thanks to this analogy between biological and artificial networks.

Neural networks play an important role in language processing in addition to visual perception. For instance, Google's Neural Machine Translation (GNMT) system achieves translation accuracy that is almost human by identifying semantic patterns in text. The notion that artificial models can replicate biological systems in language understanding is further supported by the intriguing finding that bilingual people's MRI scans reveal that human language processing stimulates similar neural networks [171]. These results imply that similar structural and functional principles govern the information processing of neural networks, whether they are artificial or biological.

Biological neurons interact with the external environment and other parts of the body through dendrites, which function as input receivers, gathering signals from other neurons or external sources. Dendrites play a key role in receiving and processing these signals, transmitting them to the neuron's body. This stage is the first in the process of information processing in biological neurons. When the sum of received signals exceeds a certain threshold, the neuron generates an output signal, which is then transmitted along the axon to other neurons or muscle fibres. This interaction mechanism allows neurons to transmit information within the nervous system. Signals transmitted through dendrites can be electrical, causing a dendritic potential. Previously, it was believed that the AP was mainly formed at the initial segment of the neuron's axon, which has a low threshold of sensitivity. However, recent scientific findings, based on accumulated data over the past decades, indicate that AP can also occur in dendrites [172]. This type of dendritic AP, to distinguish it from axonal AP, has been termed the "dendritic spike". Dendritic spikes represent a phenomenon where spontaneous APs arise locally in the dendrites of many types of neurons in the CNS under conditions of focal dendritic stimulation. There are three main types of dendritic spikes, differing by the class of active conductors underlying them: Na^+, Ca^{2+} (or plateau spikes), and N-methyl-D-aspartate (NMDA) spikes. These spikes, although they have different electrical properties and dendritic morphologies, are similar to axonal APs in their basic properties. As a nonlinear phenomenon, dendritic spikes can effectively resist the influence of other synapses, hindering the integration of additional synaptic impulses. This occurs through the local summation of synchronised clusters of incoming signals to the dendrite. Dendritic spikes can be generated either in isolation from the soma (local spikes) or coincide with back-propagating axonal APs. If a dendritic spike is strong enough, it can reach the neuron's soma and lead to the generation of a somatic-axonal AP or a burst of APs. The existence of dendritic spikes expands the computational repertoire of the neuron, providing a functional association of local incoming signals. They allow the enhancement of distant synaptic impulses, which otherwise would not affect the somatic potential, and also influence the stimulation of synaptic plasticity.

In the late 1950s and early 1960s, it was believed that dendrites of neurons were passive extensions that merely integrated incoming impulses. However, isolated evidence, particularly from Chinese neurophysiologist Hsiang-Tung Chang, who discovered that dendrites could generate APs, raised doubts about this theory. Despite this, most scientists continued to believe that APs were generated only in axons. Subsequently, in the mid-1950s, during the recording of hippocampal neuron responses, new evidence of dendritic spike generation was obtained. The detection of electrogenic responses in dendrites supported this idea. However, most scientists remained critical of these results. It was only in the late 1980s and early 1990s that scientists began to accept the view that dendrites not only transmit information but also modify and store it. The work of Stuart and Sakmann, who used whole-cell patch-clamp electrodes, provided convincing evidence for the existence of dendritic spikes and voltage-gated ion channels in dendrites. Since the 1990s, observations of dendritic spike recordings in hippocampal and neocortical neurons, both in slices and *in vivo*, have made their ability to generate regenerative spikes via voltage-gated sodium (Na^+) and calcium (Ca^{2+}) channels increasingly clear [173]. Simultaneous recordings of somatic and dendritic currents revealed the generation of dendritic spikes in layer 5 pyramidal neurons of the cortex and hippocampus. Dendritic spikes occurred before somatic ones and propagated in both types of neurons through strong synaptic excitation. Purkinje cells of the cerebellum, hippocampal interneurons, mitral cells of the olfactory system, and retinal ganglion cells also exhibited dendritic spikes separate from somatic APs. Previously, these spikes might have gone unnoticed due to their ability to be masked by the somatic spikes they induce. Typically, the generation of dendritic spikes is associated with the active conductance of Ca^{2+} or Na^+ channels or both. Depolarisation of the distal apical dendrite activates Ca^{2+} conductances, leading to the generation of Ca^{2+} spikes. In layer 5 pyramidal neurons of the cortex, a unique influence of dendritic Ca^{2+} spikes on the propagation of synaptic potential and their active impact on the soma has been discovered. Many synapses are located far from the soma, and without a compensatory mechanism, they would not affect the somatic potential. However, even active currents can significantly attenuate the incoming synaptic signal in these neurons. There is a special feature in the pyramidal neurons of the upper cortical layer—a dedicated area on their dendritic branches where spikes can occur. This is necessary for the complex computational tasks they perform. This dendritic zone contains many calcium channels, allowing them to generate a special type of spikes known as calcium spikes. These calcium spikes affect the signal transmission process within the neuron. When they are activated, they can lead to repetitive discharges and thus influence the excitation of neighbouring regions of the neuron. This is especially important for neurons in the cerebral cortex, as it can be a key mechanism for transmitting information to other neurons. In Purkinje cells, the generation of dendritic spikes is simpler because they use different types of channels. They generate dendritic APs controlled by calcium channels, allowing them to create broader spikes. Another important feature is the ability to generate spikes in any part of the dendrites, which is different from axons, where spike generation occurs only at specific points. This makes it possible to influence the accuracy of signal transmission and the use of information within the neuron. Thus, understanding the generation of dendritic spikes is crucial for revealing the intricacies of the brain's computational functions.

There is a consensus among researchers about the existence of dendritic spikes, but debates about their role and impact on neural signalling continue. Some studies indicate that Na^+ and Ca^{2+} spikes occur in dendrites but often remain within the dendrites, having a limited strength. Recordings from the axon, soma, and apical dendrite show that the axonal potential always precedes the somatic one, even if the dendritic spike occurs before the somatic potential. This suggests that dendritic spikes do not always reliably reach the soma and axon, and the depolarisation of the soma by dendritic spikes is usually small [175]. However, in some cases, dendritic spikes, together with coincident EPSPs, can cause enough depolarisation to generate an AP in the axon, while in other cases, they cause only a small depolarisation in the soma and axon, limited to an isolated dendritic spike.

Unlike the "all-or-nothing" APs in the axon or soma, dendritic spikes vary in strength and are regulated by prior synaptic activity. Some researchers suppose that dendritic spikes influence the generation of APs in the axon but are not a decisive factor. They also believe that dendritic spikes play a role in synaptic integration, but the final impact on AP generation should be attributed to the axon. Dendritic spikes differ from those occurring in the part of the neuron called the axon. Dendritic spikes can detect simultaneous signals in different parts of the neuron, aiding in information processing. There are various types of dendritic spikes, which arise due to different ion channels, such as Na^+ and Ca^{2+}. These spikes affect signal processing in the dendrites and interaction with other parts of the neuron. In pyramidal neurons, which are crucial for processing complex information, it has been found that dendritic spikes are generated upon depolarisation, with a slightly higher threshold than regular signals. Dendritic spikes occur through the activation of ion channels such as Na^+ and K^+. Their generation requires strong depolarisation since dendrites have fewer Na^+ channels than axons. The activation of NMDA receptors, which conduct current only during depolarisation, is essential, leading to the generation of dendritic spikes. Dendritic spikes can influence signals in the dendrites and even extend to the soma, enhancing the efficacy of incoming signals. Even if they do not reach the soma, they can amplify synaptic potentials and accelerate the growth of somatic spikes, ensuring precise and synchronised output signals.

Backpropagation AP (bAP) is a signal that travels from the neuron's body to its dendrites. It plays a significant role in the interaction between parts of the neuron and the influence of axonal AP on dendritic spikes and vice versa. bAP helps coordinate neuronal activity and enhances the impact of synaptic signals on the neuronal impulse. The extent of bAP spread in dendrites varies for different types of neurons and even among the dendrites of a single neuron. For instance, in hippocampal pyramidal neurons, bAP can spread in the dendrites with little or no reduction in amplitude. In Purkinje cells in the cerebellum, bAP spreads passively [176]. The density of ion channels, particularly Na^+, in dendrites also affects bAP propagation. Blocking these channels reduces the amplitude of bAP. The shape and branching of dendrites also play a role in bAP propagation. Therefore, bAP performs a crucial function in the internal interaction of the neuron, ensuring the coordination of its various parts and optimising the influence of synaptic signals. bAP is important for propagation in dendrites due to the width of the somatic AP, which affects its attenuation during propagation. For example, neurons in the substantia nigra have a broader somatic AP, which allows them to effectively release dopamine through backpropagation in their dendrites. This is also characteristic of mitral cells in the olfactory bulb. The passive electrical properties of dendrites play a significant role in the propagation or attenuation of bAP. Reducing membrane resistance can decrease the amplitude of bAP. Inhibitory neurons can also affect bAP by suppressing it and influencing dendritic function. bAP impacts synaptic potentials and their integration, and high-frequency potentials can promote the generation of dendritic Ca^{2+} spikes, affecting synaptic plasticity. bAP can also interact with synaptic potentials, influencing their amplitude and integration in the dendrites. As bAP penetrates the dendritic tree, it increases intracellular calcium concentration through the activation of Ca^{2+} channels. This can affect synaptic integration by modulating NMDA receptors and activating dendritic K^+ conductances that impact dendritic function. bAP can temporarily halt synaptic integration by returning the dendritic membrane potential to its original state or fixing excitatory synaptic conductances. In neurons without bAP or where it cannot propagate to distant dendrites, local dendritic spikes can influence synaptic integration, affecting the somatic potential. Along with bAP, the axon becomes an active participant in integrative processes, influencing not only the target neuron but also its own activity. This opens up a unique interaction between excitatory and inhibitory neurons in the CNS. The backpropagation of somatic AP has a significant impact on dendrites and the neuron as a whole, providing them with complex internal feedback mechanisms. This expands the understanding of the neuron and its role in networks.

Despite the discovery of Ca^{2+} spikes, the problem remained of how distant synaptic impulses could overcome the threshold for generating such spikes. It was found that the initial segments in cortical pyramidal cells could interact through the apical dendrite, aiding signal propagation. This allows even weak signals to contribute, as once they reach the local threshold, they activate other initial segments.

The synchronous coincidence of distant synaptic inputs in the dendrites with a bAP within an appropriate interval (20–30 ms) lowers the threshold for generating a dendritic Ca^{2+} spike and can induce a burst of APs in the soma. This phenomenon has become known as "BAC firing" (backpropagation-activated Ca^{2+} spike firing).

The generation of a burst of APs is based on the fact that a dendritic Ca^{2+} spike can depolarise the axon's initial segment, leading to the generation of complex dendritic potentials. This mechanism characterises "BAC firing", which requires precise temporal synchrony similar to the timing precision of spike timing–dependent plasticity (STDP) at distant synapses. "BAC firing" can be blocked by an inhibitory input lasting approximately 400 ms [177]. This indicates the possibility of using inhibition to control the connections between the zones of near and distant spike generation and to suppress burst discharges associated with "BAC firing".

Unlike conventional neural networks, the layer 5 pyramidal neuron of the cortex has a unique structure that allows it to process two streams of information separately and then combine them. This reduces network complexity. Due to the special features of pyramidal neurons, they can be used to associate different information pathways within a few milliseconds. Matthew Larkum hypothesised that the association of external and internal data in the cortex might occur at the cellular level with the help of pyramidal neurons, avoiding complex neural networks and their self-organisation. In the cortex, signals to pyramidal neurons come from various sensory regions, but the "BAC firing" mechanism is associated with a new type of neural code that helps detect the coincidence of input signals. According to this model, information from the external world first reaches neurons to establish a baseline spike discharge rate and then interacts with internally originating information through feedback. Larkum's model suggests that the associative function of the cortex can be realised through the simple structure of pyramidal neurons, helping to solve the homunculus problem. The ability of dendrites to generate spikes, which facilitates the integration of synaptic inputs in the CNS, is also important.

Dendrites in neurons play a crucial role in information processing. It has been found that the propagation of APs in dendrites depends on their type. Furthermore, the geometry of dendrites influences the spread of APs. For example, some types of dendrites facilitate the efficient propagation of APs with a small number of ion channels. An important factor is the number of branching points, which affects the efficiency of propagation. Changes in dendritic morphology, such as branching, can occur due to various processes, such as development, plasticity, and learning. Modelling and research confirm that dendritic morphology is a key factor in understanding the differences in AP propagation among different types of neurons. This is important not only for understanding the process of AP propagation itself but also for the overall understanding of how neurons process information. Furthermore, when creating ANNs, it is important to consider dendritic morphology for effective modelling of brain activity.

In chemical synapses, when a signal reaches the terminal branch of a dendrite, the release of chemical substances, neurotransmitters, into the synaptic cleft occurs. These substances cross the synaptic cleft and interact with receptors on the membrane of the next neuron, transmitting the signal. If the signal is electrical, a dendritic potential occurs—a brief change in the electrical potential on the surface of the dendrites. A strong signal can cause a dendritic potential, which, if it reaches the threshold, will generate an AP. The AP propagates along the axon, representing short-term changes in the membrane potential and transmitting the signal along the neuron. When the AP reaches the end of the axon, neurotransmitters are released in the synapse, affecting the dendrites of other neurons and transmitting the signal further into the nervous system. Thus, an incoming signal that reaches the threshold value causes the generation of an AP and signal transmission along the axon, allowing neurons to interact and transmit information in the nervous system.

An electrical impulse, known as an AP, propagates along the axon of a neuron at high speed when the output signal reaches the threshold level in the trigger zone. This impulse results from the opening and closing of ion channels, which allow ions to move across the axon membrane. The AP is an irreversible process and moves quickly, allowing information to be transmitted rapidly along the nervous system [178]. After reaching the end of the axon, the AP causes the release of neurotransmitters in the synapse, affecting the dendrites of other neurons and causing signal transmission. After the AP propagates, the axon membrane returns to its initial state through repolarisation, preparing the neuron for a possible next impulse. This process allows the nervous system to interact effectively with the environment and respond to various situations. When the AP reaches the end of the axon, it triggers the release of chemical mediators, neurotransmitters, into the synapse—the contact point between the axon of one neuron and the dendrites of another. The synaptic gap, the space between the end of the axon and the dendrites of the next neuron, fills with neurotransmitters that transmit the signal. When the AP reaches the end of the axon, it causes synaptic vesicles containing neurotransmitters to open. These substances are released into the synaptic gap and interact with receptors on the dendrites of the next neuron. This causes changes in the membrane potential of the second neuron. The neurotransmitters are broken down by enzymes or are restored in synaptic vesicles for future use. This process allows information to be transmitted from one neuron to another, coordinating the functions of the body.

After being released into the synaptic gap, neurotransmitters cross the synaptic cleft and interact with receptors on the recipient's dendrites, located on the membrane of the next neuron. These receptors are sensitive to specific neurotransmitters, and binding to these chemicals changes the membrane potential of the second neuron. If these changes reach a critical level, they can trigger a new dendritic potential or impulse, which propagates along the neuron. This new dendritic potential becomes a signal that is transmitted along the neuron, and this process of signal transmission from one neuron to another is repeated. The receptors on the recipient's membrane have specific sensitivities to certain neurotransmitters, ensuring the precision of signal transmission and the perception of specific information in the nervous system. This interaction process plays a key role in signal transmission and supports the functions of information perception and processing in BNNs.

After release into the synaptic gap, neurotransmitters cross the synaptic cleft and interact with receptors on the dendrites of the recipient's next neuron. These receptors respond to specific neurotransmitters, and their binding with the chemicals causes changes in the membrane potential of the second neuron. If these changes reach a critical level, they can trigger a new dendritic potential or impulse, which propagates along the neuron. This new dendritic potential becomes a signal transmitted along the neuron, and this process of signal transmission from one neuron to another is repeated. The receptors on the recipient's membrane are specific to certain neurotransmitters, ensuring the accuracy of signal transmission and the perception of specific information in the nervous system [179]. This interactive process plays a key role in signal transmission and supports the functions of perception and information processing in BNNs. When a neuron reaches a certain level of excitation, it generates an AP, which propagates along the axon. During this process, the neuron's dendrites integrate incoming signals, and if the cumulative effect reaches a threshold value, an AP is triggered. This occurs on the dendrites or cell body, where signals are integrated. After generating an AP, an electrical charge propagates along the axon due to the movement of ions and changes in electrical potential. When the impulse reaches the end of the axon, it stimulates the synaptic ending adjacent to the next neuron or effector organ. At the synaptic ending, neurotransmitters are released, crossing the synaptic cleft and interacting with receptors on the membrane of the next neuron. This can cause the generation of a new dendritic potential in the second neuron, continuing the signal transmission in the nervous system. This process of signal transmission along neurons and through synapses is a fundamental basis for the functioning of the nervous system and information processing in the organism. The generation of an AP occurs when a threshold level of excitation is reached, which may be the result of summing dendritic potentials coming from other neurons or external stimuli. The AP is transmitted along the axon, forming a wave front that carries information. Upon reaching the synapse, the AP causes the release of neurotransmitters into the synaptic cleft. These neurotransmitters cross the synaptic cleft and interact with receptors on the membrane of the next neuron, causing the emergence of a new dendritic potential. Dendritic potentials are integrated into the receiving neuron, and if their sum exceeds the threshold, a new dendritic potential is generated. APs are transmitted along axons, transmitting information through synapses to other neurons and triggering responses in muscles, learning, the regulation of internal organs, and other functions. This key mechanism ensures the coordination of various functions and responses to external stimuli, making it important for the perception, processing, and transmission of information in the organism.

Signals are transmitted through neuronal networks through synaptic connections, forming a continuous cycle of information transmission and processing. When an AP covers the distance to the synapse—the connection point between the axon of one neuron and the dendrites of another, an important process occurs. Here, the release of chemical mediators, such as neurotransmitters, into the synaptic cleft is stimulated. These neurotransmitters cross the synaptic cleft and communicate with receptors on the membrane of the dendrites of the recipient neuron, belonging to the next neuron. If the interaction at the synapse leads to the generation of a new dendritic potential in the receiving neuron, this can trigger an AP in this new neuron. The resulting AP travels along its axon, and this process can repeat, transmitting the signal to other neurons in the system. At each synapse, information is processed and integrated. Depending on the nature of the signals and the characteristics of the synaptic connections, both amplification and inhibition of signals can occur. This interaction between neurons and the transmission of signals through synapses allow the creation of complex neural networks that perform various functions in the body. This cycle of signal transmission and processing through neural networks responds to external stimuli, coordinating the body's actions and reactions.

Transmission and processing of information in BNNs are characterised by their dynamics and plasticity, which means their ability to change according to experience and learning. The dynamics of neural networks are manifested in their ability to adapt to changes in the environment or the internal states of the body. For example, neurons can change their sensitivity to stimuli depending on the context. Neural network plasticity, particularly synaptic plasticity, allows them to change through experience or learning. With repeated use or training, the strength of synaptic connections can change, affecting signal transmission. Neurotransmitters and other chemical signals play a key role in modulating the function of neural networks. These substances can influence the sensitivity of neurons to signals, facilitating adaptation to changes in the environment or stimuli. A broader view of neuron interaction indicates that they form systems where interaction can generate emergent properties and synergy. This makes BNNs adaptive and capable of performing functions in various conditions and contexts.

The complex system of neural networks in the nervous system allows the body to adapt to environmental changes and perform various functions, including response to stimuli, memory retention, and regulation of different aspects of the organism. This highly developed network of neurons not only ensures rapid and coordinated responses to external influences, helping the body adapt to changes in the environment, but also ensures survival. Neural networks play a key role in storing and reproducing information accumulated through experience and learning. This allows the organism to use accumulated knowledge to make effective decisions in various situations. Neural networks are also responsible for coordinating various body functions, including movement, homeostasis, hormone secretion, and other vital processes [180]. The ability of neural networks to learn and adapt allows the organism to interact with changing conditions and adjust its behaviour and physiology. These networks enable the perception, processing, and integration of information from different sources, creating a holistic understanding of the surrounding world. In summary, these functions form a highly efficient and adaptive system that ensures the most critical life processes in the body.

BNNs in the brain play a key role in the perception, processing, and transmission of information in the body. Neurons, as the basic building blocks of these networks, interact through synapses, creating a complex process of signal transmission. From the perception of signals through dendrites to the generation of dendritic potentials, transmission of APs along axons, and synaptic transmission through neurotransmitter release, this process forms the basis of nervous system function. Particularly important are the dynamics and plasticity of BNNs. These properties allow the system to adapt to environmental changes and perform various functions such as responding to stimuli, retaining memory, and coordinating body functions. Examining the functioning of neural networks has the potential to enhance understanding of the principles of the nervous system and improve strategies for treating disorders related to this aspect of biology. However, further research is needed to uncover more details about the processes in BNNs and their impact on human physiology and psychology. When analysing the effects of hypnosis on BNNs, changes in brain area activity and neuron interaction are observed. The hypnotic state can lead to reduced activity in areas responsible for analysing stimuli and thoughts, which aids in concentration. Changes in pain perception, memory impact, and other functions are also characteristic of hypnosis. However, the exact mechanisms of these changes require further investigation, as the response to hypnosis is individual. Hypnosis is used in medicine and psychotherapy to treat various conditions, but its effectiveness and precise mechanisms remain the subject of ongoing scientific discussion.

BNNs in the human brain are remarkably diverse and possess unique power and flexibility. They support several essential functions, making them a key element in information processing and interaction with the environment. Sensory processing is an important aspect, encompassing the complex process of handling information from various senses, such as vision, hearing, and touch. Sensory organs, such as the eyes, ears, and touch system, convert external stimuli into neural signals, which are transmitted to the brain for processing. This process includes the handling of information through neural networks, integrating diverse information, and issuing responses. The brain integrates information from different sources, creating a comprehensive perception of the environment. Sensory processing is also linked to cognitive aspects such as attention and perception, allowing a person to identify important information. Furthermore, it determines motor responses, such as movements and actions, performed according to the perceived information. All of this makes sensory processing a key element in forming a complete perception of the world and adapting the body to changes in the environment.

Motor coordination refers to the ability to control body movements to achieve specific goals. This complex process involves the interaction of various systems in the body, such as the nervous, muscular, and joint systems. The successful execution of physical and daily tasks requires good motor coordination. Developing this skill is important at all stages of life, especially in early childhood. Games, sports exercises, and other activities can help improve motor coordination. Training includes enhancing the accuracy, speed, and strength of movements. Disruptions in motor coordination can arise from various causes, such as neurological diseases, injuries, or developmental disorders. However, training and rehabilitation can help restore this skill in individuals with such issues. In summary, motor coordination is a key aspect of an active and healthy lifestyle, helping to effectively perform various physical and motor tasks.

Memory in the brain performs important functions: storing, retaining, and retrieving information. This process can be divided into three stages. The first stage, encoding, involves transforming external information, such as words while reading, into an internal code for storage. The second stage is storage, where the encoded information is held in different parts of the brain. There are short-term and long-term memory, each retaining information for different durations. The final stage is retrieval (recovery), when the brain recalls previously stored information for use in current tasks or situations. A highly developed memory enables effective information retrieval. Memory is classified based on the duration of storage:

- sensory memory (short-term storage of sensory impressions, such as vision or hearing);
- short-term memory (limited retention of information for a short period, for example, a phone number for a few seconds);
- long-term memory (prolonged storage of information, sometimes for a lifetime).

Memory capacity depends on various factors, such as age, health status, and stress. Engaging in mental exercises, maintaining a healthy lifestyle, and participating in developmental activities can improve memory.

Learning and adaptation represent the critical ability of systems and organisms to change based on experience and interaction with the environment. This universal process encompasses several key aspects. Sensory organs gather information from the surrounding world, which the brain processes to recognise patterns and create perceptions. The ability to perceive, process, and understand information forms the basis of learning processes, encompassing the retention of new knowledge in short-term and long-term memories. The ability to learn includes enriching memory with new information, retaining necessary knowledge, and using it in appropriate situations. An essential aspect is the ability to adapt to the social environment, which involves developing communication skills, emotional intelligence, and the ability to interact with other members of society. The ability to perform various actions and movements, which is part of learning and adaptation, integrates the development of motor skills in the brain and muscles [181]. Continuous interaction with the environment allows systems to adjust their activities based on received information and experiences. Learning and adaptation are integral parts of the development of organisms and systems, providing them with the ability to function effectively in a changing environment.

Emotions and mood regulation embody a vast epic of the human inner world, where each moment is a delicate scene in the dance between brain structures, neurotransmitters, and psychological factors. The emergence of emotions is the result of complex information processing by the brain, including the evaluation of events and situations. Brain mechanisms such as the amygdala, hypothalamus, and cerebral cortex interact, creating grand emotional responses. Neurotransmitters, including serotonin, dopamine, norepinephrine, and others, play key roles in transmitting signals between neurons, regulating mood and emotions. Physiological changes, such as facial expressions, body posture, and voice modulation, serve as expressions of these emotions, whether conscious or automatic reactions. Mood regulation involves strategies and mechanisms that help alter or maintain emotional states. This encompasses psychological and behavioural strategies as well as social interactions. External stimuli, cultural factors, and communication with others influence this grand spectacle of emotions and mood regulation, shaping how this process forms within the human body. This unique epic becomes an essential component of mental health and adaptation to the surrounding world.

In the depths of the body's internal regulation lies a complex array of autonomous functions, encompassing a highly organised system that regulates key processes such as breathing, thermoregulation, and heartbeat. These reflections of physiological phenomena are structured by a complex interconnection of neurons located in various segments of the nervous system, forming an extensive network to ensure precise and coordinated functioning. Breathing, as an expressive function of the delicate respiratory centre in the medulla, aptly manages the rhythm and amplitude of respiratory movements [155]. This centralised neural network responds to the parameters of blood gas composition, activating the activity of respiratory muscles to ensure efficient gas exchange. Thermoregulation is governed by the activity of the hypothalamus, located in the brain. This highly specialised regulator interacts with information about temperature changes in the environment and the internal state, initiating necessary mechanisms to maintain a stable thermal balance. Heartbeat is subjected to detailed neural control, where the sympathetic and parasympathetic branches of the autonomic nervous system closely cooperate. Sympathetic activation accelerates heart contractions, while parasympathetic activation regulates their rate. Other autonomic functions, such as intestinal peristalsis and urination, are meticulously governed by a vast neural network, embodying precise coordination of physiological responses. This interaction between systems and their central control structures in the brain determines the exact rhythm and diversity of autonomic reactions, reflecting the organism's natural needs at any given moment.

Social interaction is a highly organised process that involves recognising and responding to social signals and actively engaging with other individuals. This critical aspect of interpersonal interaction is based on the integration of various components of the nervous system and socio-psychological mechanisms [156]. Facial recognition, performed by brain structures specialised in processing visual information, is a key stage in interaction. The recognition of emotions and facial expressions triggers active responses in the corresponding neurons of the brain. The theory of understanding others, which involves comprehending the thoughts and feelings of others, engages various parts of the brain. This includes analysing information about behaviour and relationships. Empathy, which is expressed in the ability to perceive and respond to the emotions of others, activates relevant brain areas responsible for emotional reactions and sensations. Communication, an important component of social interaction, depends on the work of the brain centres for speech and speech perception, which are responsible for the effective processing of verbal information. Social cognition, including the perception and understanding of social situations and norms, involves the work of highly developed brain regions responsible for decision-making and social behaviour. Coordinated interaction between these aspects and brain systems is key to effective social interaction, which is determined by the interdependence and dynamics of these components.

Creativity and intuition, as complex cognitive processes, are rooted in the functioning and interaction of different aspects of the nervous system. Creativity, on the one hand, is associated with the ability of different brain regions to interact with each other. Associative memory, which allows connections to be made between different ideas and concepts, is a key factor in generating new combinations and solutions. On the other hand, the functioning of the frontal lobes of the brain is important for creative tasks, as they are responsible for performing complex tasks, planning, and decision-making. Neuromodulators, such as norepinephrine and dopamine, affect the emotional atmosphere and can play a role in the creative process. A certain emotional state can stimulate or inhibit creativity. Regarding intuition, it may arise from the rapid perception and analysis of information by the subconscious. The posterior lobes of the brain may be key to this process [158]. The ability of neural networks to learn and adapt is important for creative problem-solving. Learning promotes the development of new strategies and approaches to creative challenges. In summary, creativity and intuition are determined by the interaction of various brain systems that work in concert to generate new ideas and solve problems.

Researching these aspects in the context of neuroscience allows uncovering more about their nature and mechanisms of functioning. These functions interact to form an integrated action. Despite this flexibility, there are limitations such as limited resources (energy and brain size) and the speed limits of signal transmission through neural networks. Understanding these limits and constraints remains an active area of research in neuroscience.

3.4 Examining the Influence of Neural Networks on Various Aspects of Cognition

Research on the impact of neural networks on various aspects of cognition is defined by the complex efforts of science aimed at uncovering and understanding how neural networks function in the context of cognitive processes. The work of Bain (1873) and James (1890) was an important evolutionary point in the development of the concept of neural networks. In their experiments, they examined mental activity, considering it as the result of interactions between neurons in the brain. Bain proposed the idea that any activity leads to the activation of a specific set of neurons, and with its repetition, the connections between them are strengthened, contributing to the formation of memory. His theory was initially met with scepticism due to the likelihood of an excessive number of neural connections. However, modern understanding confirms that the brain is an extremely complex structure capable of performing multiple tasks simultaneously. James' theory was similar to Bain's but also noted the assumption that memory formation occurs through the passage of electrical current between neurons, without the need to create new connections for each act of memorisation or action. Physiologist Sherrington conducted experiments in 1898 to test James' theory by passing electrical current along the spinal cord of rats. The results of his experiments played an important role in the development of the theory of habituation. In 1949, Donald Hebb proposed the hypothesis that neural connections can be improved every time they are activated. This is known as "Hebb's rule". His idea is that groups of neurons tend to excite together when perceiving stimuli. This joint excitation is fixed in memory as a group of synchronised neurons, supporting the concept of the dynamic and variable nature of neural connections.

Cognition, which includes perception, thinking, memory, problem-solving, and other brain functions, has become a focal point for scientists exploring how neural networks contribute to these functions [155]. Some aspects of this research include ***analysing the role of specific neural pathways and synaptic connections in the formation and storage of memory.*** Neural networks play a crucial role in creating memories and learning new information. Other studies focus on ***how neural networks process information and respond to various stimuli, including problem-solving and decision-making***. Additional research areas cover the ***interactions between different brain regions and their impact on cognitive functions***. Specifically, scientists are investigating which parts of the brain are responsible for particular aspects of cognition and how they interact to support complex cognitive functions. These studies contribute to a deeper understanding of how neural networks work in relation to cognition and may influence the development of new strategies for improving memory, thinking, and other cognitive processes.

The influence of neural networks on cognition can be exemplified by the phenomenon of bilingualism and cognitive reserve. Research studies have indicated that bilingual individuals frequently demonstrate greater cognitive reserve, which may result in a delay in the onset of neurodegenerative diseases. The management of two linguistic systems has been shown to enhance executive control and induce structural brain changes, including increased grey matter density in the inferior parietal lobule. Research has demonstrated that bilingual individuals exhibit enhanced flexibility in cognitive processing, as their brains constantly switch between languages. This phenomenon is attributed to the enhanced synaptic plasticity and strengthened neural pathways that result from this continuous cognitive exercise. Consequently, bilingual individuals appear to demonstrate a greater degree of resilience to age-related cognitive decline. Furthermore, functional MRI studies have revealed that bilingual individuals engage broader neural networks during language tasks. This finding suggests that these individuals possess an augmented capacity for cognitive flexibility and problem-solving. The ability to maintain and manage multiple linguistic systems in parallel suggests that neural networks play an active role in reinforcing memory functions, executive control, and cognitive adaptability. These findings are consistent with broader research on neural plasticity, which suggests that frequent engagement in complex cognitive tasks can lead to long-term structural changes in the brain, improving its ability to process and retain information efficiently [174].

Research on the role of neural pathways and synaptic connections in memory formation and storage involves a thorough analysis of how brain neural networks interact to create and maintain memory. During information memorisation, neurons receive signals through their dendrites from other neurons or the external environment, initiating APs when the stimulus reaches the excitation threshold. Studies of BNNs related to memory use various methods and approaches to uncover how brain neurons process and store information. Electrophysiological methods, such as electroencephalography (EEG) and single-neuron recordings, allow researchers to examine brain activity patterns during memory formation. Functional magnetic resonance imaging (fMRI) identifies active brain areas during task performance. Imaging methods of neural circuits, such as confocal and two-photon microscopy, help examine the structure of synapses and live neurons. Genetic methods, including calcium tracing and synaptic markers, provide insights into the activity and structure of neurons during memory formation. Molecular-biological studies allow the analysis of the molecular mechanisms of memory, including changes in synaptic proteins and signalling pathways [154]. Each of these methods has its advantages and limitations, and their combination provides a more comprehensive understanding of how neurons process and store information in BNNs.

In BNNs, the formation of dendritic potentials is a complex and multifaceted process that includes a range of mechanisms at different levels of neuron organisation. A neuron initiates the formation of dendritic potentials by receiving input stimuli from other neurons or the external environment. Dendrites, branching out from the neuron's body, function as receiving antennas that perceive signals. On the membrane of dendrites, ion channels regulate the flow of ions, particularly sodium (Na^+) and potassium (K^+). These channels open under the influence of incoming signals initiated by stimuli from neighbouring neurons or the external environment. If the input stimuli lead to a sufficient flow of ions through the dendritic membrane, a dendritic potential can be generated. This is a short-term change in membrane potential that can be excitatory or inhibitory. If the dendritic potential reaches the threshold, an AP is generated on the neuron's membrane. This impulse is transmitted along the axon to the synapses, where synaptic transmission occurs. At the synapses, the AP stimulates the release of neurotransmitters into the synaptic cleft. These chemical signals cross the cleft and interact with receptors on the membrane of the next neuron. Signals received from different synapses are integrated into the receiving neuron's dendrites. If the integrated signal exceeds the threshold, a new dendritic potential may arise. In synaptic connections, calcium ions play a crucial role in neurotransmitter release. The influence of calcium ions can modulate the strength of synaptic connections and affect the formation of dendritic potentials. These processes occur at the molecular level and determine how neurons form dendritic potentials, which is a critical step in the transmission and processing of information in BNNs. Research into these mechanisms involves various methods, including electrophysiological studies, live brain imaging, and biochemical analyses.

During synaptic transmission, which is an integral part of information exchange in BNNs, an AP triggers a complex chain of events. When the impulse reaches the synaptic zone, neurotransmitters such as acetylcholine, serotonin, or glutamate are released into the synaptic cleft. These chemical signals cross the cleft and interact with corresponding receptors on the membrane of the next neuron [155].

It is important to note that the effects of synaptic transmission can be excitatory or inhibitory. When receptors are stimulated, an electrical signal in the form of a dendritic potential is initiated. This signal transmission along the dendrites and the neuron's body can lead to the generation of an AP on the axon, which sends information to other neurons. Various methods are used to analyse synaptic transmission, including electrophysiological techniques such as EEG and electrophysiological recordings, which allow tracking the brain's electrical activity and the responses of individual neurons. fMRI and positron emission tomography (PET) are used to identify active brain regions associated with synaptic events and memory. Biochemical and molecular studies allow for the analysis of the composition and functions of neurotransmitters, as well as the molecular mechanisms underlying memory processes. Genetic studies, particularly genetic modification of neurons, have become a powerful tool for understanding the impact of genetic factors on synaptic transmission and memory. Interactions between neurons, considered in the context of network dynamics, examine how neurons interact in networks to process and transmit information. These studies, which combine various methods and approaches, aim to uncover the deep mechanisms by which the nervous system perceives, processes, and retains information.

At the molecular level, complex changes occur that allow for memory retention in BNNs [167]. This process encompasses a number of key aspects that scientists investigate to understand how the nervous system stores and retains information, as well as to identify potential links to pathologies associated with memory loss. Synaptic plasticity is one of the key aspects of the molecular mechanisms of memory, encompassing changes in the strength and efficiency of synapses between neurons. Research on these processes includes analysing the role of various proteins that influence the stability and strength of synaptic connections. The molecular mechanisms of LTP and LTD of synapses are analysed to uncover how neural networks store and retain information over long periods. In the search for molecular markers of memory, scientists focus on specific biomolecules, such as proteins and RNA, which may indicate certain changes in synapses associated with memory and learning. Genetic studies aim to uncover the role of specific genes in the stability and changes in neuronal circuits, affecting memory. Research on the influence of chemical signals, such as neurotransmitters, on synaptic connections and memory expands the understanding of the chemical processes occurring in neural networks [159]. These comprehensive studies utilise high-precision methods, such as electrophysiological recordings, molecular biology, brain imaging, and genetic manipulations. The results of such studies can not only reveal the fundamental principles of neural network function but also have practical applications in treating memory disorders and neurological diseases.

In the study of long-term memory formation in BNNs, various approaches are used, encompassing electrophysiological, molecular, genetic, and brain imaging methods. Electrophysiological studies focus on uncovering the secrets of synaptic plasticity, which includes adaptations in synaptic connections under the influence of long-term memory. Recordings of APs and dendritic potentials allow for the analysis of neuron activity in the context of long-term memory retention. Molecular biology and genetics reveal deep aspects of molecular mechanisms. Studies of gene expression and genetic mutations help identify key genes associated with the stabilisation of synapses and other processes that determine long-term memory. Brain imaging methods, such as fMRI and PET, provide information about the functioning of different brain regions during the formation and use of long-term memory. Biochemical analyses examine changes in the concentration of neurotransmitters and other chemical signals, regulating synaptic processes in the context of long-term learning and memory retention [175]. These approaches interact synergistically, providing a deeper understanding of the high-level and detailed aspects of the molecular, genetic, and physiological processes underlying the formation of long-term memory in neurons.

The analysis of levels of various chemicals in synaptic spaces, such as neurotransmitters, which play a role in signal transmission between neurons, is crucial for understanding the mechanisms of long-term memory formation in BNNs. Research in this area employs various methods and techniques to analyse the functioning of BNNs at the molecular, cellular, and systemic levels. The analysis of molecular mechanisms includes examining the levels of different molecules that interact at synaptic connections, altering their state and function. The study on genetic aspects aims to uncover the role of specific genes in synaptic stabilisation and other processes that determine long-term memory. Brain imaging methods, such as fMRI, allow for identifying active brain regions during the formation and use of long-term memory. Biochemical analysis techniques include examining changes in the concentration of neurotransmitters and other chemical signals that regulate synaptic processes in the context of long-term learning and memory retention. These different approaches collectively help reveal the precise mechanisms and interactions occurring in neural networks related to memory [154]. Understanding these processes has potentially significant implications for treating memory disorders and neurodegenerative diseases in the future.

Some scientists use techniques such as EEG and fMRI to analyse brain activity during memory processes. These studies contribute to understanding the mechanisms underlying memory formation and how they can be used to develop methods for improving memory and cognitive functions. The impact of BNNs on information learning is explored in many scientific fields, such as neurobiology, cognitive science, psychology, and neuroscience. Research in these fields helps to understand how BNNs influence learning and memory processes in humans and other organisms. Studies in neurobiology aim to uncover the physiological and chemical aspects of the functioning of neural networks in the brain. Scientists investigate how signals are transmitted between neurons, how information is stored and removed, and how synaptic plasticity affects learning. In the fields of cognitive science and psychology, the study focuses on how BNNs affect cognitive processes such as perception, attention, speech, problem-solving, and decision-making. Research encompasses understanding how the brain processes information and uses it for learning. Learning neuroscience focuses on exploring how neural networks influence learning processes and memory formation. The research includes analysing the role of synaptic plasticity and the mechanisms that allow the brain to adapt to new information. Some studies use computer models to simulate BNNs and examine their impact on learning. This allows scientists to experiment with different parameters and conditions for a better understanding of learning principles [155]. The overall goal of this study is to uncover how BNNs adapt to the information they receive and how this knowledge can be used to improve learning methods and the development of artificial intelligence. Research on the influence of BNNs on information learning is conducted by numerous scientists in various scientific fields.

Prominent Austrian–American neurobiologist Eric Kandel, Nobel Prize laureate in Physiology or Medicine in 2000, was born on November 7, 1929, in Vienna, Austria. His research focuses on the neuronal mechanisms underlying long-term memory and learning. Kandel investigated the morphology and functions of neurons, particularly in the glial cells of marine slugs. His work revealed changes in synapse structure that occur during learning and memory formation. His contributions to the fields of neurobiology and psychiatry advance the understanding of how the brain stores and retrieves information, and his research is important for understanding the neuroscientific foundations of memory, learning, and cognitive brain functions. Kandel's findings highlight key aspects of the molecular and physiological mechanisms that determine the formation and retention of memory. His work uncovers the molecular basis of memory, identifying the role of specific molecules such as the protein CREB and synaptic changes in long-term memory [177]. These discoveries expand the understanding of how molecular mechanisms influence neural circuits and form memory. Kandel's research further confirms the concept of "plastic memory", emphasising that changes in synapse structure and activity can be dependent on experience and learning. An important contribution is also the confirmation of the concept of neuroplasticity, which indicates the brain's ability to change its structure and functions in response to experience and learning. Kandel's results stimulate further studies in the field of neurobiology, contributing to the discovery of new approaches to understanding the mechanisms of learning and memory in neurons.

The distinguished British neuroscientist Susan Greenfield is a leading researcher in the investigation of the impact of BNNs on cognitive processes and learning. Born in 1950, she earned a doctorate in physiology and neurobiology and has gained recognition for her contributions to understanding the interaction between the brain and cognition. Susan Greenfield's research in neurobiology focuses on the molecular and cellular aspects of neuron interactions in the brain, particularly the examination of synapses. She directed her attention to determining how these structures influence the formation and retention of memory, which is key to understanding neurobiological mechanisms. Greenfield's research into the molecular aspects of synaptic transmission includes the analysis of various biomolecules, such as neurotransmitters, receptors, and other proteins that interact in the processes of signal transmission between neurons. Detailed elucidation of these molecular details aids in understanding the precise mechanisms of the nervous system. Susan Greenfield also examines the cellular aspects of neuron interactions in memory formation. This includes the analysis of structural and functional changes that occur in neurons and their connections under the influence of learning and experience [167]. A detailed understanding of these processes is crucial for comprehending the foundations of memory formation and retention in the brain. Overall, Susan Greenfield aimed to analyse the molecular and cellular mechanisms that ensure effective synaptic transmission and influence memory formation processes in neurons. Alongside her scientific activities, Greenfield investigated the impact of technology on brain activity and cognitive functions. She issued warnings about the prolonged use of modern technologies, especially among younger generations, and advocated for their mindful and responsible use. In conclusion, Susan Greenfield's research has significantly contributed to understanding the interaction between the brain and cognition, paving the way for further neurobiology studies and examining technology's impact on the human brain and memory.

John O'Keefe is a renowned British neurophysiologist awarded the Nobel Prize in Physiology or Medicine for his significant discoveries in the field of neurobiology. His research identified cells that form spatial maps in the brain, which are crucial for understanding learning and memory processes. One of O'Keefe's notable achievements was the discovery of "place cells" in the hippocampus that determine spatial orientation and location of organisms in their environment. These findings play a key role in understanding learning and memory processes and are fundamental to the study of brain functions. John O'Keefe worked in leading scientific institutions, notably at University College London, where his contributions to neurobiology are recognised as a substantial advancement in scientific knowledge about brain functions and their impact on learning and memory processes.

British neurobiologist Richard Morris is known for his research in learning and memory, focusing on the role of the hippocampus in these processes. One of the key areas of his research is the examination of the molecular and cellular mechanisms underlying memory formation and retention. Morris is actively engaged in exploring processes of synaptic plasticity that determine changes in synaptic connections between neurons under the influence of experience and learning. An important part of his research is the interactions between different brain regions, particularly the hippocampus, and its influence on learning and memory formation. Richard Morris's work made a significant contribution to understanding how the brain processes and stores information, revealing fundamental mechanisms of learning. In addition, Morris is known for his active role in the field of neuroscience and his publications in scientific journals, which contribute to the advancement of this scientific domain. His contributions to neurobiology are recognised as crucial for expanding scientific knowledge of brain functions and learning mechanisms.

Michael Merzenich, a prominent American neurobiologist born in 1942, significantly contributed to the understanding of neuroplasticity and the impact of learning on brain structure. After obtaining his higher education and a doctorate in physiology, Merzenich actively researched the mechanisms of neuroplasticity—the brain's ability to change its structure and functions in response to experience and activity. One key aspect of Merzenich's research is how stimulation and activity can lead to changes in neural connections. His work reveals the immense potential of neuroplasticity, allowing the brain to adapt and reorganise throughout life. Merzenich's studies also emphasise the influence of learning and stimulation on brain structure and functions. He extensively investigated how various types of learning and training can lead to changes in neural connections, enhancing cognitive functions. His research provides insight into how the brain can adapt to injuries and loss of functions [164]. These findings have become the basis for developing interactive programmes and brain training methods to improve cognitive functions. Michael Merzenich is not only a scientist but also a practitioner, actively advancing methods for brain rehabilitation and training. His conclusions profoundly impacted the understanding of how learning and experience shape brain activity and functions, with significant implications for the development of educational and rehabilitation practices.

At the level of neurophysiology, scientists examine the activity of specific neurons and their roles in decision-making. Experiments recording neuron activity in various brain regions help identify which areas are responsible for specific aspects of decision-making. The field of neurophysiology has seen many impressive achievements, enhancing the understanding of the nervous system and brain function. In the 19th century, physiologist Johannes Müller and physicist Hermann von Helmholtz made key contributions to the analysis of nerve conduction, discovering the basic principles of how nerve impulses are transmitted through nerve fibres at certain speeds. This discovery laid the foundation for further neurophysiology and neurology research, advancing the nervous system. Johannes Müller's research spanned various aspects of physiology, but he showed a particular interest in the nervous system and its functions. One of his key discoveries was that different nerve fibres carry distinct functions [173]. For example, he determined that some fibres are responsible for transmitting pain signals and others for transmitting temperature signals. German physicist and physiologist Helmholtz, collaborating with Müller, expanded the research on nerve conduction. Together, they examined nerve stimulus reaction times and determined nerve conduction speed in different parts of the nervous system. They showed that signals are transmitted along nerve fibres at a certain speed, which was an important step in understanding the principles of the nervous system's operation. These discoveries indicated that there is a physical impulse transmitted along nerve pathways, forming the foundation for further research aimed at uncovering the mechanisms of nerve conduction, the physiology of nerve cells, and the interactions between nervous structures in the brain and spinal cord.

In the history of neuroscience, several scientists have investigated the structure and function of the brain, including developing methods to determine the shape and structure of individual neurons. One example is Santiago Ramón y Cajal, who used the Golgi staining method to examine neuron structure. He received the Nobel Prize in Physiology or Medicine in 1906 for his work in this field. With the development of methods for recording the electrical activity of neurons and the use of electrodes, more detailed studies of neuron interactions and understanding the electrical activity in neural networks became possible. This opened the way for further discoveries in neuroscience and helped to unravel many mysteries related to brain function [154]. Regarding the study of synapses, one of the key scientists in this area was Sir Charles Sherrington. In collaboration with colleagues, including Albert Young, Sherrington conducted significant research on neuromuscular synapses and the structure of the PNS. Sir Charles Sherrington and Albert Young jointly examined the nature of signal transmission in the nervous system and, in 1932, received the Nobel Prize in Physiology or Medicine for the discovery of the "functions of the neuromuscular junction and the regulation of muscle activity through the nervous system". Their research helped uncover and describe many key aspects of synapse functioning, namely the mechanism of impulse transmission between neurons and muscle fibres through chemical connections. This was an important step in the development of scientific knowledge about the nervous system and its functions. Prominent physiologist Ivane Beritashvili analysed the mechanisms of activation of brain structures, particularly the accentuation of the reticular formation. These achievements paved the way for further research and the development of methods for studying neurophysiology, contributing to a deeper understanding of the functioning of the human brain and nervous system.

Neuropsychologists explore how different brain disorders or mental conditions can affect a person's ability to make decisions. For example, research on patients with damage to the frontal part of the brain (prefrontal cortex) can highlight the role of this area in decision-making processes. The field of neuropsychology, which explores the interaction between the brain and the mind, has achieved numerous notable advancements. In 1861, French scientist Paul Broca made an important discovery in the field of neurophysiology, known as "Broca's area". During his research, Broca found that damage to a specific part of the left frontal lobe of the brain, now known as Broca's area, could lead to a loss of the ability to produce speech while not affecting the ability to understand it. Patients with damage to Broca's area may have difficulty with speech production, particularly in forming words and phrases. This region of the brain plays a key role in the motor aspects of speech, including movements of the lips, tongue, and other speech organs. The identification of this area and its role initiated the analysis of functional localisation in the brain and the development of the idea of language function localisation within a specific brain region. The discovery of Broca's area marked a new stage in understanding how the brain controls various aspects of cognitive functions and is one of the first examples of establishing a connection between a specific brain region and a specific function.

The analysis of the connection between the brain and memory in the 1950s to 1960s proved to be crucial in understanding the functions of the hippocampus and memory formation processes. A major part of these studies involved patients who had damage to the hippocampus and its surrounding structures [172]. The most famous case is that of patient Henry Molaison, known as H. M., who underwent surgery that resulted in significant damage to his hippocampus. As a consequence, he lost the ability to form new declarative memories (memories that can be expressed in words). These studies confirmed the essential role of the hippocampus in forming declarative memory, especially episodic memory (memory of specific events). They also pointed to the existence of different types of memory and the distribution of their functions in various parts of the brain. Thus, these studies were an important step towards understanding the mechanisms of memory and determining the role of different brain structures in these processes.

The development of modern methods, such as PET and magnetic resonance spectroscopy (MRS), allows examining visual thinking and cognitive functions of the brain [169]. The implementation of technologies like fMRI has enabled scientists to observe the activity of different brain regions during various tasks. This has significantly facilitated the examination of the functional organisation of the brain.

The application of principles learned in neuropsychology to the development of deep learning algorithms and artificial intelligence has become a key direction in information technology. These achievements indicate ongoing progress in understanding the interaction between the brain and the mind, opening up new possibilities for the development of diagnostics and treatment of brain disorders.

Cognitive scientists explore cognitive processes related to decision-making, such as attention, memory, thinking, and perception. Research in this area includes analysing different decision-making strategies and their interaction with cognitive mechanisms. The field of cognitive science encompasses various studies and achievements covering the understanding of cognitive processes such as perception, attention, memory, language, thinking, and decision-making. The development of information theory and the ideas of cybernetics in the mid-20th century laid the foundations for understanding how information is processed in the brain and artificial intelligence systems. The notion that the brain operates like a computer spurred the development of models for the computational nature of cognitive functions, such as information processing and problem-solving [159]. Research on memory helped identify the mechanisms of storing and retrieving information, as well as understanding the differences between various types of memory. The emergence of cognitive psychology allowed for a deeper exploration of thinking processes, attention, and other cognitive functions using experimental methods. The application of modern neuroimaging methods, particularly fMRI, has enabled direct observation of the activity in different parts of the brain during task performance and the study of their interactions.

The analysis of the cognitive aspects of language has propelled the development of various theories and approaches, including Noam Chomsky's generative grammar and the concept of semantic networks. Noam Chomsky, an American linguist, developed the theory of generative grammar in the mid-20th century. He proposed that human language is based on an internal linguistic competence that allows people to generate an infinite number of natural language expressions. According to generative grammar, a linguistic act is not just produced but interpreted within the framework of an internal linguistic system. The concept of semantic networks is related to the representation of semantic information and connections between concepts in the form of a graph or network [166]. This approach is based on the idea that the meaning of words is determined by their relationships and connections. Semantic networks can be used to model language understanding and concept interaction. These theories reflect an interdisciplinary approach, combining linguistics, cognitive science, and other fields. They have played a significant role in shaping the understanding of how language interacts with cognitive processes and have contributed to the development of modern approaches to studying the nature of language and its role in thinking. The development of the idea that the brain can change and adapt throughout life has led to a new field of research focused on neuroplasticity. These advancements deepen understanding of how the human brain works and have important implications for the development of artificial intelligence and information processing technologies. These studies interact and complement each other, enhancing understanding of the fundamental mechanisms behind decision-making in biological systems and the development of new approaches to creating intelligent systems.

Antonio Damasio is a distinguished American neurobiologist and neuropsychologist whose specialisation focuses on the decision-making processes and the relationship between brain activity, emotions, and behaviour. Damasio has made significant contributions to understanding emotions by introducing the concept of "somatic markers". He assumes that physical sensations, such as skin conductance response or heart rate, can serve as somatic markers for evaluating situations and making decisions. Antonio researched how emotions influence decision-making [158]. His studies confirm that emotions play a key role in decision-making processes and are an important factor in shaping people's actions. Antonio Damasio examined patients with brain damage, particularly those with lesions in the ventromedial prefrontal cortex. He discovered that this can lead to impairments in decision-making processes and a lack of appropriate emotional responses. In his famous book "Descartes' Error: Emotion, Reason, and the Human Brain", Damasio explores the relationship between emotions and the mind, highlighting the importance of the emotional component in decision-making. Antonio is the author of numerous papers and books in which he examines various aspects of brain function, including its impact on behaviour and personality formation. Antonio Damasio continues to deeply explore the role of the brain in interacting with emotions and how this influences choices and decision-making. His work contributes to the development of cognitive science and understanding the interconnections between brain activity, emotions, and cognition.

In cognitive science, Daniel Kahneman, a distinguished American psychologist and Nobel laureate, is recognised for his contributions to economics, psychology, and cognitive sciences, particularly the analysis of decision-making. In one of his seminal works, Kahneman introduced the concepts of "System 1 and System 2". System 1 is a fast and automatic mode of thinking, whereas System 2 is a more deliberate and conscious mode. He researched how these systems interact in decision-making [161]. Kahneman conducted numerous experiments and studies demonstrating that people often make decisions not based on rational principles but influenced by emotions and stereotypes. His work laid the foundation for understanding the psychology of decision-making. Kahneman identified the concept of heuristics—simple rules or shortcuts that simplify decision-making processes but can lead to systematic errors. He analysed various types of heuristics and their impact on decisions. In his book "Thinking, Fast and Slow", he presented his ideas to the general public, revealing the principles of decision-making underlying economic processes and behavioural economics. Daniel Kahneman received the Nobel Prize in Economics in 2002 for "his development of ideas concerning decision-making under uncertainty and risk". Kahneman continues his active work, consulting and publishing papers in psychology and economics. His contributions to understanding decision-making and the interplay between the mind and emotions remain significant.

The results of studies on the impact of neural networks on various aspects of cognition demonstrate considerable progress in understanding brain functions and mental processes. Scientists' work provides unique insights into the mechanisms of learning, memory, decision-making, and social adaptation. Neural networks have been found to play an important role in the formation of cognitive functions, contributing to the development of memory, perception, and problem-solving. Studies on neuroplasticity open up prospects for enhancing brain function through learning and adaptation. The combination of biological principles with the application of ANNs points to the potential for improving the imitation of cognitive processes in the future (Fig. 19). These studies open new perspectives in the field of artificial intelligence and the development of innovative technologies that contribute to the advancement of modern society.

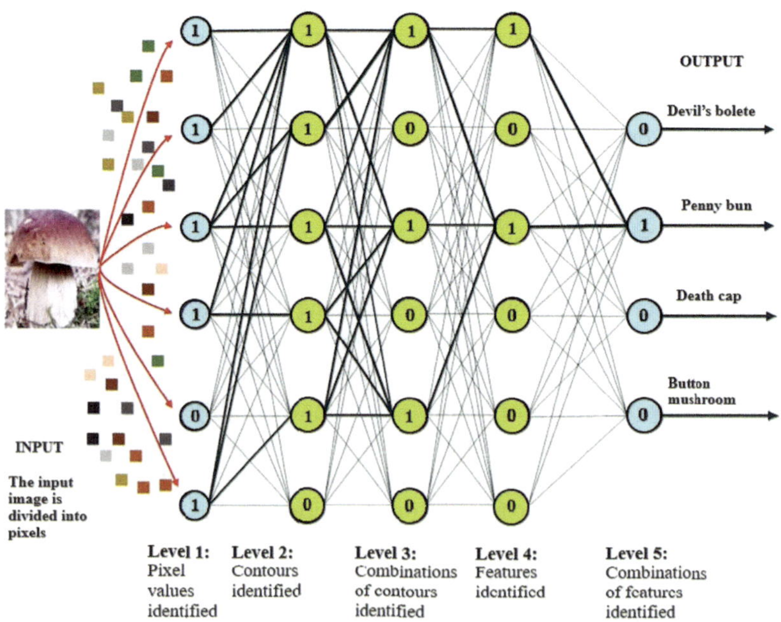

Figure 19 Example of a neural network in image classification.

3.5 Use of ANNs

In today's world, the use of ANNs has become an integral part of daily life. These innovative technologies have a significant impact on modern society and promise further development prospects. ANNs represent a complex branch of information technology developed to mimic BNNs to solve various artificial intelligence tasks. The history of this field can be divided into several key periods. The early years, particularly the 1950s and 1960s, were marked by the introduction of mathematical models of neurons, which laid the foundation for the development of both biological and ANNs (Fig. 20) [182]. It is important to note the work of Frank Rosenblatt, who, in 1957–1958, created the perceptron—a simple ANN capable of solving certain classes of tasks. His underlying idea was to create a simple artificial neuron that could serve as the basis for developing ANNs capable of solving specific classes of problems. The perceptron he developed was a fundamental element that modelled and mimicked the operation of a neuron in the biological brain. The perceptron is designed to model the principle by which a human neuron works, consisting of several key components. Each input is accompanied by a weight that reflects the importance of a particular signal. An adder computes the weighted sum of all inputs. This sum is then subjected to an activation function that determines the neuron's output signal. The core idea of the perceptron is that it can learn weights to automatically adapt them to solve specific tasks. Learning is based on adjusting weights according to the obtained results. While a simple perceptron is a linear classifier, more complex neural networks, such as multilayer perceptrons, can model more intricate interactions and solve tasks that linear models cannot effectively address. Many modern deep learning technologies are based on advanced and optimised versions of perceptrons. This model has proven quite effective in solving simple classification and pattern recognition tasks. However, it is important to note that perceptrons eventually showed limitations in solving complex problems, leading to criticism, particularly from Marvin Minsky and Seymour Papert in 1969. They demonstrated that the perceptron could not effectively solve tasks such as the XOR logical operation, causing a decline in interest in ANNs during that period. Despite the limitations of the perceptron, its work laid the groundwork for further research in the field of ANNs.

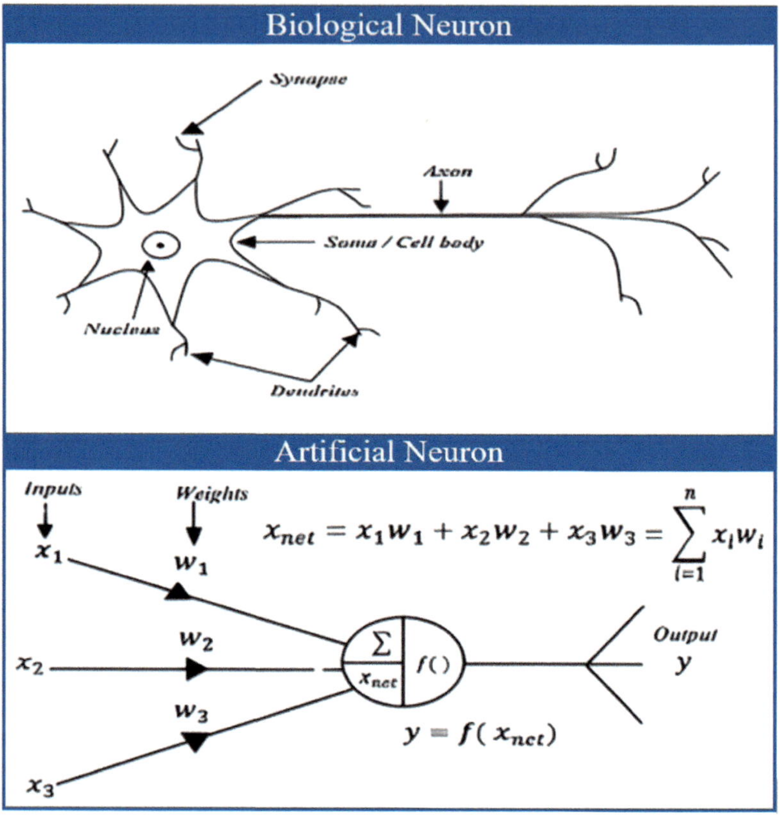

Figure 20 Comparison of biological and artificial neural networks.

The 1980s marked a renewed interest with the advent of backpropagation and the analysis of recurrent networks, which adopted effective learning methods and expanded the capabilities of ANNs. The 1980s witnessed a positive resurgence in the field of ANNs, highlighting renewed interest in this area. This period became a crucial stage in the development and application of ANNs, thanks to several key events and improvements. One of the most significant events of this period was the invention and implementation of the backpropagation algorithm [183]. This learning method allows for the effective adjustment of neural connection weights, reducing the error between predicted and actual results. Backpropagation marked a significant step toward improving the efficiency of ANNs and expanding their applications. During this period, recurrent neural networks (RNNs) were actively examined. RNNs are a family of ANNs that can retain internal states and process data sequences. For the first time, the detailed study and application of RNNs allowed the development of ANNs for more effective sequence processing, such as speech or time series. The growth of high-performance computing facilitated the application of ANNs in real-world tasks. They began to be used for image processing, speech recognition, forecasting, and other tasks that require complex data analysis. All these factors together defined the positive development and popularity of ANNs in the 1980s, shaping their further path for development and application in various fields.

Moving on, the 1990s and 2000s saw a significant rise in high-performance computing, which enabled the use of ANNs in large computational systems for image processing, speech recognition, and other tasks. The 1990s and 2000s were marked by a substantial rise in high-performance computing, which significantly altered the landscape of ANN applications. This period was transitional for the field and defined new opportunities and challenges for the development of ANNs. The increase in computational power allowed for the use of ANNs on a real and large scale, changing the landscape of their applications. They became an integral part of image processing systems, including facial recognition, object detection, and other tasks in computer vision. ANNs began to be actively experimented with in other fields, such as biomedicine, finance, engineering, and technology [184]. ANNs were used for predicting financial markets, analysing medical images, managing production, and many other tasks. During this period, the concept and spread of deep neural networks (DNNs) emerged. These are more complex architectures capable of automatically learning high-level features and data representations. DNNs became a key tool in deep learning. The application of ANNs in the field of natural language processing (NLP) has also experienced significant advancements. This allowed for the development of machine translation systems, speech recognition, and other NLP tasks at a new level of accuracy and performance. Overall, the 1990s and 2000s were pivotal in defining the potential and practical use of ANNs in various areas of life, shaping the further development and integration of these technologies into modern society.

In the "golden age of deep learning", which lasted between the 2010s and 2020s, deep learning and neural networks with deep architectures were developed. The use of ANNs in medicine and science was also a significant achievement of this period. During this time, the world witnessed incredible advancements in deep learning and neural networks with deep architectures [185]. Significant achievements in various fields and sources of innovation marked this period. Deep learning became a central element in many modern artificial intelligence applications. DNNs, particularly those using deep learning, have shown incredible effectiveness in solving tasks such as face recognition, image classification, voice control, and others. DNN architectures have become increasingly complex and efficient. Architectures such as deep CNNs for image processing and RNNs for sequence processing emerged and improved, laying the foundation for various applications. ANNs found widespread use in the field of medicine. They are used for analysing medical images, diagnosing diseases, predicting treatment outcomes, and other aspects, enhancing the accuracy and speed of medical diagnoses. Deep learning is used in research to analyse complex data, model intricate systems, and make predictions. The application of deep learning has facilitated significant discoveries in physics, chemistry, biology, and other sciences. The increase in the volume of available data and advancements in computational capabilities have enabled ANNs to handle more complex tasks and deliver high performance [186]. In summary, the "golden age of deep learning" was a period of emergence and stabilisation of technologies that have now become an integral part of daily life, contributing significantly to the development of various fields and improving the quality of life.

The current stage, 2021 and beyond, is characterised by attention to the ethical aspects of using ANNs and research into quantum neural networks, opening new prospects for the further development of this captivating field. At the modern stage, covering 2021 and later, the use of ANNs has become a subject of attention not only from a technical perspective but also from an ethical one. This period is marked not only by technical achievements but also by a growing awareness of responsible use and the ethical issues related to the application of artificial intelligence. As the implementation of AI in various societal areas increases, ethical questions have become pertinent. There is a need to establish standards and regulations aimed at protecting people's rights and privacy, preventing discrimination, and avoiding other potential negative consequences. One of the areas of active research is quantum neural networks. These new models are based on the principles of quantum mechanics and have the potential to solve problems that are difficult or impossible to address using classical methods. They open new perspectives for enhancing the computational power and efficiency of ANNs [187]. The growth in the capabilities of ANNs poses new challenges for society related to security, efficiency, and regulation. Research in this area is aimed at addressing these issues and ensuring the rational and responsible use of these technologies. The use of ANNs is constantly being improved and integrated with other advanced technologies, such as cyber-physical systems, the Internet of Things (IoT), and blockchain. This creates new opportunities for innovation and improvement in various industries. All these aspects define the current stage of ANN development and provide momentum for the further enrichment of this fascinating field of knowledge.

A notable example of ANNs in action is Google's DeepDream algorithm, which mimics human visual hallucinations by enhancing patterns in images using ANNs [188]. DeepDream creates dreamlike and surreal pictures by overactivating specific visual patterns. It's interesting to note that MRI research indicates that people who have visual hallucinations as a result of LSD use or schizophrenia experience a similar overactivation. This case study offers insights into both artificial intelligence and human cognition by demonstrating how ANNs may simulate and magnify perceptual events.

Thus, the use of ANNs is becoming a dominant phenomenon in the technological space, and its impact on modern society is far from exhausted, opening paths for new achievements and challenges. ANNs play a key role in various fields due to their unique properties of learning and adaptation. The essence of an ANN lies in imitating BNNs to solve artificial intelligence tasks [189]. An ANN consists of neurons that interact with each other, forming a large network. The human nervous system derives the fundamental components and principles of their functioning. The key unity is artificial neurons, which perform calculations and information transmission. The main elements of an artificial neuron include weights, an adder, an activation function, and an output. Each input signal received by a neuron has its weight, which determines its importance for performing calculations. These weights are subjected to the next processing stage, namely the adder. The adder performs the operation of summing the input signals multiplied by their weights. As a result, a weighted sum is obtained, which serves as the primary internal value for further calculations. This stage is crucial in determining how activated the neuron will be. The activation function determines whether the neuron will be activated based on the received weighted sum. This function can be varied, including step, linear, or other types of functions that regulate the neuron's output signal. The output of the neuron represents the result of the activation function's work. This value can serve as the input signal for the next artificial neuron in the network or be outputted as the final result of the neural network's work. Thus, an artificial neuron interacts with input signals, calculates their importance using weights, and generates an output signal based on the activation function [180]. ANNs vary in architecture and size, and they can be trained using machine learning algorithms. The training process involves determining the optimal weights for effectively solving a specific task.

In summary, the essence of an ANN lies in creating computer models that can automatically learn and solve tasks inspired by the human brain's BNNs. Their wide use spans areas such as image recognition and classification, speech recognition, autonomous vehicles, medicine, finance, e-commerce, voice assistants, research, virtual reality, and energy. In the first area, image recognition and classification, ANNs perform automatic face, object, and scene recognition in images, and they assist in classifying text information. In speech recognition, ANNs are used for automatically converting speech to text and responding to voice commands. Autonomous vehicles are becoming a reality thanks to autonomous control systems built based on ANNs [190]. This application solves the problem of controlling vehicles such as cars, drones, and others. In medicine, ANNs are used for diagnostics, disease recognition, and early-stage pathology prediction. In finance, their capabilities are leveraged to analyse market dynamics and forecast trends in financial markets. In e-commerce, ANNs analyse habits and provide personalised recommendations, while voice assistants interact with users through voice command recognition and execution. ANNs, which are used to model complex systems, solve mathematical problems, and analyse large volumes of data, have boosted research. In virtual reality, ANNs ensure simulation and immersion, creating realistic virtual worlds [191]. Finally, in energy, ANNs are used to create energy-saving systems, optimise energy efficiency, and predict energy consumption.

The use of ANNs in various fields is very diverse and expanding due to the continuous development and improvement of this technology. A wide range of methods is employed, including supervised and unsupervised learning, image recognition, NLP, vehicle management, medical diagnostics, financial analysis, e-commerce recommendations, voice interfaces, and modelling complex systems in research. The application of ANNs depends on the specific task and the availability of large datasets for training. Artificial intelligence technologies continue to evolve, expanding the possibilities for using ANNs in various industries. In recent years, ANNs have achieved significant successes, surpassing human intelligence in highly specialised tasks such as playing chess, video games, and medical diagnostics. However, the human brain has demonstrated superior performance in more general and abstract tasks requiring adaptive learning and cognition.

One key difference between ANNs and BNNs is the higher energy efficiency and ability to perform parallel computations in biological systems. Neurons in biological networks operate in parallel, interacting with each other and processing information efficiently. In addition, BNNs possess the ability for real-time learning, self-repair, and autonomous consciousness, allowing them to actively adapt to changes in the environment. The fundamental unit of BNNs is the neuron, which processes information by generating APs according to its membrane potential. In biological systems, signals are transmitted through synapses that connect neurons. The strength of synapses adaptively changes according to incoming signals, playing a crucial role in neuroplasticity—the brain's ability to learn and store memories. BNNs use an "all-or-nothing" principle, where a neuron fires if its membrane potential exceeds the firing threshold [192]. This simplifies information processing and makes BNNs effective in handling complex tasks.

In summary, although ANNs have made great strides, BNNs remain an incredibly powerful and efficient source of inspiration for the development of artificial intelligence. The future of ANNs encompasses several promising areas. Expanding their applications in manufacturing promises process optimisation and automation. In medicine, the use of ANNs for data analysis can support decision-making and the development of personalised treatment approaches. Intelligent systems, such as smart homes and cities, can enhance automation, ensuring more efficient resource use. ANNs can benefit robotics by creating more intelligent robots, enabling energy-saving capabilities, and optimising energy consumption. The development of learning algorithms considers the use of ANNs with limited data volumes. Ethical aspects and security are becoming increasingly important, requiring reinforced privacy measures. Finally, hybrid systems that combine ANNs with other technologies open new possibilities in the fields of quantum computing and robotics.

An ANN is a computer model inspired by the biological neural system that solves artificial intelligence tasks. The main idea is to emulate the principles of the human brain, where neurons collaborate to process information. An ANN consists of interconnected artificial neurons organised into layers that work together to solve tasks such as image recognition, data classification, or prediction. Each artificial neuron has weights for input signals, a summing junction that calculates the weighted sum of the inputs, and an activation function that determines the neuron's output signal [193]. Artificial neurons are organised into layers, interpreting input data and passing it through the network to produce output results. ANNs are used in various fields, including NLP, computer vision, medical diagnostics, finance, and many others. Thanks to the ability to learn from data, it has become a key tool for solving complex tasks and achieving significant advances in modern research and technology.

ANNs are widely used in the field of medicine to enhance diagnostic processes, treatment, and management of medical information. In medical diagnostics, ANNs automatically analyse images such as X-rays or CT scans, helping to detect pathologies early. Furthermore, they are used in dermatoscopy for recognising signs of skin diseases. In the area of therapy and treatment, ANNs are used to develop personalised treatment plans and select optimal medications for patients. They assist in processing large volumes of medical data to identify correlations and make predictions. Patient data management also benefits from the use of ANNs in electronic medical record systems. They facilitate the processing and analysis of electronic medical records, improving the ability to make effective decisions and promoting efficient information management. ANNs are also used to predict risks and disease progression, enabling the development of preventative strategies and personalised health approaches. In the field of mental health, ANNs are used to analyse changes in patient behaviour and the early detection of signs of mental disorders [182]. In addition, ANNs have found their application in signal processing, specifically in analysing electroencephalographic signals to examine brain activity and create interfaces for controlling devices through thoughts. Overall, their implementation fosters innovation and improves the quality of medical services.

The future of ANNs in medicine, biology, and psychology holds many exciting opportunities for advancement and enhancement. These technologies can be key in a number of aspects. In medicine, ANNs can facilitate personalised medicine, where they are used to develop personalised diagnostic and treatment methods, considering each patient's unique characteristics, such as genetic makeup, immunological features, and other factors. They can also find application in neurorehabilitation, where they help in restoring motor and cognitive skills. In biology, ANNs are used to delve deeper into the structure and functions of the nervous system, which aids in understanding neurobiological mechanisms and developing new research methods [184]. They can be an important tool in investigating biological processes and discovering new patterns in biological systems. In psychology and psychiatry, ANNs can be used for the diagnosis and treatment of mental disorders, as well as for creating innovative approaches to maintaining emotional stability. Biotechnological implants based on ANNs could revolutionise the development of technologies for treating various diseases and injuries, while brain-computer interfaces could find wide application in the implementation of augmented reality technologies and interaction with computing systems. In addition, the development of hybrid systems that combine ANNs with other technologies, such as quantum computing or robotics, could lead to the creation of even more powerful and effective systems in various fields. Considering these aspects, ANNs define prospects in the deep understanding and use of biological and psychological processes, the implementation of innovative treatment methods, and the exploration of new scientific fields.

ANNs can play a key role in understanding the action of hypnosis on the human brain due to their ability to model and analyse neuronal processes. Hypnosis remains a complex phenomenon, and the use of ANNs in this context can contribute to scientific disclosure and a better understanding of this phenomenon [185]. Firstly, ANNs can be used to create virtual models of brain neural networks that consider various stimuli, including those that occur during hypnosis. This allows for analysing intra-regional interactions and their impact on a person's state during hypnotic sessions. Secondly, ANNs can serve as a tool for exploring how hypnosis affects various brain functions, such as attention, memory, and perception. Analysis of changes in neuronal activities that occur during hypnosis helps to understand its impact on different aspects of brain activity. Thirdly, ANNs allow examining brain plasticity, i.e., its ability to adapt and change. This can include analysing changes in connections between neurons that may occur during hypnotic states and their impact on brain functions. In addition, ANNs can help uncover the mechanisms of emotional response to hypnotic stimuli, which helps to understand how hypnosis affects the emotional sphere of a person. Thus, the use of ANNs to examine hypnosis can open new horizons in understanding the neuronal mechanisms that define this phenomenon and promote the development of methods for influencing brain activity in medicine and psychotherapy.

Chapter 4

Hypnosis and Brain Activity

4.1 Trends in Brain Research and Hypnosis

Scientific and technological advances, which have become the foundation for ground-breaking research, present the neuroscientific and psychophysiological aspects of brain function and its response to hypnosis. Modern imaging techniques, in particular, functional magnetic resonance imaging and electroencephalography, can illuminate the activity of various brain regions. This opens up new opportunities for understanding mental processes and examining their dynamics. The use of transcranial magnetic stimulation (TMS) and brain interfaces (BIs) by a computer opens up opportunities for research and influence on brain activity. Such methods are becoming not only learning tools but also promising technologies for modifying brain functions. In hypnosis, interest in neurophysiological approaches opens up new horizons. Research focuses on uncovering the brain mechanisms of hypnotic states and their impact on a person's mental state. This chapter covers scientific research as a key tool for understanding and influencing brain function and investigating the phenomenon of hypnosis in the context of modern neuroscience discoveries. The examination of the brain and hypnosis is one of the key areas of modern science, defining new horizons for understanding human consciousness and mental processes [194].

Mechanisms of Brain Function and Adaptive Potential
Liana Spytska

ISBN 978-981-5129-91-5 (Hardcover), 978-1-003-71697-6 (eBook)
www.jennystanford.com
DOI: 10.1201/9781003716976-5

The importance of these studies is determined by their potential to review the mechanisms of perception, memory, and mental disorders, including practical use in medicine, psychotherapy, and improving quality of life. Further research aimed at uncovering and understanding the internal mechanisms of the human mind and its interaction with the phenomenon of hypnosis is of great value.

The investigation of the brain and hypnosis is currently an extremely relevant and promising field of science since their interrelationships help to identify and understand the deepest aspects of the functioning of the human mind and psyche. These studies open up the possibility of enriching scientific knowledge about brain activity and the mechanisms of consciousness. It is noted that understanding the influence of hypnosis on brain activity not only enriches the theoretical basis of the science of the nervous system but also opens up broad prospects for practical use in medicine, psychotherapy, and other areas related to improving people's quality of life. Activity in this area can lead to the development of new treatments and approaches to psychotherapy and improve the overall understanding of human consciousness. Thus, the subject of brain research and hypnosis has a substantial impact on the development of science, contributing to the discovery of new facets of knowledge of the human mind and opening up prospects for practical application in various spheres of life. An overview of key points in the development of brain research and hypnosis indicates the continuous development of these scientific fields. From the beginning of brain research to current research in hypnosis, substantial transformations have been observed. In the initial period of brain research, emphasis was placed on the anatomical structure and basic functions. However, with the development of technologies such as neuroimaging, researchers have been able to examine brain activity in detail in real time. An important trend in the field of hypnosis is the growing interest in understanding the brain mechanisms underlying hypnotic phenomena. Current research follows the relationship between hypnosis and brain activity and explores the possibilities of using hypnosis in medicine and psychotherapy. All this points to the continuous development and deeper understanding of the phenomena of the brain and hypnosis, which serves as the basis for further scientific research in these important areas.

In the history of brain research and hypnosis, there are prominent figures whose contributions have become the most valuable in the intellectual development of these areas. One of the key figures is Charles Sherrington, who won the Nobel Prize for his work on brain physiology. His studies set the stage for further research on brain activity. Another important figure is S. Freud, the founder of psychoanalysis, who made a substantial contribution to the understanding of mental processes and the use of hypnosis in psychotherapy. Modernity is defined by trends established by scientists such as Oliver Sacks in neuroscience and Antonio Damasio in neuropsychology. They focus on understanding the connections between the brain and the emotional sphere, which identifies new pathways for further research. Summarising, the contributions of such researchers as Sherrington, Freud, Sacks, and Damasio identify current trends in brain and hypnosis research aimed at a deep understanding of brain processes and their impact on mental health [195].

Current trends in brain research are determined by the use of advanced techniques and technologies that allow examining this complex organ in more detail and accurately. For example, a new industry such as ***neuroimaging*** represents an important area in the examination of brain activity and structure, which is undergoing substantial development thanks to new technologies. Functional magnetic resonance acts as a key tool in this context, allowing researchers to assess the activity of different brain regions while performing different tasks. The use of fMRI allows obtaining a detailed image of the activity of brain regions with high spatial and temporal resolution. This method allows observing changes in brain activity associated with specific tasks, which provides a deeper understanding of the functional organisation of the brain.

Interest in neurophysiological methods has expanded the field of hypnosis research. Research aims to understand the brain processes underlying hypnotic states and how they affect mental states. Finding the brain correlates of hypnosis has been made possible in large part by functional neuroimaging, especially fMRI. Hypnosis may alter pain perception at the brain level, as evidenced by a noteworthy fMRI study that showed hypnotised subjects had decreased activation in the anterior cingulate cortex, a region involved in pain perception [196]. This discovery has important ramifications for the application of hypnosis in pain management and psychotherapy, as well as strong evidence for its function in changing pain-related brain circuits.

An additional step in the development of neuroimaging is the introduction of high-resolution anatomical and functional neuroimaging tomography methods. These methods allow not only to obtain high-quality images of the brain's structural elements but also to examine their functional interaction. Current approaches to the examination of hypnosis and its brain correlates reflect trends aimed at uncovering and understanding this mysterious state of consciousness. In the field of hypnosis research, the use of neuroimaging and MRI is important. These technologies allow us to analyse brain activity during hypnotic states and determine changes in functional brain regions. Using neuroimaging, including MRI, researchers can obtain detailed images of the brain and determine its response to hypnosis. The activity of various brain regions can be considered to understand how hypnosis affects neurophysiological processes. In addition, establishing changes in functional brain regions and their relationships in hypnotic states is an important step in uncovering the mechanisms of hypnosis and its effective use in treatment and other practical aspects. Electroencephalography and magnetoencephalography are used to record the electrical activity of the brain during various phases of hypnosis. These methods allow analysing changes in the frequency of waves and magnetic fields associated with states of hypnosis. The EEG registers electrical signals generated in the brain, while MEG measures magnetic fields generated by electrical activity. These methods help to identify the electrophysiological basis of hypnosis and determine its impact on various aspects of brain activity. Structural magnetic resonance allows assessing structural changes in the brain, such as the size and volume of individual regions, in the context of hypnosis. This method examines the relationship between the structural aspects of the brain and the depth of hypnotic exposure, revealing possible changes in brain anatomy associated with entering a state of hypnosis and its effect on structural components [197].

The cognitive-neuroscience approach involves using cognitive-neuroscientific paradigms to examine the relationship of attention, memory, and other cognitive functions with hypnosis. This approach involves analysing the effects of hypnosis on information processing and decision-making and identifying the mechanisms underlying changes in cognitive function during hypnotic states. These approaches interact to uncover deeper aspects of hypnosis and its relationship to brain activity, making a substantial contribution to the development of this mysterious field of science. A detailed review of the methods used in brain research includes various techniques that allow examining brain activity at various levels, from neural networks to molecular processes [198].

Genetic research is aimed at identifying the role of genes in the formation of brain structure and functions. This approach allows identifying genetic factors that affect brain development and may be important for understanding its functioning, including mechanisms associated with hypnosis [199].

Biochemical studies are aimed at determining the chemical composition of brain tissue and identifying molecular pathologies. This approach allows analysing biochemical processes in the brain, which helps in revealing the features of the brain functioning in hypnotic states and identifying changes in its chemical composition.

An industry such as ***molecular neuroscience*** is determined by the application of advanced molecular biology techniques to examine the genetic and biochemical aspects of the functioning of neurons and their networks. Molecular neuroscience involves the examination of the genetic and biochemical processes that may be associated with hypnosis. A detailed analysis of the effects of neuromodulators and genetic variations allows for determining the relationship between specific molecular factors and the occurrence and retention of hypnotic states. This contributes to understanding the molecular basis of hypnosis and opens up opportunities for developing new approaches to the investigation and application of hypnotherapy. This area of research allows identifying the deep molecular mechanisms that underlie the physiology and pathology of the nervous system. Using the latest molecular biology techniques, such as *genomic sequencing, edge gene editing, and high-speed biomolecule analysis*, allows scientists to analyse the genetic composition of neurons and examine their expression and interaction with other cells. Molecular neuroscience aims to uncover the role of specific genes, proteins, and other molecules in the functioning of the nervous system. This is important for understanding the molecular mechanisms of learning, memory, and the development and functioning of neural networks [200].

Electrophysiology and integration of data are key components in modern brain research, providing an opportunity to examine its functional activity in detail and understand the relationships between different aspects of physiology. Electrophysiological techniques, such as electroencephalography and magnetoencephalography, allow assessing the electrical activity of the brain in real time. The EEG measures the electrical signals generated by neural activity, while MEG registers the magnetic fields that accompany this activity. These methods provide an unsurpassed opportunity to examine the dynamics of brain activity at different levels. Another important component is the integration of data from different sources. It involves combining electrophysiological data with genomics, metabolomics, and structural data to create a more complete picture of brain function. This approach allows examining the interactions between genetic factors, metabolic processes, and structural features, contributing to a deep understanding of brain activity and the relationships between different levels of biological organisation.

Innovative methods for increasing resolution include using advanced technologies to examine brain structures in detail with the highest possible resolution. One of these methods is *optical cochlear tomography*. Optical cochlear tomography uses high-resolution optical techniques to investigate the microstructures of the brain. This method allows viewing brain regions with exceptional resolution, revealing details at the nanometre level. This allows researchers to get a more accurate picture of the brain's architecture and functions, which is important for a deep understanding of how it works. The use of such high-resolution methods allows identifying even the smallest details of brain structures, which opens up new opportunities for analysing the microarchitecture and functions of the nervous system. These trends in brain research indicate the continuous development and improvement of research methods, leading to a deeper understanding of this fundamental organ [201].

Brain–computer interfaces (BCIs) are an active area of study that aims at creating technologies that allow people to directly use brain activity to interact with computers and other systems, without using muscles or language. This is an extremely important area that can revolutionise the way we communicate and control. BCIs use electroencephalography techniques and other technologies to record the brain's electrical activity and then translates these signals into commands for electronic devices. This can be especially useful for people with disabilities who may lose the ability to move or speak. Integration of BCIs with *artificial intelligence* (AI) extends the capabilities of these systems. AI can improve brain signal recognition, make predictions, and even interact with the user at a higher level, making brain-technology interaction more efficient and convenient. The integration of brain signals with AI systems is an active area of research aimed at expanding the cognitive capabilities of people. Researchers are exploring the possibility of combining content derived from brain signals with state-of-the-art AI algorithms to improve information recognition, analysis, and interpretation. This allows expanding the functionality of BCIs and creating new opportunities for interaction between humans and technology. The use of AI in the analysis of brain signals helps to effectively interact with the user, predict their needs and respond to them. This area of research may have substantial implications in the field of technology development to support people with various limitations and create interfaces that allow them to interact more effectively with digital systems using only brain activity [202].

Hypnosis research represents an important area in modern science, where hypnotherapy is considered a method of treating a variety of conditions, including pain, stress, and anxiety disorders. Recently, there has been a growing interest in the latest trends in the use of hypnosis, such as its effect on physiological processes, its use in cognitive behavioural therapy, and the development of personalised approaches to patients. A deep understanding of these aspects can open up new perspectives in hypnotherapy and improve treatment methods, considering the individual needs and characteristics of each patient.

Neurophysiological approaches to hypnosis are aimed at examining the brain mechanisms that underlie hypnotic states. These trends indicate continued progress in analysing the brain and understanding its functions. Investigating the interaction of hypnosis with neuronal activity opens up new opportunities for developing effective treatments and improving cognitive abilities. This approach allows integrating modern discoveries in neuroscience with practical applications in hypnotherapy, contributing to the further evolution of this scientific field.

Current research on the effects of hypnosis on the brain identifies important discoveries, establishing the mechanisms of this phenomenon and its impact on different levels of brain activity. Studies on functional changes in the brain using functional magnetic resonance imaging indicate that hypnosis can lead to changes in the functional activity of certain brain regions, such as the anterior cortex, which is responsible for decision-making and control. This suggests the possibility of hypnosis interacting with important functional aspects of the brain and its impact on higher cognitive functions. Examinations of the brain's electrical activity using electroencephalography show that hypnosis can affect the frequency and synchronisation of electrical brain waves. This may be due to a state of deep relaxation and altered consciousness, indicating electrophysiological correlates of hypnotic states. Examinations of molecular processes and genetic aspects show that hypnosis can influence the expression of stress-related genes, adaptation, and regulation of emotions. This indicates molecular and genetic changes that can be caused by hypnosis and affect the physiological state of the body. Research shows that hypnosis can influence pain perception by modulating brain pathways, including those responsible for pain sensations and stress responses. This confirms that hypnotic effects can have a substantial effect on pain perception and response in brain activity.

Discoveries regarding neuroplasticity of the brain under the influence of hypnosis indicate changes in the connections between neurons and the ability to reprogram certain brain functions. This indicates that hypnosis may have an impact on the structural and functional aspects of the brain through its interaction with neuroplasticity. Research also highlights the positive effects of hypnosis on mental health, particularly in reducing stress, anxiety, and symptoms of depression. This indicates the potential psychotherapeutic benefits of using hypnosis in the context of maintaining mental health. These findings help uncover the complexity of the interaction between hypnosis and brain activity and also open up prospects for using hypnosis in medicine, psychotherapy, and other fields to improve a person's physical and mental health. A review of possible applications of brain research and hypnosis in modern medicine and psychotherapy points to a wide range of promising uses for these approaches to improve physical and mental health. The pain and rehabilitation investigation highlights that hypnotherapy can be used to reduce pain, especially in patients with chronic pain syndromes. Brain research helps to understand the neurophysiological mechanisms of the analgesic effects of hypnosis, which is important for the development of effective rehabilitation methods. New trends in pain research and rehabilitation include integrating cutting-edge technologies, such as the use of virtual reality (VR) and neuromodulation, to enhance the effectiveness of hypnotherapy. For example, the use of VR during hypnotherapy sessions can create an immersive environment that promotes better analgesic response and improves patient rehabilitation [203].

New trends in the research on the treatment of mental disorders include expanding the range of applications of hypnotherapy for the treatment of various mental conditions, involving those patients resistant to standard methods and innovative approaches, for example, the use of hypnotherapy in combination with AI technologies for an individualised approach to each patient. Brain research identifies the mechanisms that may underlie successful hypnotherapy treatment of various mental disorders more deeply. The use of hypnotherapy in combination with AI technologies to create individualised treatment programmes that consider the characteristics of each patient. The studies aim to introduce hypnotherapy into new areas, such as the treatment of concomitant conditions in mental disorders and the basis of a psychotherapeutic approach for various categories of patients. These areas indicate the need to find innovative and effective methods of treating mental disorders through hypnotherapy, considering individual characteristics and expanding the scope of application of this method.

Research on the use of hypnotherapy in the treatment of forbidden habits and addictions is supported by some new trends. Integration of hypnotherapy with cognitive behavioural techniques. Research shows that combining hypnosis with cognitive behavioural therapy strategies can provide more effective outcomes in treating addictions. The use of VR technologies. Some studies explore the possibility of using hypnosis in combination with VR to create special scenarios that promote the rejection of bad habits. These areas of research point to the prospects for the development of effective methods of treating addictions and forbidden habits through hypnotherapy, integrated with other psychotherapeutic approaches and the use of the latest technologies. Hypnotherapy is used to treat various forms of addictions and forbidden habits. For example, a smoking treatment study used hypnotherapy in combination with cognitive behavioural therapy, which led to an improvement in the effectiveness of treatment programmes [203]. Brain research in this context helps uncover the neurophysiological mechanisms associated with habit formation and control.

In terms of treating mental disorders, hypnotherapy is also effective in treating depression, anxiety and post-traumatic stress disorders (PTSDs). Brain research points to changes in the functioning of emotional and memory areas that support the success of hypnotherapy interventions. In addition, hypnotherapy can be used to break bad habits, such as smoking or alcoholism. Addiction research confirms that hypnotherapy can help shape and maintain positive changes in the mental health and behaviour of people with addictions [204]. Hypnotherapy is effective in changing behaviour and cognitive processes. For example, the use of hypnosis can affect learning skills and improve cognitive function, which is supported by brain research results. Hypnotherapy is also used to deal with fears and phobias, which helps improve patients' quality of life. Brain research in this context allows expanding the understanding of the mechanisms underlying the effects of hypnosis on mental processes. It indicates changes in the activity of various brain regions that contribute to the adaptation and improvement of a person's cognitive functions.

Recent trends in hypnosis research in the field of medical support and training highlight the expansion of its application and improvement of techniques. New studies provide further evidence for the effectiveness of hypnosis in reducing anxiety before surgery. Examples point to the successful use of hypnosis to facilitate the treatment process, including regulating patient insurance responses. These new approaches emphasise the potential of hypnosis to improve patient well-being and optimise medical procedures. Recent trends in hypnosis research point to its promising possibilities in managing dental pain and anxiety. Hypnotherapy is considered an alternative method for reducing pain and stress during dental procedures. Brain research is aimed at the mechanisms of analgesic effects of hypnosis in this context, which allows understanding the physiological aspects of its effectiveness and developing new methods of pain management in dental practice. New trends in hypnosis research highlight its potential in improving sleep quality and treating insomnia. For example, hypnotherapy is used to modulate sleep structures, and brain research indicates the mechanisms of this influence, which opens up new opportunities for developing effective approaches in the field of sleep and insomnia.

Brain science and hypnosis can be classified according to various criteria, considering specific aspects of these areas.

4.1.1 Classification of Brain Research

Structural neuroscience. New trends in brain anatomy studies include the use of advanced technologies to examine different departments and structures in more detail. For example, the use of three-dimensional magnetic resonance (3D MRI) and VR allows for the accurate and amazing analysis of the anatomical details of the brain and examination of them in the context of various physiological conditions. One example of cutting-edge research is a project that uses a combination of 3D MRI and VR to assess changes in the structure of the hippocampus, an area of the brain responsible for memory and learning. This allows researchers to not only review the anatomy in detail but also understand how changes in this area can affect brain function and behaviour. Such innovative methods not only identify new aspects of the brain's anatomy but also provide opportunities for a deeper understanding of the relationships between its structure and function in different settings.

Functional neuroscience. Recent research in the field of analysing the activity of various brain regions under normal and specific conditions demonstrates great achievements. Studies conducted using functional magnetic resonance imaging allow examining the activity of various brain regions during the performance of specific tasks or in response to specific stimuli. New analysis methods allow obtaining more accurate and detailed data on the brain's response to various influences. For example, a study examining brain activity during concentration tasks reveals specific brain interactions related to attention and cognitive processes. Such studies can have important implications for understanding the mechanisms of attention and opening up new opportunities for its improvement. In addition, new trends in the use of machine learning and large-volume data analysis in neuroscience allow expanding horizons in understanding complex brain processes. Examples of such research include developing models that predict brain responses to various stimuli or interactions that occur at different levels of the neural network.

Psychophysiology and psychology. One of the latest trends is the use of deep learning and neural networks to analyse functional magnetic resonance. The researchers identified that this approach allows for more accurate identification of brain regions that interact during specific cognitive tasks, expanding the understanding of mental processes. In addition, the use of innovative real-time methods, such as functional electroencephalography, helps to identify accurate temporal characteristics of brain responses to various external stimuli. Considering these new approaches opens up the possibility of a deeper understanding not only of the physiology of the brain but also of its relationship to mental processes, which has important implications for psychiatry and neuroscience.

Brain stimulation and interfaces. New trends in brain activity research include the use of advanced techniques such as TMS. This method allows to safely influence brain activity using magnetic fields, which opens up new opportunities for research and treatment of various mental and neurological disorders. Another prospect is the development of interfaces for brain interaction, such as brain–computer interfaces (BCIs). These technologies allow direct reading and interpretation of signals from the brain, which opens up opportunities for creating new communication methods and therapies for people with various disorders. For example, the use of TMS in combination with functional magnetic resonance imaging may allow accurate identification of areas of the brain that are responsible for specific functions. In the case of BIs, research aims to develop more accurate and efficient interfaces that can be used to control machinery or virtual environments using brain signals. These areas of research point to broad prospects for developing technologies that can improve the quality of life and treatment of patients in the future.

Brain simulation and artificial intelligence. One of the new trends in brain research is the development of models that mimic brain processes and their use in the field of artificial intelligence. Research focuses on creating algorithms and systems based on the principles of the functioning of the human brain to improve artificial intelligence. For example, the use of neural networks that mimic the structure and function of neurons can improve the efficiency of machine learning systems and solve complex problems by reproducing the natural principles of information processing in the brain. These studies are aimed at creating more efficient and flexible AI systems.

Possible ***prospects for using research results*** in practical aspects include:

Individualised hypnotherapy techniques. New trends in hypnosis research point to the importance of considering individual characteristics of brain activity. This allows developing more effective and personalised hypnotherapy techniques for each patient, targeting specific problems or conditions, which can improve outcomes and provide a better clinical effect. For example, applying personalised approaches may include considering individual responses to different types of hypnosis and adapting techniques to the specific needs of each patient.

Medical integration of hypnosis. New trends in hypnosis research point to the possibility of introducing this method into medical practice for the treatment of pain, stress, and psychosomatic diseases. For example, research shows that hypnosis can be an effective tool for reducing pain symptoms, coping with stress, and improving patients' mental well-being. This paves the way for the development of comprehensive therapeutic strategies that use hypnosis as an important component of treating various diseases.

Development of new therapeutic programmes. New trends in hypnosis research are focused on understanding the mechanisms of its effects on the brain, which may open the door to developing new therapeutic applications. For example, research focuses on the use of hypnosis to improve cognitive function in patients with neurological diseases or experiments using hypnotherapy to treat chronic pain syndrome. These areas of research allow developing innovative methods aimed at improving both physical and mental health.

Improve learning and cognitive development. Modern research in the field of hypnosis is aimed at developing and improving the learning process. The use of hypnosis in this context may be supported by new trends in the study of the effects of hypnotic states on cognitive function and memory. For example, research focuses on the possibilities of hypnosis in improving the ability to remember and understand information, which opens up new prospects for the use of hypnosis in educational processes.

Application of hypnotherapy in rehabilitation. New trends in brain research highlight the importance of introducing hypnotherapy into rehabilitation programmes to facilitate adaptation and recovery of physical and mental health. For example, considering the results of research can contribute to the development of innovative approaches to hypnotherapy that consider the individual characteristics of the patient and their brain activity. Such personalised techniques can maximise the effectiveness of rehabilitation programmes, promoting optimal function recovery and improving the quality of life.

Using innovative approaches to medical training. Recent research shows new trends in the use of hypnosis in medical training, in particular, the implementation of expansion through new methods. Scientific discoveries in the field of neuroscience and neuroimaging provide an opportunity to better understand the role of the brain in shaping the psycho-emotional state of patients before medical procedures. For example, examining brain activity during hypnotherapy can lead to innovative approaches to individual medical training, providing optimal conditions for each patient.

Development of technological tools for monitoring hypnotic states. Recent research points to the development of technological tools for monitoring hypnotic states based on the study of brain processes. These technologies can greatly facilitate objective measurement and analysis of the depth and nature of hypnosis in patients, opening up new opportunities for accurately determining the effectiveness of hypnotherapy and understanding its impact on brain activity.

These prospects open up a wide scope for improving the methods of applying hypnosis and introducing innovative approaches to the practice of medicine and psychotherapy, which will help ensure effective and individualised treatment.

4.1.2 Classification of Hypnosis Research

Research on psychophysiological aspects of hypnosis. This study examines the physiological aspect of the hypnotic state, analysing the impact of this state on various systems and organs of the human body. One of the key trends in this direction is the use of advanced technologies of generated data recording, such as functional magnetic resonance imaging and electroencephalography, for a more detailed review of the internal processes that occur during hypnosis. Research shows that hypnosis can affect the activity of different parts of the brain. For example, during a hypnotic state, there are changes in the activity of the anterior cortex, which is responsible for higher mental functions such as attention and decision-making. In addition, physiological changes in the pituitary gland and hypothalamus, which regulate hormone production, are also explored in the context of hypnosis. The latest technologies are also used to examine the autonomous functions of the body in a hypnotic state. Studies using cardiovascular monitoring and electrodermal activity show that hypnosis can affect the regulation of heart rate and the level of electrical activity of the skin. One example of modern research is the work conducted by a group of researchers in the field of psychophysiology. They used functional magnetic resonance imaging to analyse brain activity in volunteers who were in a hypnotic state while performing certain tasks. The results showed that hypnosis helps activate certain areas of the brain responsible for performing tasks and can improve functional connections between them. The general conclusion of such studies indicates that hypnosis can affect the physiology of the body at various levels, from brain activity to the regulation of autonomous functions. These findings open up new possibilities for the use of hypnosis in medical and psychotherapeutic practice to improve the health and well-being of patients.

Research on the neurophysiology of hypnosis. In modern neuroscience, much attention is paid to the examination of the brain mechanisms responsible for the state of hypnosis. Recent research in this area identifies a number of key aspects that expand our understanding of this phenomenon. One aspect of the study is to identify brain regions that are activated during a hypnotic state. Functional magnetic resonance imaging and electroencephalography allow the accurate determination of changes in the functional activity of various parts of the brain. For example, some studies suggest that hypnosis can be accompanied by changes in the functioning of the anterior target cortex, which is responsible for performing tasks and controlling behaviour. Another important trend in hypnosis research is to examine the role of networks involving the cerebral cortex in this phenomenon. In particular, the interaction between the anterior part of the brain and the medial part of the network of small brain hemispheres was identified. This interaction can play a crucial role in creating a hypnotic state and influence the perception and processing of information. When examining hypnosis, new methods and approaches are also emerging to reveal its mechanisms. For example, the use of immersive VR technologies can help create specific scenarios for hypnotic research and the visualisation of brain responses. Another current trend is the assessment of individual variability in the response to hypnosis. Some people may be more susceptible to hypnotic influences than others, and this may be due to genetic traits or previous experiences. These trends and the latest research not only determine the complexity of the brain mechanisms of hypnosis but also open up prospects for the development of innovative methods of using hypnosis in clinical and psychotherapeutic practice and for the development of personalised approaches to the examination of this unique phenomenon.

Research of cognitive and psychological aspects. In modern scientific literature, much attention is paid to the investigation of mental aspects, particularly attention and memory, in hypnotic conditions. Hypnosis, as a state of altered consciousness, arouses researchers' interest in its effects on cognitive function and brain activity. The latest research in hypnosis uses a variety of techniques, such as functional magnetic resonance imaging and electroencephalography, to monitor changes in brain activity during hypnosis. For example, studies using fMRI have shown that hypnosis can affect the activity of certain parts of the brain associated with attention and memory tasks. In hypnotic settings, the effect of hypnosis on peripheral and central aspects of attention is also considered. Some studies show that hypnosis can improve concentration and help increase mindfulness of certain stimuli. One example of recent studies is a paper taking on the effects of hypnosis on episodic memory. Research has shown that the hypnotic state can substantially increase the memorisation of episodes, in particular, in the context of stories or events that became the object of attention during hypnotic exposure [205]. The general trend indicates that hypnosis can affect cognitive processes, such as attention and memory, and opens up new possibilities for using hypnosis in clinical and psychotherapy practice. However, it is important to continue research to better understand the mechanisms of these influences and their potential benefits and risks for different populations.

Research on hypnotherapy. Modern medicine faces urgent challenges from mental and physical illnesses, and in the search for effective treatment methods, scientists are examining alternative approaches, particularly hypnosis. Recent years have been characterised by a deeper understanding of the mechanisms of hypnosis's influence on the mental and physiological levels and an expansion of its use in the treatment of a wide range of diseases. Recent studies have shown that hypnosis can affect brain function by activating certain neurophysiological processes. MRI and EEG show changes in the activity of various brain regions that may be associated with pain relief, stress reduction, and improved mental health. Many studies show that hypnotherapy can affect the functioning of the amygdala by regulating feelings of fear and anxiety. Hypnotherapy has been shown to be effective in treating various mental illnesses, such as anxiety disorders, depression, and PTSD. The mechanisms underlying this effect are associated with changes in mental state and psychosomatic symptoms. The latest research confirms the effectiveness of hypnosis in reducing the symptoms of anxiety disorders. Hypnosis also shows potential in treating physical illnesses such as chronic pain, asthma, insomnia, etc. The interaction of hypnosis with physiological processes can lead to relief of symptoms and accelerated recovery. Numerous meta-analyses of clinical trials show the effectiveness of hypnosis in treating chronic pain and supporting healing processes. Modern research in the field of hypnosis identifies new possibilities for its use in the treatment of mental and physical diseases. Neurophysiological mechanisms, effectiveness in the treatment of mental disorders, and physiological benefits make hypnosis a promising area for further research and implementation in clinical practice.

Research of social and ethical aspects. Hypnosis, as a psychotherapeutic method and tool of influence, is becoming the subject of growing interest in modern society. The use of hypnosis in practice is based on psychological and neurophysiological foundations, which creates extensive opportunities for influencing social relations and forming ethical issues. This review examines the impact of hypnosis on social aspects and the educational process, and analyses current trends and examples of its use in various fields. One of the most important areas of influence of hypnosis is its application in social relations. More recent research suggests that hypnotherapy can help resolve conflicts, improve communication skills, and increase empathy between people. In particular, hypnosis can be used to reduce social anxiety and develop positive interpersonal relationships. The use of hypnosis also raises ethical questions that require serious investigation and regulation. For example, it is important to consider consent and confidentiality when using hypnotherapy. In addition, the question arises of the effectiveness and safety of hypnosis in various contexts, such as medicine and psychotherapy. Current trends in the use of hypnosis cover a variety of areas, including medical procedures, sports psychology, and educational programmes. For example, hypnosis can be used to train athletes to achieve optimal psychological shape before competitions. In education, hypnotherapy is used to improve concentration and relieve stress before exams. The impact of hypnosis on social relationships and ethical issues is a substantial subject for research and discussion. The latest trends in the use of hypnosis in various fields demonstrate its potential to improve the quality of life and achieve positive changes in society. However, it is important to maintain a balance between innovation and ethical standards to ensure the safety and well-being of people who use hypnosis in different contexts.

This classification can be used to structure and organise scientific research in the areas of the brain and hypnosis. Researchers in the field of interaction between the brain and hypnosis continue their research in various directions, using modern methods of neuroscience and psychology. In the modern world, the examination of the interaction of the brain and hypnosis is an important area of research that opens up broad prospects for science to understand the complex psychophysiological processes of the human body. Being at the intersection of neuroscience, psychology, and medicine, this field of science attracts the attention of world-class scientists. The development of the latest technologies in neuroimaging allows solving the mysteries of how hypnosis affects brain activity step-by-step. The use of methods such as functional magnetic resonance imaging and electroencephalography allows obtaining unique data on changes in the work of various brain structures during hypnosis.

One of the key areas of research involves the investigation of neurophysiological correlates of hypnosis [206]. This means analysing the effects of hypnosis on different types of brain activity, especially on waves of different frequencies, which can open up new perspectives for understanding the state and functioning of the human brain. An important aspect is also the development of new methods of hypnotherapy aimed at the individual characteristics of the patient. Examining individual differences in the response to hypnosis at the brain level allows creating personalised exposure programmes that can be effective for a wide range of mental and physiological conditions. Equally important is the use of hypnosis to treat mental disorders, which opens up new opportunities for patients facing depression, anxiety, and other mental health problems. The future of brain-hypnosis interaction research promises a deeper understanding of the nature of these phenomena and the development of new, more effective methods of influencing human mental and physiological health. With each new discovery in this area, new opportunities open up for improving the quality of life and developing innovative methods of psychotherapy.

In the modern scientific and clinical environment, active development of *individualised hypnosis therapy* is observed, which is based on the unique characteristics of brain activity and the genetic profile of each patient. This innovative approach involves creating personalised hypnotherapy sessions, considering the individual characteristics of each client [103].

Using cutting-edge scientific research on brain activity and genetic traits, doctors can adapt therapeutic effects based on unique brain response patterns. This allows creating effective and personalised sessions aimed at achieving the specific goals of each client. In addition, advanced clinical studies and successful examples already confirm the effectiveness of these individual hypnosis therapies in various fields. This opens up new prospects for improving the effectiveness and personalisation of approaches in hypnotherapy.

In the field of hypnosis research, substantial importance is attached to *the analysis of ethical aspects and establishment of a legal framework for its use*, in particular, in the medical and psychotherapeutic fields. This becomes a vital element in ensuring the safety and effectiveness of hypnosis in clinical practice. Research on the ethical aspects of hypnosis use focuses on a number of issues, such as informed patient consent, privacy, integration of medical personnel, and consideration of personal moral beliefs. It is important to develop ethical standards that define clear principles for patient-therapist interaction during hypnosis sessions. The creation of a legal framework for the use of hypnosis for medical and psychotherapeutic purposes includes defining licensing of specialists, standards of practice and mandatory means of quality control of services. In addition, regulation may include setting rules for maintaining medical records and considering cases where hypnosis is used for medical procedures. These ethical and legal guidelines are important for ensuring patient confidence, safety, and the effectiveness of hypnosis use in clinical scenarios. Continuing research in this area contributes to the development of standards that guarantee a high level of professional ethics and legal responsibility.

Hypnosis, as an innovative method, is marked by the potential for application in the field of training and training. The implementation of *hypnotic techniques in educational programmes is aimed at improving students' cognitive abilities and activating their creative processes.* This challenges teachers and curricula to develop effective strategies for using hypnosis in the learning environment. The use of hypnotic methods in teaching includes the creation of special programmes that consider the individual characteristics of each student. With the help of hypnosis, it is possible to stimulate concentration, increase memory, and facilitate the process of learning new material. Hypnotic techniques can be aimed at improving cognitive functions such as logical thinking, analytical skills, and problem-solving. Using hypnosis in this context can help students better learn and understand complex concepts. Hypnosis also shows the potential to stimulate creativity and creative processes. With the help of hypnotic techniques, the unconscious part of the mind can be activated, which helps to identify new ideas and approaches in the educational process (Fig. 21). However, the use of hypnosis in training requires careful research and the development of safety standards. It is necessary to consider the individual characteristics of students, avoid possible risks, and ensure ethical standards. The overall goal is to create innovative teaching methods that maximise the potential of hypnosis to improve the quality of education and student development [207].

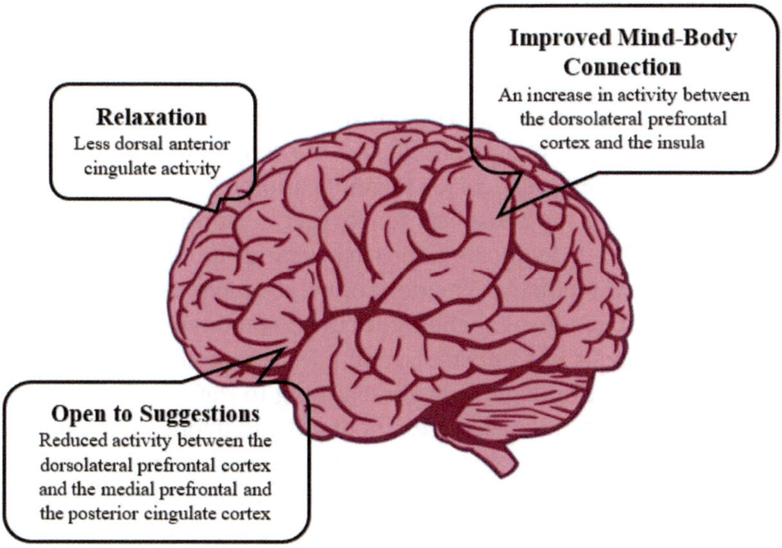

Figure 21 Diagram of the effects of hypnotic techniques on different brain regions.

Considering modern possibilities, hypnosis identifies *synergistic potential with other cutting-edge technologies,* in particular, with VR. Integrating these technologies can create unique and effective approaches to treatment, training, and development. Combining hypnosis and VR can help create immersive and individualised scenarios for therapy. VR can help reproduce situations to improve the therapeutic effect of hypnotherapy. The use of hypnosis and VR in learning can provide interactive and in-depth learning methods. This is especially useful in situations where practical experience is important, such as in medical or military training. Combining hypnosis and VR can create unique psychotherapy scenarios for dealing with phobias, anxiety, and other mental health problems. This can help patients immerse themselves in controlled virtual environments to deal with their emotions. For example, in the field of rehabilitation after injuries or operations, hypnotherapy can be combined with VR. The patient can be in a virtual environment created to restore mobility and functionality. Hypnosis can improve the perception of this environment and increase the patient's motivation to actively participate in the rehabilitation process. However, it is important to consider the ethical aspects that arise when using VR and hypnosis. Ensuring patient privacy and safety and developing standards for the use of these technologies is a critical challenge for successful synergy in these areas [208].

A general discussion of current trends and future directions of brain and hypnosis research indicates the constant development of this field and the prospects for its application in various spheres of life. Highlighting the prospects for further scientific discoveries in brain research and hypnosis has great potential for change in society, education, and medicine. The development of hypnotherapy opens up new prospects for expanding medical treatment, in particular, for the treatment of pain, stress, and psychosomatic diseases. New research and technologies are contributing to the creation of more effective and individualised treatments. Using hypnosis to manage pain provides an opportunity to reduce the need for analgesics and medications, which can be especially important when treating patients with chronic pain or during surgery. Hypnotherapy can be an effective tool for reducing stress and anxiety in patients before surgery, treatment of chronic diseases, or other medical procedures. Hypnotherapy can be useful in the treatment of psychosomatic diseases, where the patient's emotional state affects physical health. This can include conditions such as headaches, gastrointestinal diseases, etc. Understanding individual characteristics of brain activity and genetic profile allows creating personalised hypnotherapy approaches. This may be particularly important for patients with specific needs or treatment response options. The use of technologies such as neurofeedback and functional brain imaging allows for more accurately and effectively influencing the patient's brain activity during hypnotherapy. It is important to develop norms and ethical standards for the use of hypnosis in medical practices. Further research and clinical trials are needed to confirm the effectiveness and safety of hypnotherapy in various medical contexts.

The disclosure of the brain mechanisms of hypnosis opens up new prospects for the development of innovative psychotherapeutic techniques aimed at treating various mental disorders. Understanding the relationship between hypnosis and brain activity is becoming a key element in improving psychotherapy and improving its effectiveness. Knowledge of the individual characteristics of brain activity and response to hypnotherapy allows creating individual psychotherapy programmes for each patient. This is especially important in the treatment of various mental disorders, as each person is unique. Hypnotherapy can be effective in treating psychosomatic disorders where emotional state affects physical health. New techniques may be aimed at correcting the brain mechanisms behind such disorders. Understanding how hypnosis affects areas of the brain responsible for processing stress and trauma allows developing psychotherapy strategies for treating post-traumatic disorders and stress responses. Hypnotherapy can be used to correct sleep disorders and restore emotional balance. Innovative approaches may include hypnotherapy sessions aimed at improving sleep quality and regulating emotions. Using hypnosis to create virtual scenarios can help patients overcome phobias, anxiety states, and other mental obstacles. The development and introduction of VR and other technologies can improve the integration of hypnosis with other methods of psychotherapy, expanding the possibilities of treating mental disorders. The growing interest and use of hypnotherapy require ethical use and compliance with safety regulations, especially when dealing with new methods and technologies.

The use of individual brain characteristics to develop personalised therapies and psychotherapy is a promising area aimed at increasing the effectiveness and adaptability of approaches to the needs of each patient. Based on individual characteristics of brain activity, hypnotherapy approaches can be adapted to the specific needs and responses of the patient. This involves considering the variability of the brain's response mechanisms to hypnosis and adapting sessions to individual neurophysiological features. Considering the patient's genetic profile can determine their propensity for certain forms of hypnosis and influence the choice of hypnotherapy strategies and techniques. Understanding the genetic aspects allows creating individualised and genetically adapted approaches

[209]. Evaluating a patient's cognitive-emotional profile can help identify key areas of focus in hypnotherapy. Applying individual approaches to each aspect of treatment can improve outcomes and patient convenience. Personalising hypnotherapy sessions based on individual brain characteristics can lead to a greater increase in the effectiveness of treatment. Analysis of brain functional activity can indicate the optimal time and format of sessions for each patient. The use of advanced technologies, such as neurofeedback and cognitive interfaces, allows real-time adaptation of hypnotherapy to the needs of the patient's brain activity, which contributes to a high individualisation of therapy. These innovative approaches can substantially improve the quality of hypnotherapy and make it more accessible and effective for different categories of patients.

Creating new scientific disciplines that combine brain research and hypnosis is an exciting area that opens up unique opportunities for the development of the scientific community and a deeper understanding of the interaction of these two areas.

For example, the integration of hypnosis and neuroscience is called ***neurohypnology***—a promising scientific discipline aimed at the in-depth examination of the interaction of hypnosis with various aspects of brain activity and neuronal mechanisms [210]. This area of research is defined by considering a wide range of aspects that cover the analysis of neuronal mechanisms of hypnosis. The study aims to identify specific neural pathways that are activated under the influence of hypnosis. Uncovering these mechanisms will allow a better understanding of how hypnosis modulates brain function and affects its functionality. Investigation of the dynamics of brain activity during various hypnotic states to determine the features of electrical and chemical processes in the brain under the influence of hypnosis. Examination of the interaction of different groups of neurons and their responses to different methods of hypnotherapy. This allows understanding those specific parts of the brain that respond to hypnosis and how it affects mental functions. Development of models for reproducing and analysing brain scenarios that occur in a state of hypnosis. This allows predicting and determining effective methods of influencing brain activity. Using an interdisciplinary approach, combining knowledge from neuroscience, psychology, and hypnotherapy to gain a comprehensive understanding of the mechanisms of hypnosis's influence on brain activity. Neurohypnology acts as a key component for explaining and optimising the effectiveness of hypnosis in the context of neuroscience research, which opens up new prospects for the development and improvement of hypnotherapy.

Psychoneurology of hypnosis is defined as a synthesis of psychological and neurological approaches in the examination of hypnosis. This new field of science aims to deeply understand and explain the relationship between the mental phenomena of hypnosis and neurological processes in the brain. Disclosure of neurophysiological processes occurring in the brain during hypnosis. The examination of changes in the operation of specific neural networks and their impact on the mental states of a hypnotic session. Analysis of how the psychological aspects of hypnosis interact with changes in the functioning of individual neural systems. Determination of which aspects of the subject's mental state can trigger specific neurological responses. Understanding how hypnosis affects cognitive functions of the brain, such as attention, memory, and the intensity of perception. Research on the influence of hypnosis on high-level cognitive processes. The use of electroencephalography techniques to record the electrical activity of the brain in hypnotic states. Analysis of frequency-amplitude characteristics to determine specific hypnotic effects. Research on the use of hypnosis as a tool for examining and improving perception, reactions, and other psychological processes at the neurological level. Investigating how the psychoneurology of hypnosis can help develop new therapeutic approaches for various mental disorders. Psychoneurology of hypnosis opens up new perspectives for understanding, using and optimising hypnosis in the context of modern psychotherapeutic and scientific research.

Cognitive neurohypnotherapy is an innovative approach that combines elements of hypnosis and cognitive behavioural therapy to improve cognitive processes in the brain and treat mental disorders [102]. This branch of science aims to reveal the possibilities of using hypnosis to modulate cognitive functions and improve cognitive processes. Examination of the effects of hypnosis on various aspects of cognitive functions, such as attention, memory, perception, etc. Identifying specific hypnotic techniques that can improve these processes. Analysis of the influence of hypnosis on the formation and change of cognitive circuits and beliefs. Exploring the possibilities of using hypnotherapy to modify negative cognitive patterns. Developing and testing new hypnotherapy approaches for the treatment of cognitive disorders such as attention disorders, cognitive deficits, etc. Examining the role of hypnosis in improving learning and memorisation processes. Developing hypnotherapy techniques to increase cognitive performance and academic success. Research on the use of hypnosis to preserve and improve cognitive function in the elderly. Development of programmes to prevent cognitive loss. Conducting empirical studies to evaluate the effectiveness of cognitive neurohypnotherapy methods in comparison with other therapeutic approaches. Cognitive neurohypnotherapy is defined as a promising pathway in the development of therapeutic strategies aimed at improving cognitive function and treating various psychiatric disorders.

The idea of creating a new industry—***neuropsychiatry hypnosis***—is interesting and may be useful for developing innovative methods of treating patients with mental disorders. Some studies already point to the potential benefits of using hypnosis in the treatment of various neuropsychiatric conditions, such as anxiety, depression, phobias, and PTSD. Hypnotherapy can be used to direct the patient's attention to internal resources and self-medication opportunities, reduce stress levels, and improve mental well-being. However, it is important to keep in mind that hypnosis is not a universal method and may not be suitable for all patients or types of neuropsychiatric disorders. Research in this area can help to understand the mechanisms of hypnosis's effects on mental health and develop effective and safe approaches to using hypnosis in the treatment of various neuropsychiatric conditions. Research in this direction surely can lead to an expansion of the arsenal of tools for treating mental disorders and improving the quality of life of patients [211].

The synthesis of engineering and neurohypnotherapy can define a new direction, which can be called "***neurohypnotherapy engineering***". Research in this discipline can be directed to the development of technologies for optimising hypnotherapy interventions [212]. The idea of "neurohypnotherapy engineering" sounds interesting and may open up new opportunities for optimising hypnotherapy approaches through the use of engineering technologies and neuroscience discoveries. The fusion of engineering and neurohypnotherapy can lead to the creation of innovative tools and techniques to improve the effectiveness and personalisation of hypnotherapy.

Possible areas of research include the development of devices for monitoring and analysing the psychophysiological indicators of a patient during hypnosis, the use of integrated technologies in the field of VR to create special environments for hypnotherapy, or the development of machine learning algorithms for individualising hypnotherapy sessions. This may also include using neural interfaces to interact with the patient's brain activity during hypnosis and adapt real-time therapeutic strategies. This hybrid approach can help improve the effectiveness and individualisation of hypnotherapy, contributing to better outcomes for patients with various psychiatric disorders.

All this can determine a new level of integration of technology with traditional methods of treatment in psychiatry and psychotherapy, creating a promising area of research and practical application. The creation of these new scientific disciplines allows exploring more deeply the limits of the possibilities of hypnosis and its interaction with neuroscience, which leads to the emergence of innovations and methods in treatment and psychotherapy. These prospects for further scientific discoveries point to the substantial impact of brain research and hypnosis on various areas of life and provide an opportunity to implement innovative solutions in practice. Thus, the coverage of brain research and hypnosis points to the importance of this area for medicine, psychotherapy, and education. Understanding the brain mechanisms of hypnosis opens up new possibilities for treating various mental and physical illnesses, optimising learning processes, and improving the overall health of society. The importance of further research in this area is to expand our knowledge of the intricacies of brain-hypnosis interaction, develop effective therapeutic strategies, and promote the development of new technologies to support physical and mental health. Further research can contribute to the creation of innovative treatment methods, improve the quality of life of patients, and develop new scientific approaches. Thus, the deepening of knowledge in this area is an important step for further improving the practical application of hypnosis in various spheres of life. The joint efforts of scientists, doctors and psychologists in the direction of further research will reach new heights in the treatment and care of patients and contribute to the discovery of new aspects of human consciousness and brain function. The relevance of continuing the development of scientific research and its practical application lies in the constant search for innovative solutions to solve complex problems in the fields of medicine, psychotherapy, and education. Deepening knowledge in the field of brain and hypnosis is essential for improving treatment methods, developing effective therapeutic strategies, and implementing cutting-edge technologies. The practical application of research results has a huge potential for improving the quality of life of patients, developing innovative methods of psychotherapy and optimising learning processes. Continuous participation in scientific research ensures the improvement of modern approaches, which is important for ensuring effective solutions to the challenges facing modern society in the field of health and mental well-being. Efforts in this direction can open up new perspectives for treatment, psychotherapy, and training, improving people's quality of life. Striving to develop innovative methods and use cutting-edge technologies is key to addressing health and mental well-being challenges. Scientists, doctors, psychotherapists, and educators are encouraged to collaborate by sharing research ideas and results to work together to improve treatments and develop new learning strategies. It is important to create favourable conditions for integrating scientific discoveries into practice, developing innovative technologies, and providing support for the introduction of new approaches in the medical and educational spheres. This will allow using the potential of hypnosis and brain research to improve the quality of life of people and the development of society as a whole.

4.2 Analysis of How Hypnosis Can Affect Brain Activity and Consciousness

In recent years, intensive scientific progress in neurophysiology and psychology has brought the issue of the impact of hypnosis on brain activity and consciousness to the forefront. Recent studies suggest that hypnosis may not only be an intriguing phenomenon in psychology but also a key tool for exploring and understanding the functions of the human brain. Scientists are focusing on how hypnosis can affect various aspects of brain activity, including electrophysiological indicators, functional magnetic resonance imaging, and other brain imaging techniques. The results of these studies indicate that hypnosis can contribute to changes in the functioning of specific brain regions, affecting sensations, attention, and other aspects of cognitive activity (Fig. 22). However, the detailed mechanism of the impact of hypnosis on brain activity remains the subject of active debate among scientists. Some researchers argue that hypnosis can cause changes in the redistribution of brain resources, while others speculate about possible changes in neural connection networks. Despite these discussions, research in this area opens new horizons for understanding the nature of consciousness and brain activity. In this context, the analysis of recent studies aims to uncover and comprehend the interaction between hypnosis, brain activity, and consciousness to reveal the potential of this phenomenon for the development of modern science and practice.

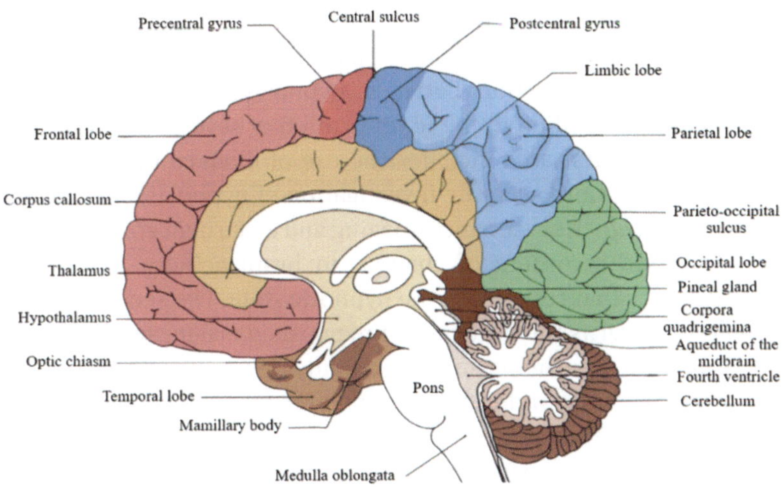

Figure 22 How the brain responds to hypnotic states.

Hypnosis, which has remained on the periphery of scientific interest for a long time, is now becoming the subject of increasingly intensive research, highlighting its significance in contemporary science. Expanding knowledge about brain activity and consciousness through the latest research methods allows for a deeper understanding of how hypnosis affects the functions of the human brain and psyche. One of the key aspects of the significance of hypnosis is its potential role in therapeutic applications, such as reducing pain, overcoming stress and depression, and improving cognitive functions. Modern studies indicate that hypnotherapy can be effective in treating various medical conditions, opening new possibilities for integrating hypnosis into contemporary medical practice. In addition, hypnosis serves as a tool for examining the nature of consciousness itself, allowing scientists to explore different aspects of cognitive function and mental processes. This opens up opportunities for understanding and enhancing mental health and psychotherapeutic approaches. To summarise, hypnosis in contemporary studies appears not only as a phenomenon that arouses curiosity but also as an important tool for expanding knowledge about human brain activity and consciousness. Its importance is evident in developing therapeutic practices and understanding fundamental aspects of mental processes, opening new perspectives for further research and applications in various fields of science and medicine.

In recent years, researchers have increasingly focused on analysing the interaction between hypnosis, brain activity, and consciousness, leading to the identification of several key trends in this field. The first area includes the use of modern neuroimaging methods for a more accurate investigation of hypnosis. Specifically, functional magnetic resonance imaging and electroencephalography allow for the recording of changes in brain activity, identifying specific areas that respond to the hypnotic state. The second area of research concerns the influence of hypnosis on cognitive functions. Work in this area helps to understand how hypnosis can affect attention, memory, and other aspects of cognitive activity. Overall, these trends in the investigation of hypnosis in the context of brain activity and consciousness open new horizons for understanding this phenomenon and its potential applications in science and practice.

Despite the diversity of studies on the impact of hypnosis on the brain, scientists are particularly focused on its effects on memory. This subject generates a significant amount of debate, discussion, and divergent conclusions within the scientific community. Researchers are interested in how hypnosis can influence various aspects of memory, including the mechanisms of encoding, retrieval, and forgetting of information. A large number of studies aim to understand how the hypnotic state can modify the perception and storage of information in the brain. However, the unconscious processes that occur during hypnosis remain a subject of active discussion among scientists. Some researchers support the idea that hypnosis can enhance the ability to remember and retrieve information, while others believe that the effects of hypnosis on memory may be limited or even contradictory. These differences in conclusions may be explained by the diversity of research methods, variability in people's responses to the hypnotic state, and the complexity of the memory phenomenon itself. Thus, despite a wealth of scientific literature and research in this area, the exact mechanisms of hypnosis's influence on memory remain a subject of further investigation and discussion within the scientific community.

The findings of the London Taxi Driver Study have been shown to provide a relevant analogy to the study of hypnosis, in particular with regard to experience-driven changes in brain structure [213]. In a manner analogous to how intensive spatial navigation training has been demonstrated to physically reshape the hippocampus, it is hypothesised that hypnosis may also be capable of inducing structural and functional changes in the brain regions associated with cognition, memory and perception. By elucidating the mechanisms underlying neuroplasticity, researchers can further explore how hypnosis may serve as a cognitive tool for modifying brain activity, supporting mental resilience, and even influencing therapeutic outcomes. This perspective reinforces the importance of continued research into hypnosis as a scientifically grounded phenomenon with significant implications for neuroscience, psychology, and clinical practice.

Hypnosis has long been associated with the potential to influence memory, making a significant contribution to clinical and legal practices. Researchers note phenomena such as hypnotic hypermnesia, pseudomemories, and amnesia related to memory and hypnosis. The analysis of the relationship between hypnosis and memory plays a crucial role in understanding memory and behaviour. Hypnosis creates a favourable environment for examining social, neuropsychological, and cognitive changes associated with memory. Proposed prospective research areas can expand the understanding of memory and its underlying mechanisms. Hypnosis opens new possibilities for investigating two main areas of memory: false memories and the influence of hypnosis on them. The first area is marked by the potential for creating false memories and unwarranted confidence in their accuracy, which has clinical and forensic implications.

Hypnosis, as a historically conditioned phenomenon, attracts researchers' interest due to its potential impact on memory and self-identity. One of the old beliefs about hypnosis is its ability to enhance memory, rooted in the practices of animal magnetism from the early 19th century, particularly in the observations of the Marquis de Puységur. These observations showed that somnambulist patients subjected to hypnosis could reproduce past events with details they could not recall in their ordinary state. Phenomena such as hypermnesia, amnesia, and the belief in hypnosis's potential to unlock hidden memories became subjects of research and are intrinsically linked to the history of hypnosis and memory. In the late 19th century, hypnosis played an important role in the development of experimental psychology. Many psychologists, such as William James, Wilhelm Wundt, Théodule Ribot, and Pierre Janet, examined hypnosis and noted the impressive memory capabilities exhibited by hypnotised individuals. Introduced by Ribot, the term "memory elevation" indicated hypnosis's ability to alter the way people remember and recall information. This period of observations of hypnotic patients also prompted the development of theories on dissociation, multiple personalities, unconscious cognition, and emotions.

Despite accounts of hypnotic hypermnesia, post-hypnotic amnesia, and post-hypnotic suggestion, some researchers, notably H. Bernheim [210], expressed doubts about the effectiveness of hypnosis and its potential impact on memory and self-identity. Hypnosis, as a method of therapy and research, has sparked lively discussions and debates among professionals throughout the history of psychology. These conversations encompassed questions about the effectiveness of hypnosis as a tool for retrieving traumatic memories and its impact on memory formation. Some researchers and practitioners regarded hypnosis as a means to help recover suppressed or forgotten memories, including those associated with traumatic events. Others questioned its efficacy and the legality of its use, especially in the context of legal practice.

The issue of whether traumatic memories emerging during hypnosis could be the result of suggestion and influence, and whether they might be false, prompted further research and discussions. This debate produced various perspectives, with some supporting the idea of hypnosis's effectiveness in memory retrieval and others highlighting the potential for creating false memories during hypnotic sessions. Research and experimental literature, particularly the works of J. R. Laurence and C. Perry, reveal various aspects of this issue, including the influence of suggestion and belief on memory formation [211, 212]. Debates among professionals in both scientific and legal arenas reflect differing approaches to the use of hypnosis and its potential impact on memory. Ultimately, the development of forensic hypnosis, which was proposed as a means to recover important but inaccessible conscious memory information, led to new debates and criticism. Some researchers questioned the successes claimed by hypnosis advocates and expressed doubts about its legality and ethicality.

Thus, the history of discussions surrounding hypnosis highlights the complexity of psychological and legal issues related to its use, effectiveness, and potential risks. It underscores the need for further research and careful consideration to understand how hypnosis can affect memory and testimony and how it can be used in medical and legal practice.

In the 19th century and today, matters have become more convoluted and complex with the renewed interest in multiple personalities, order, and the belief that not only does hypnosis allow unconscious memories to surface, but also that the recovered memories are often embedded in various "personalities" overall. The use of hypnosis by police has raised questions about whether suggestion leads to the introduction of misinformation, while the accounts of patients with different personalities have highlighted that suggestion can create entirely new (and false) memories and crystallise fabricated or fictional personalities. The importance of beliefs in shaping memory narratives is understood through the notion that hypnosis partially influences the formation of memory reports. Consequently, what people remember during hypnosis depends on their beliefs about the impact of hypnosis on memory. One common belief is that during hypnosis, participants recall events more thoroughly and accurately compared to those who do not use hypnosis. This belief can be traced back to the clinical work of Janet, who was one of the first therapists to use hypnosis to help patients recover memories of traumatic events. The roots of early practices like Janet's work and the belief that hypnosis can restore hidden memories of traumatic events have permeated many therapeutic approaches. Regarding the dissemination of these beliefs today, they remain significant in psychological and psychotherapeutic disciplines, although they are largely viewed critically and considered in the context of modern research and methods [214].

Studies conducted by M. D. Yapko [215], D. A. Poole et al. [216], E. Legault and J. R. Laurence [217], and J. Ost et al. [218] revealed high levels of endorsement by psychotherapists regarding hypnosis as a means of memory enhancement. For instance, a survey of over 850 psychotherapists in independent practice conducted by M. D. Yapko [215] indicated that 75% of respondents agreed with the statement that hypnosis allows people to accurately remember things they otherwise could not. Similar results were obtained in other studies, such as those conducted by D. A. Poole et al. [216], E. Legault and J. R. Laurence [217], and J. Ost et al. [218], which examined the opinions of psychologists and psychotherapists from various countries. These studies suggest that although a substantial amount of scientific literature has presented data contradicting the beliefs about hypnosis and memory, a significant number of professionals and laypeople still believe in these ideas. For example, the studies by W. G. Whitehouse [219] found that 93% of participants in the experiment believed that hypnosis could improve memory retrieval, whereas M. E. Johnson and C. Hauck [220] recorded that 88% of respondents agreed with the idea that hypnosis could help people remember things they otherwise could not.

A study conducted by J. P. Green et al. [221] covered the beliefs of students about hypnosis in four countries: Australia, Germany, Iran, and the USA. Despite the lack of reported cross-cultural differences in responses to statements about hypnosis and memory, approximately 90% of participants agreed with these statements. The influence of beliefs and expectations on memory, whether with or without hypnosis, has clinical and legal implications. Belief in memory shapes not only personal memory during hypnosis but can also lead to reports of memory recovery, even creating new identities. According to a survey conducted by clinicians Legault and Laurence, 17.5% of women who came to therapy without any real memories of abuse "recovered" these memories during therapy, regardless of whether hypnosis was used. Therapists who reported using many "recovery" techniques, including hypnosis, reported a higher number of recovered memories. The interaction of beliefs, expectations, and the recollection of life narratives remains an open field for research. The transmission of false beliefs (and expectations) and their potential consequences is an important aspect of understanding memory in action.

Regarding the question of whether hypnosis improves memory, empirical evidence is mixed. While some studies support hypnotic hypermnesia, controlled studies have shown how inaccurate memory reports during hypnosis can be. These studies also demonstrate that the focus on memory can distract from other important aspects, such as the accuracy of recall and memory errors. For decades, researchers evaluated memory in terms of the quantity of remembered items, but little attention was paid to accuracy and errors in memory. Modern research highlights that suggestions made during hypnosis and other similar methods can affect memory recall and recognition, often guiding them in specific areas.

Hypnotic hypermnesia has been a subject of research since the early 1970s. This term refers to extraordinarily vivid and expressive memories that arise during or after hypnosis. Primarily, these are memories that may relate to autobiographical events or even involuntarily evoke stories beyond personal experience, such as film scenes or conversational stories. One method of exploring this phenomenon is age regression, which involves asking participants, including patients, to "go back in time" and "relive" past events. However, controlled studies do not confirm the possibility of accurate regression to earlier stages of development or the unequivocal correctness of memories obtained through regression.

M. T. Orne [222, 223], as early as 1951, warned that hypnotic regression could lead to confabulations—the creation of fabricated events or memories. Researchers also emphasise that the historical accuracy of memories related to past lives or early childhood events is often difficult to independently verify. Future research plans include investigating the impact of hypnosis on memory, both in laboratory settings and in forensic cases involving eyewitness testimony. H. Relinger's [224] research results indicate that hypnosis can facilitate the reproduction of meaningful material, such as live or recorded scenes, but it may not have such an impact on other types of information, such as lists of words or meaningless syllables.

M. H. Erdelyi's [225, 226] research provides important conclusions regarding the influence of hypnosis on memory. They confirm that segmenting studies by type of memory test and types of stimuli used aids a more accurate assessment of the hypermnesic effect of hypnosis on memory. Studies focused on recognition demonstrate the absence of hypnotic hypermnesia for recognition, regardless of whether the material is meaningful or nonsensical. In studies comparing the memorisation of nonsensical stimuli with meaningful ones, the absence of hypermnesia for nonsensical stimuli is confirmed. However, for meaningful stimuli, most studies confirm the presence of a hypnotic hypermnesic effect, which exceeds normal recall. A caveat is that some studies do not consider that simple repetition can lead to increased memory retention without hypnosis. Studies that employ the typical "sleep hypermnesia design" produce ambiguous results, sometimes confirming the absence or insignificance of hypermnesia. Some studies suggest that hypnosis can lead to an increase in memory errors, which can pose a risk of false memory formation. In addition, enhanced recall following hypermnesia suggestion during hypnosis may simply reflect normal (non-hypnotic) hypermnesia. Therefore, the research confirms the potential for effective use of hypnosis to enhance recall but also demonstrates its potential risks in creating false memories.

Contemporary cognitive research indicates that memory is a fallible and flexible construct, heavily dependent on current beliefs, emotions, expectations, images, and assumptions about past events. Many researchers support the idea that memories, even the most ordinary, consist of a combination of accurate and inaccurate elements. In this context, it is not surprising that suggestive methods, including hypnosis, can distort the content of memories. Initial observations in this area were made in the late 19th century when H. Bernheim [210] introduced a false story to a patient, which was then remembered and reproduced under hypnosis. Even after the session ended, patients remained convinced of the accuracy of their memories. In the 1970s and 1980s, substantial evidence emerged about memory distortion during hypnosis. Empirical studies, such as those conducted by Laurence and Perry, showed that participants under hypnosis could create false memories under the influence of suggestion, even if they had no real basis. This demonstrates that memory can be easily distorted by external influences, and people can remain confident in the accuracy of their memories, even if they are incorrect.

Early studies on memory indicate its vulnerability to suggestive influences, including hypnosis and various suggestive procedures, such as dream interpretation and imagery. For example, researchers like P. Malinoski et al. [227] investigated the impact of hypnosis on participants' memories and showed that after hypnotic induction, a significant portion of participants recalled memories that were distant in time, compared to non-hypnotised participants. Studies also suggest that warnings about the possibility of false memories during hypnosis can reduce their occurrence, but the effect of these warnings remains limited. Some studies, including those conducted by C. A. Burgess and I. Kirsch [228], confirm a link between hypnosis and increased confidence in responses, regardless of their accuracy. This can lead to memory distortions, which may become apparent during or after hypnosis when witnesses misinterpret events or memories due to altered criteria. However, there are also studies showing the limited impact of hypnosis on enhancing memory accuracy.

In summary, research indicates the complexity of the impact of hypnosis on memory, particularly regarding the accuracy of memories and confidence in them, and the need for further studies to understand these processes. In experiments using hypnosis to assess confidence in memory accuracy, confidence ratings in hypnotised subjects exceeded those expressed without the use of hypnosis. This phenomenon has surprised memory researchers, as it contradicts the usual theory that reducing the criterion shift should lead to a hypermnesic effect. The findings support the idea that hypnosis can cause memory illusions and affect memory attribution. Further studies confirm that visual imagination and source monitoring may play a key role in this process, as they can lead to confusion between visual images and actual memories. Other studies confirm that hypnosis can make new memories appear very similar to old ones, even when this does not correspond to reality. The uncertainty of scientific evidence, the lack of theoretical justification for using hypnosis to improve memory, and the problem of excessive trust in hypnosis in enhancing witness memories indicate the unsoundness of using this method in legal practice.

For instance, a disorder such as functional amnesia, characterised by sudden memory loss due to psychological trauma and not associated with brain damage, is a key aspect of dissociative disorders, which include amnesia and identity disorder. This phenomenon attracts the attention of scientists due to its spontaneity, rapid onset, and complexity. Posthypnotic amnesia (PHA), in particular, involves the inability to recall certain material after hypnosis without a specific cue. Both of these phenomena pique the interest of scientists and raise questions about the relationship between memory, psychological mechanisms, and emotions, although their spontaneity complicates systematic research. Typically, functional amnesia is observed in highly hypnotisable individuals, where explicit memory is disrupted. This leads to discrepancies between implicit and explicit memory and reversibility, as noted by A. J. Barnier [229–231]. Researchers documented functional amnesia concerning various elements studied or experienced during hypnosis, such as lists of words in J. F. Kihlstrom's study [232–237], information received before hypnosis in the work of A. J. Barnier et al. [230], and past personal experiences in A. J. Barnier's experiment. The similarities between functional amnesia and PHA, namely the disruption of explicit memory and the dissociation between explicit and implicit memory and reversibility, have prompted researchers to use PHA as a laboratory paradigm for modelling and examining personal memory disorders. For example, researchers R. E. Cox and A. J. Barnier [238–240], R. E. Cox and R. A. Bryant [241], and A. Mendelsohn et al. [242] suggest that PHA can be used to model functional amnesia. A. J. Barnier [229–231] conducted a programmatic examination of PGA as a model of functional amnesia, focusing on personal, emotional, and everyday events, not limited to simple materials, such as word lists, or less important experiences, such as hypnosis events. For example, in A. J. Barnier's study on high and low hypnotism, participants recalled their early days in high school and university. After hypnotic induction, half of the participants received a PGA offer for the first day of school, and the other half received a PGA offer for the first day of university. Before applying the reverse signal to cancel PGA, the researcher evaluated the participants' implicit memory for events and their ability to explicitly recall events repeatedly. The results showed discrepancies between explicit and implicit memory for people with high levels of hypnotism but not for those with low levels of hypnotism. For individuals with high levels of hypnotism, their explicit memory returned to its original level after the reversibility signal. Thus, A. J. Barnier's research has created a useful method for examining the temporary and reversible forgetting of autobiographical memories.

R. E. Cox, R. A. Bryant, and A. J. Barnier point to the usefulness of methods that allow examining changes in memory availability and comparing them with clinical disorders such as dissociative identity disorders and PTSD, which are often accompanied by obsessive memories of trauma, states C. R. Brewin [243]. During hypnosis, the researchers asked participants with low levels of hypnotism to recall memories of their first romantic relationships according to 10 replica phrases. Then, they applied a psychological amnesia procedure targeting either all memories from a specific period or specific events within that period. Following the application of PHA and after reversing its effects, researchers assessed explicit memory by recalling five key phrases presented during exposure, as well as five "new cues" that had not been previously presented. It was discovered that PHA had the greatest impact on highly hypnotisable individuals, causing them to forget old cues when the suggestion was aimed at the entire period rather than specific events. Highly hypnotisable individuals also displayed dissociation between explicit and implicit memories, as determined by social judgement tasks. According to the authors, the pattern of results aligns with the predictions derived from M. A. Conway [244] autobiographical memory model, which organises memory into three levels of interaction: lifetime periods, general events, and specific events. In summary, a PHA suggestion targeting the entire period had a greater impact on recall efficiency than one targeting specific events, supporting M. A. Conway and C. W. Pleydell-Pearce's [245] autobiographical memory model.

In their study, A. Mendelsohn et al. [242] used functional magnetic resonance imaging to investigate the impact of hypnotic suggestions on brain activity associated with episodic real-life memories of watched films and the context in which they were viewed. Participants were compared based on their level of hypnotisability, with some able to experience a hypnotic amnesic effect and others not. After watching a documentary and being hypnotised with the instruction to forget some of its details, participants were tested on their memory of the film and its context. It was discovered that the group experiencing PHA forgot more details than those who did not experience PHA. After the recall instruction was reversed, memory in both groups became equivalent, but the PHA group retained context memory even with impaired film memory. Researchers observed differences in brain activity between the groups, suggesting that the activation observed in PHA may reflect a form of rapid memory inhibition through increased prefrontal cortex activity. They also noted that the results might highlight important aspects of memory and encourage researchers to consider differences between these studies and studies on functional amnesia.

Hypnotic suggestions have also been used to model identity changes, where some participants, under hypnosis, perceived themselves as individuals of the opposite sex. This led to changes in their perception of information and selective attention, which was confirmed by post-experiment studies. R. E. Cox and A. J. Barnier [238–240] also demonstrated that hypnotic influences could lead to the creation of more specific autobiographical memories corresponding to the participants' new identities. Thus, research indicates the profound potential impact of hypnosis on memory, identity, and perception, opening new avenues for exploring these phenomena in the brain and human psyche.

In a series of studies aimed at creating laboratory analogues of real-life delusional beliefs (such as the mirror misidentification illusion or alien control delusion), A. J. Barnier et al. [231] convincingly showed that hypnotic illusions could help understand the social and cognitive factors involved in these illusions and ultimately provide intriguing insights into how they might be eliminated or treated. These experiments also highlight the role that autobiographical memory plays in sustaining these disorders.

Hypnosis can also play an important role in the examination of mood, autobiographical memory, and self-identity. Some researchers, such as F. MacCallum et al. [246], have shown that specific moods induced during hypnosis affect the type of autobiographical memories reported in response to positive (happy), negative (sad), and neutral cues. Specifically, while low-hypnotisability moods were not affected by induction, highly hypnotisable individuals responded differently to various cues in a sad state. When asked to recall specific personal memories for each cue provided by the experimenter (e.g., "smile" or "anger"), highly hypnotisable individuals had less access to specific positive memories during sadness. This finding is consistent with previous research indicating that people experience greater difficulty retrieving positive memories when depressed, as described by L. Goddard et al. [247]. In this case, hypnosis helped researchers understand the relationship between the valence of emotional states and memory performance, showing that hypnotic mood induction can be a useful procedure for exploring the connection between mood and autobiographical memory. This suggests that hypnotically induced delusions can influence information processing and may be very relevant for understanding the potential links between delusional beliefs, autobiographical memory, and the subjective experiences of individuals suffering from delusions.

Research on hypnosis explores the possibility of its application for modifying personal identity through memory transformation. Although hypnosis is used in clinical practice to create pseudo memories, there is little scientific evidence of its use for forming a positive and functional self-identity. The phenomenon of functional amnesia is considered not only as a means of memory modification but also as a tool for understanding intentional and automatic forgetting. Disruption of normal retrieval strategies has been observed in experiments using PHA, indicating the complexity of its mechanism of action. Amnesia induced by hypnosis can be either targeted or general, depending on the experimental conditions. Intentional forgetting, presented in the form of directed forgetting (DF) and memory suppression, is compared with amnesia induced by hypnotic suggestions; however, the results of studies regarding their similarity are mixed. The difficulty in comparing both forms of amnesia is highlighted due to differences in instructions and memory task requirements for research participants.

Studies comparing functional amnesia with procedures inducing hypnotic amnesic effects have found differences in their effectiveness, as noted by J. F. Kihlstrom [232–237]. Compared to functional amnesia, PHA induced greater amnesia, with fewer items being recalled, and a reversal cue prompted the return of most memories, especially in highly hypnotisable participants. On the other hand, studies by B. H. Basden et al. [248, 249] showed that functional amnesia had a very pronounced effect on forgetting and on enhancing memory access during reversal. Although contexts may be similar, it is challenging to investigate the role of hypnosis, hypnotisability, or amnesia within the framework of functional amnesia. However, these studies suggest the possibility of a forgetting mechanism that escapes the attention of functional amnesia researchers, especially in highly hypnotisable individuals. Voluntary suppression, as in the Think/No-Think task, is another intriguing area of research where the use of hypnosis and PHA can extend the understanding of the mechanisms involved in attempts not to think about something, as noted by M. C. Anderson and C. Green [250]. This task resembles the paradigm of ironic effects, where trying not to think about something often leads to the paradoxical emergence of that very thought or word, according to Wegner. The studies by K. S. Bowers and E. Z. Woody [251] discovered ironic effects of voluntary thought suppression among people with varying levels of hypnotisability, including the suggestion of PHA. According to the results, only highly hypnotisable participants who received PHA showed a significant reduction in the number of intrusions. Studies on the effectiveness of PHA compared to intentional suppression, such as the Think/No-Think task, have not yet been published. However, results from several unpublished studies by the first author indicate that highly hypnotisable individuals perform better in inhibiting target words in this task compared to participants with medium and low hypnotisability. These studies confirm the possibility of suppression without hypnosis, suggesting that hypnotic induction is not necessary to provoke forgetting through inhibition. A. Mendelsohn et al. [242] noted that amnesia induced by posthypnotic suggestion affects the pre-monitoring process, which is conducted before deciding whether to continue searching. This process may play a role in the difference between inhibition when comparing functional amnesia and PHA. Studies like those by A. Mendelsohn et al. can help understand the cognitive processes associated with functional amnesia, PHA, and intentional suppression.

Regarding memory accuracy, some researchers have considered the possibility of using hypnosis to improve memory accuracy. Studies have shown that suggestions made during hypnosis can increase false memories and enhance confidence in them. However, it is possible to manipulate expectations and demand characteristics to reduce false memories. Research by A. Oeberst and H. Blank [242] demonstrated that the misinformation effect can be altered through a "clarification" procedure, which improves source monitoring by testing the contribution of the source to the final memory. G. F. Wagstaff et al. [253, 254] and A. Oeberst and H. Blank [242] explored the potential of hypnosis to influence participants' memory and behaviour. They point to the possibility of using hypnosis to enhance metacognitive monitoring and to distinguish between true and false memories. The use of hypnosis may be significant for both memory research and forensic and clinical studies. Special attention is given to "emerging" involuntary autobiographical memories that arise in both ordinary and hypnotic contexts. Researchers examine brain activity during hypnosis and compare it to default mode activity. New methodologies for examining involuntary memories provide opportunities for a better understanding of this phenomenon. The research also highlights the concept of post-hypnotic suggestion as a form of prospective memory, which can be explored to understand the mechanisms of involuntary memory retrieval. Overall, the use of hypnosis in memory research opens up broad possibilitics for advancing knowledge about this complex process. Hypnosis becomes an important model for understanding memory and related processes that play a role in everyday life and clinical practice.

The second most important and frequently studied area is the impact of hypnosis on pain. Hypnosis is used as a psychological method for pain management, involving relaxation, focused attention, and targeted verbal influence to alter the perception and response to pain, as described by M. P. Jensen and D. R. Patterson [255]. Hypnosis is considered an effective and low-risk treatment method, as it has few or no side effects and can be administered in audio-recorded format without the presence of a medical professional, based on research by M. P. Jensen et al. [256]. Recent studies by A. Del Casale et al. [257] have shown that hypnotic influence causes changes in the activity of key brain areas responsible for pain perception and regulation. This opens up possibilities for using hypnosis as an analgesic method. However, exaggerated claims about the analgesic properties of hypnosis have led to some degree of scepticism about its effectiveness among scientists like M. Larkin [258]. Therefore, rigorously controlled studies are necessary for an objective evaluation of its role in pain relief. The results of a meta-analysis by T. Adachi et al. [259] indicated that hypnosis might be an effective method for reducing pain. However, these results are primarily based on a limited number of studies and require further scientific validation. In addition, there are certain methodological limitations in these studies that need to be addressed in future research to more accurately determine the effectiveness of hypnosis in pain control and its impact on different types of pain.

Pain affects up to 1.5 billion adults worldwide, as noted by researcher F. Yaqub [260], and significantly diminishes their quality of life. Furthermore, pain is a leading cause of years lived with disability (GBD Causes of Death Collaborators), and it also imposes a substantial economic burden. Healthcare costs and productivity losses related to pain amount to up to $635 billion annually in the USA alone, according to J. Gaskin and P. Richard [261], which exceeds the costs for cardiovascular diseases, cancer, or diabetes. The increasing concerns about the side effects, addiction risks, and costs of opioid medications highlighted the urgent need to identify non-pharmacological pain management methods that are effective, safe, and accessible. One of the popular psychological methods for pain management is hypnosis, which typically involves relaxation, attention, and targeted verbal suggestions to alter perception and behaviour, as indicated by studies by M. P. Jensen and D. R. Patterson [255]. Hypnosis is easily administered, has few or no side effects, and is inexpensive when delivered in pre-recorded formats (e.g., audio recordings), which do not require the presence of a practitioner, based on research by M. P. Jensen et al. [102, 201]. Recent studies have shown that hypnotic suggestions cause changes in the activity of key brain areas involved in pain regulation, which may underlie potential analgesic effects. Nevertheless, exaggerated claims about hypnotic analgesia have led to scepticism regarding its efficacy, criticises researcher Larkin, and a stringent evaluation of controlled studies is needed to properly assess and quantify its effectiveness in pain reduction.

One meta-analysis conducted by T. Adachi et al. [259] led researchers to conclude that hypnosis is effective for pain relief. However, this conclusion was primarily based on a subgroup of 4 studies that compared hypnosis with standard treatment, which were scarcely replicated in other subgroup comparisons. The results of individual studies were conflicting, likely due to differences in pain conditions, control comparisons (e.g., usual care, no intervention), and the hypnotic susceptibility of the study samples. Thus, this meta-analysis does not provide a clear overall picture of the analgesic benefits of hypnosis. Other reviews pointed to the beneficial effects of hypnosis on pain during childbirth in studies by K. Madden et al. [262] and on fibromyalgia in studies by K. Bernardy et al. [263], but all concluded that the supporting clinical evidence is of low methodological quality.

The impact of hypnosis on pain has also been investigated through experimental studies to provide a level of methodological control that is difficult to achieve in clinical settings. A meta-analysis of 18 studies, including 12 experimental and 6 clinical studies, by G. H. Montgomery et al. [264–266] revealed a significant moderate analgesic effect of hypnosis. While this is an important finding, several critical limitations should be noted, primarily due to the lack of available data. Firstly, it is challenging to determine the level of significant analgesia from hypnosis, given the absence of a measure to reflect meaningful clinical changes. Secondly, the assessment of somatic analgesia was complicated by substantial heterogeneity in the control comparators. Thirdly, factors such as hypnotic susceptibility and the use of direct suggestions, which may be crucial for treatment success, as noted by M. P. Jensen and D. R. Patterson [255], could not be adequately assessed. Since this meta-analysis, conducted almost 20 years ago, many experimental studies have been published, opening new opportunities for a more reliable assessment of the effectiveness of hypnosis in reducing pain and evaluating potential beneficial factors.

To fill gaps in the current understanding of the effectiveness of hypnosis for pain relief, a meta-analysis was conducted comparing hypnotic interventions with no-treatment controls in studies using experimental pain models in healthy participants. The specific aims included obtaining precise estimates of the magnitude of hypnotic analgesia on both standardised and non-standardised scales and the extent to which the effectiveness of the intervention depends on both hypnotic suggestibility and the inclusion of direct suggestions for pain relief. The conclusion indicated that hypnosis represents a promising psychological method for reducing pain. The results of the studies confirm that hypnotic interventions can be effective in pain management, especially in patients with chronic conditions and in experimental pain models. The impact of hypnosis on pain is analysed in both clinical and laboratory settings. Despite positive results in some studies, it is important to consider methodological limitations and the diversity of approaches to the use of hypnosis. It is also crucial to consider individual differences in response to hypnotic interventions, including potential side effects. Further development in the field of hypnosis and pain requires additional research that carefully controls methodological aspects and examines the mechanisms of action of hypnosis on different types of pain. The growing interest in alternative pain management methods highlights the importance of further research in this area to develop effective and safe approaches to managing pain with hypnosis.

Scientists investigated the effectiveness of substance abuse involving various substances such as alcohol, nicotine, narcotics, and excessive consumption of unhealthy food [267]. This issue requires serious investigation due to its widespread impact on health and society as a whole. The study on alcohol abuse effectiveness pertains to its effects on physical and mental health, including its social consequences. For example, heavy alcohol consumption can lead to mental disorders, problems with internal organs, accidents, and violence. The analysis of the effects of nicotine is also significant, given its association with the development of addiction and various lung and heart diseases. Narcotic drugs, in turn, can have a devastating impact on users' mental and physical health, leading to addiction and even fatal outcomes [132]. Unregulated eating is another issue that requires exploration. Excessive consumption of unbalanced foods can lead to obesity, cardiovascular diseases, diabetes, and other chronic conditions. All these aspects necessitate a comprehensive approach to understand the causes, consequences, and effective strategies for preventing and treating substance abuse and unregulated eating. Such studies help develop better prevention and intervention programmes to improve the overall health and well-being of society [268].

Substance use and abuse are costly problems for both individual users and society, with substance-related deaths continuing to rise year on year. Approximately half of those suffering from substance use disorders also suffer from co-occurring mental health disorders. Unfortunately, only half of the treatment programmes offer mental health services. Relapse rates remain high, and the likelihood of multiple treatment episodes is substantial. This is because many treatment centres do not address the underlying cause of substance use, which is a symptom of a more serious issue.

Much has been written about the causes of substance abuse, and leading experts in the field link pain and trauma to this problem. The higher the rating of post-traumatic stress, the higher the risk of developing substance addiction later in life. This information is sufficient to show that even if a person who abuses substances does not have a co-occurring mental health disorder, they have past pain and trauma leading to their addiction. This underscores the necessity of addressing past traumas and pain for the effective treatment of substance use itself. Substance use impacts the formation of neural connections in the brain, which alters the perception of use as necessary for survival. The reinforcement of old neural connections occurs through highly emotional memories of use, making it challenging to stop this process by creating new neural connections [269, 270]. Repeated substance use leads to the formation of deep neural pathways in the brain and radically alters the body's level of homeostasis. Hypnotherapy is a method that allows for the healing of past traumas and the introduction of positive changes in brain activity, thereby reprogramming negative thought patterns and behaviours. Hypnosis enables access to information such as belief systems, emotions, and memories stored in the subconscious mind. The subconscious mind retains over 95% of an individual's state, including information that triggers the symptoms of the problem. In the state of hypnosis, a certified specialist can understand and remove the pain and trauma that lead to substance use. In this state of suggestion, the client can accept new proposals for a life without substances. Most people are susceptible to at least a light hypnotic state, making this approach valuable for use. A key factor in hypnotherapy is the use of visualisation [271, 272]. The practice of using visualisation to stimulate feelings forms new neural pathways in the brain. Creating new pathways will be based on the client's visualisation of life without substances. The negative effects of substance use on the brain highlight the importance of using hypnotherapy to reprogram neural pathways in the brain. The number of studies on the use of hypnotherapy for substance use treatment is limited, indicating the need for more research in this area. Available studies show that hypnotherapy has helped with cessation and the reduction of cravings, which are two symptoms of substance use disorders that need to be managed to maintain sobriety [273, 274]. Abstinence is the goal for most substance users, and studies show that hypnotherapy can help achieve this goal. Other negative symptoms have also been reduced through this approach, making it useful in both the field of addiction and mental health. Research in neuroscience indicates that reprogramming the brain through visualisation can change the brain in a positive way.

Recent trends in the analysis of the interaction between hypnosis, brain activity, and consciousness are shaped by the application of advanced neuroimaging methods. Functional magnetic resonance imaging and electroencephalography have become key tools for gaining a deeper understanding of the effects of hypnosis on brain activity. These methods allow real-time visualisation of changes in brain activity during the hypnotic state. Analysis of the latest technologies in neuroimaging highlights specific brain areas where activity changes in response to hypnosis, revealing the precise mechanisms of this phenomenon. It is noted that these new technologies not only allow observation of changes in brain activity but also determine their functional consequences for consciousness. Such in-depth analysis provides a better understanding of the relationship between hypnosis, brain activity, and states of consciousness, opening new perspectives for further research and potential clinical applications.

Recent research in the field of hypnosis and its effects on brain activity focuses on identifying regional changes in the brain. This research area opens new opportunities for understanding how hypnosis interacts with specific brain areas and what functions of these areas may be altered. One of the key trends is the use of functional magnetic resonance imaging to determine regional changes in brain activity during hypnosis. Studies enable the identification and mapping of specific zones where significant changes in activity are observed, indicating the influence of hypnosis on various functional systems of the brain. One interesting finding is that hypnosis can induce changes in regions associated with sensations, attention, and emotions. For example, changes in brain activity are observed in the anterior cingulate cortex, which is responsible for task execution and attention control. Further analysis of the latest research in this field can contribute to uncovering the specific mechanisms of hypnosis's impact on brain activity, which, in turn, can have practical applications in areas such as medicine and psychotherapy.

Recent studies examining the role of the anterior and posterior parts of the brain during hypnotic states reveal intriguing insights into the interaction between these regions and the influence of hypnosis on their functionality. In-depth scientific investigations increasingly define the role of the brain's anterior regions, particularly the prefrontal cortex, in hypnotic states and their impact on cognitive functions. This aspect, significantly involving the prefrontal cortex, is considered in the context of hypnosis without focusing on specific studies. Further research confirms that the anterior parts of the brain, especially the prefrontal cortex, play a crucial role in hypnotic states. Paying special attention to this area allows for a more detailed understanding of how hypnosis affects internal plans and behavioural strategies. For instance, studies have used functional methods, such as functional magnetic resonance imaging, to observe activity in the anterior brain during hypnotic states. It has been found that this region shows increased activity, particularly when individuals engage in tasks related to planning and coordinating actions.

On the other hand, analysis of electroencephalogram data indicates that hypnosis causes changes in the electrical signals of the brain's anterior regions, reflecting their increased involvement in regulating internal plans and behavioural strategies. Therefore, a comprehensive understanding of the interaction between the anterior parts of the brain and hypnosis is vital for further research. This would highlight the profound mechanisms that may define hypnotic states and their consequences for cognitive activity. Such studies lay the groundwork for understanding how hypnosis can modify activity in the brain's anterior regions, enhancing or altering the ability to plan and execute tasks. These findings are crucial for understanding the mechanisms of hypnosis's impact on cognitive functions and behavioural control.

Studies covering the impact of hypnosis on the frontal cortex provide deeper insights into how this key brain region regulates high-level cognitive functions such as decision-making and emotional control. The frontal cortex, particularly the dorsolateral prefrontal cortex (DLPFC), is recognised for its crucial role in cognitive processes and emotional regulation. Studies using functional magnetic resonance imaging demonstrate that hypnosis can induce changes in the activity of the frontal cortex, especially the DLPFC. For example, during hypnosis, there may be an increase in DLPFC activity, which may be associated with an increased ability to solve cognitive problems and support better decision-making functioning. Some studies also suggest that hypnosis can affect the number of activated neurons in the frontal cortex, which can determine the level of cognitive reserves and control over emotions during a hypnotic state. This effect can have practical applications, for example, in improving cognitive function in patients with various disorders or in optimising training to improve cognitive performance. Thus, research on the role of the frontal cortex in the hypnotic state provides unique opportunities for understanding and improving cognitive functions through hypnosis.

Studies on the role of the posterior parts of the brain in the hypnotic state allow uncovering the complex mechanisms of interaction of these regions with hypnosis. For example, functional magnetic resonance imaging has opened up new horizons in examining the activity of the parietal cortex during hypnosis. This area of the brain, which is responsible for processing sensory information and spatial perception, detects changes in perception under the influence of a hypnotic state. Changes in pain or temperature perception indicate the effect of hypnosis on the processing of sensory stimulation by the back of the brain, expanding understanding of the effects of hypnosis on sensory experience. The role of the parietal cortex in hypnosis is considered in the context of visual information processing. Numerous studies show that during a hypnotic state, a transformation in the perception of faces, colours, and objects can occur. Observations of changes in responses to visual stimuli indicate the ability of the brain to adapt to new ways of perception, solving the mysteries of the effect of hypnosis on visual perception. Understanding these regional changes allows not only to determine how hypnosis affects specific parts of the brain but also to highlight possible practical aspects, such as the use of hypnosis in medicine to treat diseases or in psychotherapy to change perceptions and responses. Such discoveries are key to the development of scientific and practical aspects of the use of hypnosis in the modern world. These results suggest that hypnosis can cause regional changes in the brain that affect high-level cognitive processes, attention, and the perception of sensory information. It is important to consider individual variations in responses to hypnosis, but these studies open up prospects for understanding the mechanisms of hypnotic phenomena and their effects on different parts of the brain.

Recent research on the interaction of hypnosis with the limbic system, which is responsible for emotional processes in the brain, has attracted considerable attention from scientists. The analysis of the latest research in this direction points to several interesting aspects. Studies using functional magnetic resonance imaging show changes in the activity of limbic structures, in particular, the amygdala and hypothalamus, during hypnosis. This suggests that hypnosis can affect emotional processes in the brain. The latest research points to the possibility of using hypnosis to regulate emotions. Hypnotherapy can be used to reduce stress levels and work with emotional responses, which can have a positive impact on mental health.

Research on hypnosis as a therapeutic tool demonstrates its success in treating mental disorders related to emotional aspects. For example, hypnotherapy can be effective in treating PTSD or anxiety conditions. Scientific work shows that hypnosis can affect pain perception and stress responses, in part by modulating the limbic system. This indicates the potential use of hypnosis in dealing with pain and stress.

Psychological therapy aimed at treating PTSD is effective, especially if memories of traumatic events are actively processed. For example, cognitive behavioural therapy for trauma is a well-proven treatment method, according to J. L. Hamblen et al. [275]. Despite the success of such therapy for many patients, a substantial number of people continue to experience residual symptoms of PTSD as mentioned by researchers P. P. Schnurr and C. A. Lunney [276] and M. A. Schottenbauer et al. [277]. T. Michael et al. [278] claim that research into new intervention methods that could improve the efficacy of trauma-focused cognitive–behavioural therapy (CBT) is crucial. PTSD is characterised by disrupted emotional states and memory, leading to the continual re-experiencing of traumatic events in the present, avoidance of trauma reminders, and a constant sense of threat, according to A. Ehlers et al. [279].

PTSD occurs when the brain forms strong, unpleasant memories after a traumatic event. During such an event, people may associate ordinary events with danger. Whenever they encounter these things again, they may re-experience traumatic memories and feel threatened. This can lead to anxiety and insecurity. PTSD therapy tries to help individuals re-experience and process these memories. It takes place in a safe environment where patients can recall the trauma without real danger. This helps to reduce the fear and stress associated with the memories. Periodically, this therapy also helps individuals focus on what they learned during the traumatic event to better understand and control their emotions. Recent studies show that sleep after learning can enhance the perception and retention of new information. In particular, slow-wave sleep promotes the processing and retention of memory [280]. This may be because deep parts of the brain are activated during sleep, facilitating the transfer of memory from short-term to long-term storage. Some studies suggest that slow-wave sleep helps to process and strengthen fear memories. However, there are other opinions that rapid eye movement (REM) sleep is also important for memory retention. In addition, sleep can help generalise knowledge from the learned situations to similar but slightly different scenarios. Some scientists suppose that the impact of sleep after experiencing trauma could enhance the treatment efficacy of PTSD. However, studies on the effect of sleep on the retention and generalisation of trauma memory have not yet been conducted. Nonetheless, scientists have considered sleep hypnosis as a means to improve sleep quality and increase the duration of slow-wave sleep, which is associated with memory consolidation. U. Halsband and T. G. Wolf [281] showed that sleep hypnosis can improve subjective sleep quality and the duration of slow-wave sleep. This is particularly interesting as slow-wave sleep is linked to memory consolidation and the alleviation of fear memories. One study investigates whether this consolidation occurs through slow-wave sleep or REM sleep. Based on this, studies were conducted to examine the impact of sleep hypnosis on sleep quality and the retention and generalisation of trauma memory. Participants in the experiment were subjected to a fear extinction technique, followed by either sleep or sleep hypnosis under controlled conditions. The next day, a test on memory retention and generalisation and an assessment of fear levels were conducted. The results showed improvements in sleep duration and quality after sleep hypnosis, as well as the generalisation of trauma memory [282, 283]. Previous data also indicated a reduction in intrusive memories and ruminative thoughts after sleep hypnosis.

One interesting study aimed to determine how hypnosis affects brain activity using various imaging techniques. Scientists examined how different levels of hypnotisability affect brain activity compared to the normal state. The results were complex due to the application of different imaging methods. Specifically, it was discovered that placebo therapy in highly hypnotisable individuals led to changes in brain activity associated with a reduction in the sensation of pain. Those with higher levels of hypnotisability exhibited different brain activity patterns during hypnosis compared to those who were less hypnotisable. High levels of hypnotisability also influenced the amplitudes of various brain waves, including changes in amplitudes in different brain areas during different states, such as rest, hypnosis, and emotional states [284, 285]. Thus, the study showed that hypnosis affects brain activity, and this influence varies depending on the level of an individual's hypnotisability.

In another study, scientists explored how hypnosis affects people on a molecular level. They discovered that oxytocin, produced in the brain, can reduce the sensation of pain during hypnosis, especially in individuals who are highly susceptible to hypnosis. This occurs through the activation of a special system in the brain responsible for pain reduction. Regarding the results of functional brain examinations using fMRI, the study revealed that the activity of certain parts of the brain decreases during hypnosis compared to the waking state, and the connectivity between some brain regions changes. People with high susceptibility to hypnosis exhibited greater activity in certain brain areas associated with pain sensation [286]. Therefore, the research indicates that hypnosis can affect the human brain, reducing pain sensation, especially in those who are easily hypnotised. In addition, positron emission tomography (PET) analysis indicates significant activation of specific brain areas during hypnosis, particularly the right supplementary visual area and the anterior cortex. This activation is associated with experiencing pain and discomfort during hypnosis. Changes in regional cerebral blood flow during hypnosis were also observed, reflecting different parts of the brain. Overall, this research aims to understand how hypnosis affects the brain in the context of dental applications. The results indicate changes in brain activity, particularly in areas associated with pain sensation and emotional response, which may have important implications for the practical use of hypnosis in medical practice. Generalising the results is complicated due to the application of different imaging methods and various contexts of hypnosis use. Due to the heterogeneity of studies, the numerous imaging methods used, and the lack of comparability, it is recommended to have a more uniform description in the design of future hypnosis studies investigating brain activity using imaging techniques [287]. Hypnotisability does not play a role in individual treatment with hypnosis, but is used as a criterion in many studies and is assessed using validated questionnaires. In the future, it is recommended to examine not only highly hypnotisable individuals but also those with medium and low levels of hypnotisability to provide a better assessment of the effectiveness of hypnosis for the entire population.

In summary, the latest research provides evidence of the impact of hypnosis on the limbic system and confirms its role in the regulation of emotions and the treatment of mental conditions. This opens up new possibilities for the use of hypnosis in clinical practice and the development of psychotherapeutic approaches. Recent trends in the analysis of hypnosis include the integration of cutting-edge technologies, notably the use of VR. Analysing this integration allows for a better understanding of how technologies can enhance or modify hypnotic effects [101].

VR has become a platform for increasing hypnosis immersion. During VR hypnotherapy sessions, a virtual environment can be created that helps the patient delve deeper into the hypnotic state. The use of VR in hypnosis allows therapists to create specific virtual scenarios for working with patients. This can include situations for overcoming stress, phobias, or dealing with traumatic experiences, creating an immersive environment for therapeutic purposes.

The integration of biofeedback into the virtual environment allows for the monitoring of physiological indicators during hypnotic sessions. This enables therapists to adapt the hypnotic approach based on the patient's real physiological responses. The use of hypnosis in combination with VR can be beneficial for performing medical procedures where reducing pain, stress, and anxiety is important. This can open up new possibilities for comfortably conducting procedures. The use of VR allows for the creation of interactive hypnotic scenarios where the patient can actively participate in their own hypnotic exploration or therapy. These recent trends in the integration of VR and hypnosis demonstrate the potential for further development of therapeutic methods and enrichment of the hypnotic experience. However, it is important to consider the ethical and psychological aspects of using such technologies in clinical practice. The use of VR in hypnotherapy opens up new opportunities for improving outcomes and ensuring a more effective impact on patients. For example, recent studies illustrating the use of VR in hypnotherapy include the treatment of phobias such as arachnophobia or aerophobia. Patients are immersed in a virtual environment where they can confront objects that trigger their phobia in a controlled and safe space. Studies have shown successful results in reducing anxiety levels and improving patient functionality. The use of VR allows for immersive interaction with phobic objects, which can enhance the effectiveness of hypnotherapy.

The use of VR to provide an analgesic effect during hypnotherapy for pain management is another example where patients are immersed in natural or relaxing virtual environments, which can help reduce the perception of pain. Studies have shown that the use of VR in conjunction with hypnosis can decrease pain intensity and improve overall patient comfort. This indicates the potential effectiveness of such an approach in clinical practice. The use of virtual environments to simulate anxiety-inducing situations and subsequent hypnotherapy exposure allows patients to confront realistic stressful scenarios, supporting their work with anxieties. Studies show a reduction in anxiety levels and improved adaptation to stressful situations following VR hypnotherapy. This can be significant for patients with anxiety disorders.

These examples highlight the potential of using VR to enhance hypnotherapy in various areas, from treating phobias to pain management and addressing anxiety conditions. The use of VR allows for the creation of customised and immersive therapeutic scenarios, which can contribute to improved hypnotherapy outcomes. Recent studies on the interaction between virtual environments and brain activity reveal new possibilities for understanding and enhancing the use of VR in various fields, including psychology, neuroscience, and medicine. Brain activity during immersive VR therapy, as shown by some studies using functional magnetic resonance imaging, has revealed activation of brain functions during VR therapy sessions for treating PTSD. The results indicate that VR therapy has a positive effect on emotional processing and stress response. Electroencephalography measures electrical brain activity during VR use to enhance cognitive functions and learning. It shows increased activity in areas responsible for attention and memory during immersive virtual environments. VR is also utilised in the rehabilitation of patients with motor function impairments following injury or surgery. Research demonstrates that the use of VR can increase motor cortex activity and promote faster recovery of movement functions. Studies on brain activity during immersive virtual research to simulate real-life situations and elicit emotional reactions indicate a high degree of activity in brain areas responsible for emotional experiences, suggesting the effectiveness of VR in evoking strong emotions. These studies highlight the powerful impact of VR on brain activity and its potential in therapy, education, and rehabilitation. The use of advanced technologies in conjunction with brain activity analysis expands the possibilities for analysing and enhancing the application of VR in various scientific and medical fields.

Recent research in the field of hypnosis provides detailed insights into individual differences and responses to hypnotic influences. Studies confirm that individual responses to hypnosis are variable, with significant differences in the degree of susceptibility to hypnotic influences among different people. Some research suggests that the level of hypnotic susceptibility may be linked to certain personality traits, such as creativity, imagination, and openness to experience. Using neuroimaging, it has been established that individuals with high hypnotic susceptibility may exhibit unique features in brain activity, particularly in the functioning of the prefrontal cortex and the limbic system. New research indicates that genetic and epigenetic factors may play a role in determining an individual's response to hypnosis, with some traits potentially being inherited across generations. Studies show that individuals with various mental disorders may exhibit different responses to hypnosis. Hypnotherapy can be beneficial for certain groups of patients, such as those suffering from anxiety disorders or chronic pain. Expectations and beliefs play a crucial role in determining the effect of hypnosis. Personal beliefs and attitudes towards hypnosis can influence its effectiveness. The analysis of these studies underscores the necessity for an individualised approach in hypnotherapy and the understanding that the response to hypnosis is unique to each person. Future research in this area should focus on identifying specific factors that influence individual responses to hypnotic stimuli, as well as developing personalised hypnotherapy methods. Such research is a complex task due to the multitude of factors that can affect hypnotic susceptibility, including psychological, sociocultural, and biological factors. However, general and theoretical considerations regarding potential genetic components in hypnosis can be explored. There is a hypothesis that certain psychological and perceptual traits that influence hypnosis may have genetic roots. For instance, characteristics such as creativity, openness to new experiences, or the ability to focus deeply might be influential factors. Some studies suggest that the propensity for trance states, including hypnosis, may have a genetic basis. However, this data still requires further validation and specification. An important aspect is that genetic predisposition can interact with external factors, such as experience and upbringing. This interaction can determine how the genetic predisposition to hypnosis is expressed.

The growing capabilities in the field of genetic research allow for studies at the level of specific genes that may be associated with susceptibility to hypnosis. Nevertheless, this remains a subject of active scientific debate.

Overall, the genetic aspects of hypnosis remain a subject of research and theoretical reflection, and further scientific investigations are needed to elucidate this complex issue. It is important to consider that hypnosis is a multifactorial phenomenon, and understanding its genetic determinants requires the integration of various scientific approaches. Recent studies focus on understanding the role of prior experience and personality traits in an individual's response to hypnotic influences. An analysis of these studies reveals significant aspects that impact the effectiveness of hypnotherapy and opens avenues for further research.

Research confirms that individuals respond to hypnotic influences differently. Prior experience and personality traits can determine how susceptible a person is to hypnosis and what specific effects they may experience. Personality traits, such as the degree of openness to new experiences, the propensity for realistic thinking, and the level of hypnotic susceptibility, significantly influence the success of hypnotherapy. Individuals with significant prior experience with hypnosis are more likely to achieve deeper hypnotic states and respond more effectively to therapeutic influences. Some studies suggest that a personalised approach to hypnotherapy, considering personality traits and prior experience, can improve therapy outcomes. Belief in hypnosis and the perception of the hypnotherapist also play crucial roles. People who believe in the possibility of hypnosis and have a positive perception of the therapist may show better results.

Despite these conclusions, more in-depth research is needed to understand the specific mechanisms that determine the interaction between prior experience, personality traits, and hypnotic influences. Some studies also emphasise the importance of considering ethnic and cultural characteristics in the analysis of hypnosis and its effects on different populations.

In summary, recent research highlights the importance of considering individual characteristics and prior experiences when examining responses to hypnotic influences. This paves the way for the development of personalised and effective hypnotherapy methods.

Recent studies in the field of mental disorders and hypnotherapy define modern approaches to using hypnosis as an additional tool in treating various mental conditions. A number of studies investigate the effectiveness of hypnotherapy in reducing anxiety and stress levels, as well as improving patients' ability to cope with stress. Some research indicates positive changes in patients' mental well-being, including a reduction in anxiety symptoms and an improvement in quality of life. Hypnotherapy is being explored as a potential treatment for depressive disorders, aimed at reducing symptoms and improving patients' mood. Results suggest that hypnotherapy can be beneficial as part of a comprehensive treatment for depression, helping to reduce symptoms and restore emotional well-being.

Experiments in the field of hypnotherapy for PTSD focus on how hypnosis can help reduce symptoms and improve the functioning of affected individuals. Some findings point to the potential benefits of hypnotherapy in decreasing the intensity of PTSD symptoms and enhancing patients' quality of life. There is also research into the use of hypnotherapy for managing chronic pain and reducing the need for analgesics. There is some evidence that hypnosis can affect pain perception and improve the physical condition of patients with chronic pain.

Studies are examining how the integration of hypnosis with other treatment methods (e.g., CBT) can improve outcomes and the duration of treatment. It has been shown that combining hypnosis with other therapeutic methods can be effective in treating mental disorders. Reviewing these studies, it can be concluded that hypnotherapy is recognised as a promising method in the comprehensive treatment of various mental disorders. However, many aspects, such as individual responses, mechanisms of action, and optimal treatment protocols, require further research and clarification. Recent research in hypnotherapy indicates promising results in treating anxiety, depression, and PTSD. Studies confirm that hypnotherapy can be effective in treating anxiety, depression, and PTSD. A series of clinical trials showed a significant reduction in symptoms in patients receiving hypnotherapy compared to those undergoing traditional treatment methods. Hypnotherapy can be effectively integrated into a comprehensive approach to treating anxiety and depression. The combined use of hypnotherapy with other psychotherapeutic methods, such as CBT or other approaches, can improve treatment outcomes. The use of VR in combination with hypnotherapy can be effective in treating PTSD. Patients who received therapy incorporating virtual scenarios during hypnosis reported a reduction in PTSD symptoms and an improvement in overall condition. The effectiveness of hypnotherapy may depend on individual patient characteristics. Personalised approaches, considering the patient's specific traits and response to hypnosis, can ensure better treatment outcomes.

Research into the neurophysiological aspects of hypnosis helps to understand the brain mechanisms underlying its effects. Neuroimaging studies show changes in brain activity during hypnosis, indicating its impact on neurobiological processes. The general trend is an increasing awareness of the effectiveness of hypnotherapy in treating anxiety, depression, and PTSD, as well as the development of new techniques and approaches to improve treatment results. Research in this field will continue to expand the understanding and practical application of hypnosis in psychotherapy and clinical medicine.

Recent research in the field of mental health medicine confirms the effectiveness of hypnotherapy in treating various mental disorders and psychotherapeutic challenges. Some studies affirm the efficacy of hypnotherapy in reducing symptoms of anxiety disorders. Hypnotherapy can promote relaxation and reduce anxiety by providing patients with self-regulation tools and coping strategies. Experiments confirm that hypnotherapy can be beneficial in alleviating symptoms of depression. Hypnosis can help change negative beliefs and improve mood, impacting overall mental health.

Research supports that hypnotherapy can be effective in managing pain and stress. Hypnosis can alter pain perception, giving patients a means to cope with physical and psychological discomfort. Using hypnotherapy in treating PTSD can lead to a reduction in the intensity and frequency of symptoms. Hypnosis can assist patients in processing traumatic experiences and enhancing their psychological resilience. Hypnotherapy can be beneficial in addiction treatment programmes, helping patients reduce cravings and change behaviours. Hypnosis is used to modify stereotypical thought patterns and influence motivation for a healthy lifestyle. These research findings highlight the significant contribution of hypnotherapy to modern mental health medicine. A deeper understanding of the mechanisms of hypnosis and further research will help optimise this method and expand its applications in treating various mental conditions.

Recent studies in the field of hypnosis and cognitive functions elevate this subject to a new level, revealing complex interactions between hypnosis and cognitive processes. Hypnosis can positively affect a range of cognitive functions, including attention, concentration, and memory. Experiments have shown that hypnosis can enhance focus on certain stimuli, improve concentration, and foster positive cognitive strategies. Hypnosis is accompanied by changes in brain activity, which may be linked to improved cognitive functions. Neuroimaging studies indicate the activation of specific brain regions, particularly those responsible for attention and information processing, during hypnotic states. There is evidence that hypnosis can enhance creative cognitive processes. Some studies have shown that hypnosis can foster a "hypnotic effect" that supports creativity and innovation. Hypnosis can influence decision-making processes and the formation of cognitive attitudes.

Experiments have demonstrated that hypnosis can modify cognitive schemas and beliefs, impacting how decisions are made.

There are variations in how hypnosis affects cognitive processes, which can depend on individual characteristics and the depth of the hypnotic state. The mechanisms through which hypnosis impacts cognitive processes include reprogramming perceptions, altering sensory information filtering, and reducing criticality. Individual differences in response to hypnosis can determine how effectively it influences cognitive functions. Patients exhibit varying levels of susceptibility to hypnotic effects, which can affect changes in their cognitive functions. The general trend is that hypnosis shows significant potential for influencing cognitive processes, but the mechanisms behind this influence require further elucidation. New research in this area will help deepen the understanding of the interaction between hypnosis and cognitive functions and develop more precise methods for using hypnosis in clinical and practical contexts.

Numerous experiments investigate changes in cognitive processes during hypnosis. Participants were subjected to hypnosis, during which they were instructed to focus on a specific task. Increased attention and concentration were recorded in participants during the hypnotic state, resulting in a higher number of correctly completed tasks. Participants used hypnosis to retrieve memories or reduce negative memories. It has been shown that hypnosis can affect memory processes, improving the recall of certain memories or, conversely, aiding in the forgetting of negative details. Participants used hypnosis to enhance their intellectual abilities or creative thinking. Some studies indicate that hypnosis can boost creativity and provide new approaches to problem-solving. Participants were subjected to hypnosis to explore its effect on their ability to evaluate time or perform quick reactions. Some experiments show that hypnosis can influence time perception and alter reaction time parameters. Participants used hypnosis to change the way sensory information is filtered. Hypnosis can affect sensory information filtering by changing the way the brain processes and integrates incoming stimuli. These experiments demonstrate the variety of aspects in which hypnosis can influence cognitive processes. However, it is important to consider that individual reactions to hypnosis may vary, and some effects may be more pronounced in certain individuals. Strengthening these findings in the future may contribute to the development of hypnotherapy and its use in various clinical and non-clinical contexts.

Recent research into the interaction between the brain and hypnosis reveals a number of key findings that enhance the understanding of the impact of hypnosis on brain activity and consciousness. Hypnosis demonstrates the ability to modulate brain activity, especially in the limbic system and frontal cortex areas. Changes in these areas can affect emotional reactions and cognitive processes. Individual responses to hypnosis vary among individuals, underscoring the heterogeneity of hypnosis effects. Some people may be more susceptible to hypnotic influences, while others are less vulnerable. The posterior parts of the brain play an important role in sensory information perception and processing visual stimuli during hypnosis. The frontal parts of the brain are activated in regulating executive functions and metacognitive processes. Hypnosis can induce changes in consciousness and transform perception, including the ability to change pain perception, alter visual perceptions, and react to external stimuli. The use of advanced technologies, such as VR, neuroimaging, and biofeedback, allows for the enhancement and intensification of hypnotic effects, providing new opportunities for research and practical application.

In summary, research into the interaction between the brain and hypnosis confirms the complexity of this process and highlights the potential of hypnosis in modifying brain activity and consciousness. These discoveries contribute to the development not only of a fundamental understanding of hypnotic phenomena but also of the practical application of hypnosis in medicine, psychotherapy, and other fields.

The field of hypnosis and brain activity remains dynamic and promising for future exploration. Understanding the variability of individual responses to hypnosis and identifying the factors that influence this response. Developing personalised approaches to hypnosis, considering individual characteristics and reactions. Identifying specific brain mechanisms responsible for hypnotic phenomena and effects. A deeper understanding of the neurobiology of hypnosis and the development of targeted methods to influence brain processes. Maximising the use of advanced technologies, such as neuroimaging, VR, and artificial intelligence, to enhance hypnotic techniques and their study. Expanding the potential applications of hypnosis in medicine, psychotherapy, and other fields through the use of cutting-edge technologies. Increasing the number of clinical studies to determine the effectiveness of hypnotherapy in treating various mental and physical disorders. Developing standards and recommendations for the use of hypnotherapy in clinical practice. Developing ethical standards in the use of hypnosis, particularly in the context of advanced technologies. Ensuring the safety and effectiveness of hypnosis, as well as understanding its potential side effects. A detailed study of hypnotic states and their variations to develop more accurate and standardised descriptions of hypnosis. Developing new theoretical models and classifications of hypnotic states to improve scientific clarity and mutual understanding. The overall perspective is that future research in the field of hypnosis and brain activity aims to expand knowledge, improve clinical applications, and enrich scientific understanding of this unique phenomenon.

People react to hypnosis individually, which complicates the generalisation of research results. Individual differences in response to hypnosis may arise due to different degrees of susceptibility to hypnosis, belief in the effectiveness of hypnotherapy, individual brain characteristics, and other factors. Some aspects of brain activity can only be measured with limited accuracy using neuroimaging techniques. The accuracy of measurements may vary depending on the chosen method (e.g., functional magnetic resonance imaging or electroencephalography). The use of advanced technologies in brain research raises ethical issues, particularly regarding the confidentiality and security of data obtained from neuroimaging studies. The concept of hypnosis remains subjective and difficult to measure. Different definitions of hypnosis can lead to ambiguity in determining its effects on brain activity.

Modern technologies, such as fMRI and MEG, provide a very precise image of brain activity, allowing researchers to study in detail how hypnosis affects specific brain regions. Recent research allows expanding the understanding of the mechanisms of hypnosis, including changes in the functional and structural aspects of the brain during hypnotic states. Examining brain activity during hypnosis allows for the integration of data from different methods, creating a more complete picture of the effects of hypnosis on the brain.

The understanding provided by recent research can contribute to improving the clinical application of hypnosis, helping to develop more effective treatment methods and aiding in understanding patients' individual reactions.

Overall, despite the complexities, recent research on the brain and its interaction with hypnosis opens up new perspectives for the scientific understanding and clinical application of this phenomenon. In the future, researchers may expand their understanding of the brain's relationship with hypnotic states using advanced technologies and approaches. The development of advanced neurointerfaces capable of reading and interacting with brain activity could help study how hypnosis modifies various aspects of neuronal activity. This could open up new possibilities for controlling hypnotic states and individualising hypnotherapy. Developing intelligent algorithms for analysing large amounts of data from neuroimaging studies can help identify complex patterns and correlations between hypnosis and different brain regions. This could help determine key factors influencing hypnotic states. Using data from hypnotic studies to develop individualised hypnosis programmes that consider the unique characteristics of each person's brain. This could help enhance the effectiveness of hypnotherapy and develop personalised approaches to managing stress, pain, or other conditions. The integration of VR with hypnotic techniques may allow researchers to create controlled environments to study the brain's response to different stimuli during hypnosis. This could be useful in clinical treatment and the analysis of psychological mechanisms. Using technologies to examine hypnosis in real time may allow researchers to observe changes in brain activity and reactions to hypnotic influences immediately. This could facilitate the precise determination of the timing and dynamics of hypnotic states. These ideas are just a modest sample of possible directions for future research. With technological advancements and a deeper understanding of brain processes, future research may reveal new horizons in investigating the interaction between the brain and hypnosis.

Studies of the impact of hypnosis on brain activity and consciousness open up interesting prospects in the field of neuroscience and psychology. Studies confirm that hypnosis can lead to changes in brain activity, reflected in electroencephalographic indicators such as changes in the frequency spectrum of waves and synchronization of different brain regions. According to research, hypnosis can affect the functioning of various brain structures, particularly the activity of the anterior and posterior parts of the cerebral cortex, responsible for information processing and voluntary actions respectively. Changes in thalamic functioning, which plays a key role in regulating consciousness and controlling important brain functions, have also been observed. Recent research shows that hypnosis can influence a person's level of consciousness and experience, making them more susceptible to the influence of subconscious processes and the identification of altered states of consciousness. Discoveries of these mechanisms are important for understanding the nature of consciousness and its modification capabilities. The overall conclusion from the research shows that hypnosis is a powerful tool for analysing brain mechanisms and investigating the nature of consciousness. Considering the complexity of these processes, further scientific investigations in this area have the potential to reveal new aspects of the human mind's functioning and its interaction with the environment.

Chapter 5

Practical Applications and Future Research

5.1 Examination of Hypnosis, Its Possible Benefits and Risks for the Brain

Hypnosis research is an important branch of science and covers various aspects. Researchers examine the neurophysiological mechanisms of hypnosis, its effectiveness and application in the treatment of mental and physical diseases, mechanisms of influence on the human psyche and behaviour, and biological and sociocultural factors that affect its effectiveness. The latest research is aimed at developing new techniques and methods of hypnosis, assessing the ethical aspects of its use, and creating appropriate legal norms and standards. These areas of research contribute to understanding the nature of hypnosis and developing new treatments and mental health support.

Hypnosis can be very useful. It is used in medicine to reduce pain and stress, improve sleep, and treat psychological and physical illnesses such as anxiety, depression, and chronic pain (Fig. 23).

Mechanisms of Brain Function and Adaptive Potential
Liana Spytska

ISBN 978-981-5129-91-5 (Hardcover), 978-1-003-71697-6 (eBook)
www.jennystanford.com
DOI: 10.1201/9781003716976-6

Hypnotherapy (HT) also helps improve one's health by changing their views on certain problems or habits. Many people use hypnosis to improve their lives, develop themselves, and achieve personal goals. In some cases, hypnosis can help unlock the potential of memory and creativity. However, it is important to note that the results of hypnotherapy can vary depending on the individual characteristics of each person. Using hypnosis can also bring risks and limitations.

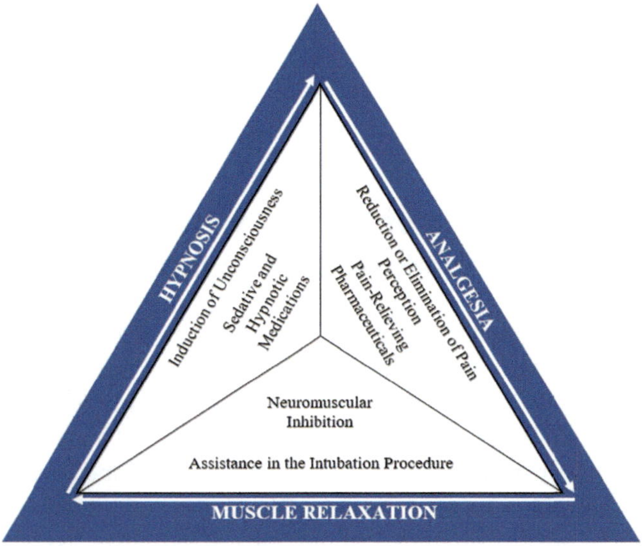

Figure 23 The concept of general anaesthesia: Hypnosis, analgesia, and muscle relaxation.

For example, during hypnosis, unpredictable reactions can occur, such as panic or loss of consciousness, especially in people with mental disorders. There may also be a negative impact on mental health in cases of unskilled hypnosis use or in vulnerable individuals. Some people may feel insecure or uncomfortable in a state of hypnosis, while others may experience unpleasant effects after the session, such as a headache or dizziness. In addition, there may be ethical issues related to identity control and data privacy during hypnotherapy. Therefore, before using hypnosis, it is important to conduct a detailed consultation with a qualified specialist and consider all possible risks and limitations.

The latest practical applications in various fields of science and technology are widely considered and developed in the modern world. In medicine, for example, new treatments, the use of artificial intelligence for diagnosis and individual treatment, and the development of nanotechnology for targeted drug delivery are just some of the areas that are rapidly developing. Another significant area of research involves BCI, which offers groundbreaking advancements in neuroscience and rehabilitation medicine.

One such example is the development of BCI technology for patients with paralysis. According to a case study by L. R. Hochberg [288], paralysed people can operate a robotic arm with just their brain activity thanks to the application of BCI technology. The patient is able to perform intricate motions without the need for physical contact due to the decoding of neural signals from the motor cortex. For people with severe movement disabilities, this breakthrough offers promise for improved autonomy and quality of life by showcasing the possibility of direct neural connectivity with external devices.

Although the scientific approaches of BCI technology and hypnosis differ, both disciplines investigate cognitive control, brain plasticity, and mind-body interactions. Through directed suggestions, hypnosis can alter brain activity, which may help with motor rehabilitation, pain management, and anxiety reduction. According to certain research, hypnosis can support neurorehabilitation by improving focus, lowering stress levels, and encouraging brain healing. By enhancing adaptability and interaction with neuroprosthetic devices, hypnosis combined with technology such as BCI may lead to new therapeutic opportunities for patients.

In the field of information and communication technologies, the latest applications include the development of artificial intelligence to automate business processes, the use of blockchain technologies to ensure data security, and the integration of large amounts of information for analysis and decision-making. In science and research, the latest practical applications cover the development of genetic engineering to develop new treatments for diseases and the development of artificial intelligence and robotics to automate and optimise various medical processes. In all these areas and many others, the latest practical applications are vital to making progress and addressing today's challenges in the world.

Hypnotherapy is used to treat a variety of mental disorders, such as phobias, anxiety, depression, and post-traumatic stress disorder (PTSD). This allows therapists to work with the subconscious aspects of the patient's personality and promote positive changes in their attitude towards themselves and their problems. Recent research in hypnotherapy on phobias, anxiety, depression, and PTSD demonstrates substantial potential for this treatment [289].

One of the directions is the assessment of the effectiveness of hypnotherapy in reducing anxiety symptoms. Some studies have shown that hypnotherapy can help people with anxiety reduce stress levels and improve their overall mental health. For example, a study by the authors K. E. Valentine, et al. [290] (University of Hartford, Connecticut, USA), which is a meta-analysis, quantifies the effectiveness of hypnosis in treating anxiety. Researchers investigated the effect of hypnosis on reducing anxiety compared to the control state. They used an intersubjective or mixed model to compare the effectiveness of hypnosis with other psychological methods. After analysing 399 papers, the authors identified only 15 studies that met the necessary criteria. After completing the scientific analysis, the researchers made an important discovery: participants who were treated with hypnosis had an average reduction in anxiety levels by 79% compared to the control group. This effect remained noticeable even with the longest follow-up; the condition of participants who participated in seven trials improved by 84% compared to the control group. However, the researchers' findings showed that hypnosis was more effective in reducing anxiety when used in combination with other psychological techniques rather than as a stand-alone treatment. This study is interesting for several reasons. Firstly, it explores the effectiveness of hypnosis in treating anxiety, which is a relevant and important subject in psychological science. The authors used a meta-analysis to combine data from different studies and draw more objective conclusions. The results of the study showed that participants who received hypnosis in combination with other psychological methods showed a substantial reduction in anxiety levels compared to the control group. This is an important discovery that may have practical implications for psychotherapists and patients looking for effective treatments for anxiety. Therefore, this study attracts attention for its relevance, methodological validity, and practical importance of the results.

In another paper by author A. Trumm [291] (Middlesex University, UK), the effectiveness of hypnotherapy was evaluated based on the results of studies on hypnotherapy for the treatment of anxiety disorders, which were conducted from 2007 to 2018. Overall, 320 participants took part in the study, including children and adults with any anxiety disorder. They received various types of hypnotherapy and other types of therapy. Hypnotherapy was identified to be beneficial for people with PTSD and anxiety disorders. However, the research results are not completely convincing due to methodological problems, such as insufficient randomness in the sample. Therefore, the author expresses the need for better research to accurately determine the effectiveness of hypnotherapy for the treatment of anxiety disorders. Therefore, this study is important in the context of understanding the effectiveness of hypnotherapy for the treatment of anxiety disorders and indicates the need for further research to confirm its results.

In the case of phobias, hypnotherapy has also proven effective in reducing the intensity and frequency of panic attacks and phobic reactions, helping people with phobias cope with their fears. Phobias are considered anxiety disorders that are characterised by excessive and unacceptable fear or avoidance of certain objects, situations, or actions that may be safe in other circumstances. Phobias can be classified according to various criteria, including the object or situation they cause or by their origin. For example, phobias occurring due to fear of specific objects or situations, such as spiders (arachnophobia), heights (aerophobia), crowded places (agoraphobia), etc.

J. Hinze et al. [292] conducted a study using functional magnetic resonance imaging that focused on specific phobias, particularly spider phobia (arachnophobia). They identified various brain regions that are responsible for processing fear in patients with such phobias. Functional and structural differences in the brains of people with phobias were established, indicating the biological roots of these disorders. Research has also confirmed that both conscious and subconscious fear processing play an important role in phobic disorders. Cognitive behavioural therapy (CBT) is recognised as an effective treatment for certain phobias, and its effects can be tracked using state-of-the-art neuroimaging techniques. In addition, research shows that new approaches, such as the use of virtual reality (VR) or augmented reality (AR), can be just as effective as traditional "*in vivo*" treatments and can expand treatment options for a variety of medical or individual scenarios. The general conclusion is that further research is of great importance, as it can discover new treatments and a deeper understanding of the neurobiological mechanisms of phobias and their treatment.

K. Fuhr et al. [293] conducted a pilot clinical study aimed at assessing the effectiveness of hypnotherapy in *reducing the symptoms of agoraphobia* compared to the control group. This study also considered identifying genetic and epigenetic changes that may affect treatment outcomes. The study methodology consisted of a randomised controlled trial (RCT) that was conducted over two years. The study involved 36 patients with agoraphobia, who were randomly divided into two groups: one received hypnotherapy, and the other waited for it. Patients who received hypnotherapy underwent 8 to 12 sessions over 3 months. The study determined that hypnotherapy did help reduce the symptoms of agoraphobia compared to the control group. Patients who received hypnotherapy experienced substantial relief at the end of the course of treatment, while patients in the control group showed only minor improvement during the waiting period. The COMT Val108/158Met genotype was also determined to affect agoraphobia symptoms and COMT DNA methylation levels. This is the first study to show that hypnotherapy has been effective in reducing anxiety symptoms. The authors call for further research that would confirm the effectiveness of hypnotherapy compared to standard treatments for anxiety disorders on a larger scale and research on the relationship between hypnosis propensity and genetic characteristics that can predict the success of hypnotherapy treatment.

Many recent studies have examined the relationship between the genetic origin of mental disorders and the positive effects of hypnotherapy. This study also tracks the relationship between the genetic origin of mental disorders and the positive effects of hypnotherapy. Through an analysis of the latest research in psychology and genetics, it discusses what genetic factors can influence the development of mental disorders and the role that hypnotherapy can play in reducing their symptoms. Mental disorders such as depression, anxiety, and personality disorders are serious medical problems that substantially affect a person's quality of life and functioning. Studies show that these disorders have both genetic and environmental factors. Genetic studies show that mental disorders have a substantial genetic component. Different genes can influence the predisposition to develop such disorders, but genetics is not the only factor that determines their occurrence. Although genetic factors can determine predisposition to psychiatric disorders, the positive effects of hypnotherapy can affect the expression of genes associated with these disorders. Hypnotherapy can help reduce stress levels, which can activate genes that contribute to the development of disorders. Overall, the relationship between the genetic origin of mental disorders and the positive effects of hypnotherapy reflects the complexity of the factors that determine the development and treatment of these conditions. Additional research in this area may help to better understand the mechanisms underlying the effectiveness of hypnotherapy and its impact on the genetic dimension of psychiatric disorders. This paper serves as a further step in understanding the complex relationship between genetics and psychotherapy, which may have implications for future research and clinical practice.

Social phobias arise in response to fear of situations of social interaction. For example, *fear of public speaking (glossophobia).* O. V. Ebrahimi et al. [294] focused on public speaking phobia, which is the most common type of social phobia in the population. This condition can lead to substantial functional limitations and negative consequences for work, study, and social spheres of life. Public speaking phobia in adolescence and adulthood was identified to be associated with an increased risk of developing general anxiety disorder, which further complicates the situation. The purpose of this meta-analysis is to assess the short- and long-term effects of psychological methods on public speaking phobia and general social anxiety disorder. There is also an assessment of possible differences in effectiveness between technologically assisted methods (e.g., online therapy) and more traditional methods of providing treatment (e.g., personal therapy). It was investigated whether there is a difference in the weight of effects between different theoretical frameworks and the correspondence of self-esteem and other measures of effectiveness (physiological and behavioural). Several RCTs were conducted to assess the impact of psychological methods on public speaking phobia. 30 RCTs with 1,355 participants were included, and most of the papers examined the effectiveness of CBT. The results showed that psychological methods are effective in reducing public speaking phobia and have a moderate effect on overall social anxiety disorder. It was also established that the effect of hypnotherapy remains stable against both active and passive controls. Thus, psychological methods have proven effective in treating the phobia of public speaking. The use of technologies to apply various methods has also demonstrated effectiveness on a par with traditional methods. These findings point to the possibility of increasing access to effective treatment through online therapy, which can be implemented in schools, primary and specialised psychiatric care.

Xenophobia is defined as fear or dislike of strangers or people who are considered "different". This is a psychological condition that can have various manifestations, including feelings of alienation, hostility, rejection, or even aggression towards other cultures, races, or ethnic groups. The problem of xenophobia arises from various sociocultural and psychological factors. This can be caused by stereotypes, negative experiences, xenophobic rhetoric in the media or politics, or even be the result of cultural or historical influences. Research on xenophobia helps to understand its roots, mechanisms, and consequences for individuals and society in general. A study by N. Harrison and N. Peacock [295] includes the analysis of intergroup relationships, psychological experiments, the examination of emotional response, and other aspects. Understanding xenophobia allows developing strategies to overcome this condition, including cultural education, cross-cultural communication, developing tolerance and promoting mutual understanding between different groups. Such research helps build a more harmonious society where everyone can feel respected and safe, regardless of their culture or background. Hypnotherapy can be used to treat anxiety disorders, including phobias such as xenophobia. Hypnotherapy can help the client achieve a deep state of relaxation, which can reduce overall levels of anxiety and fear and change negative beliefs and stereotypes that can be the basis for xenophobia. This can help the client view the world from a new, less disturbing perspective. Hypnotherapy can be used to introduce a client to fictional situations where they encounter the object of their xenophobia. This allows them to experiment with their feelings and reactions in a safe environment, which can help reduce fear. Hypnotherapy can help the client feel more confident and independent in situations involving xenophobia. This may include working on positive beliefs and mantras that help increase self-confidence and reduce anxiety. It is important to note that hypnotherapy should be performed by a qualified professional with appropriate training and experience in the treatment of anxiety disorders. This can be part of a comprehensive approach to treating xenophobia, which may include other forms of therapy, such as CBT, psychotherapy, and medication, if necessary.

I. Mc Intosh [296] describes *flight-related phobias* that are common among passengers. Unexpected accidents and plane crashes make many travellers anxious and filled with fear of flying, a condition that most people overcome by rationalising and blocking thoughts; other people develop a phobia of flying. This condition responds successfully to hypnotherapy, which, however, can take a long time. Limited sessions of short, directed, selective hypnotherapy can cure overwhelming fear, impaired conditioned responses, and evasive behaviour. Comparing the effectiveness of treating phobias with hypnotherapy to other methods, including drug approaches, is complex and depends on many factors, such as the type of phobia, the individual characteristics of the patient, and the methods used by the specialist. Hypnotherapy differs in that it is usually aimed at working with the subconscious mind, where the roots of fears are laid. This approach can help identify and relax the subconscious beliefs and mechanisms underlying phobias and help change attitudes toward them. Hypnotherapy can also promote a deep relaxation state that helps reduce anxiety. Compared to medication, hypnotherapy often has no side effects that can occur from taking medication. In addition, hypnotherapy usually focuses on developing self-control skills and stress management strategies that can be useful for managing phobias in the future. However, hypnotherapy can sometimes be expensive and not effective for everyone. Some people may experience an insufficient response to hypnotherapy due to their individual characteristics or phobias. Therefore, the effectiveness of hypnotherapy in comparison with other methods of treating phobias is determined by a complex of factors, and the decision on which method is best suited for a particular patient should be made individually and under the supervision of qualified specialists.

Another study by V. Papagianni and Y. Kotera [297] determined that from 3 to 15% of people in their lives experience certain phobias that can be long-lasting and cause distress. Usually, the treatment of such phobias can be long and unpleasant. The study examined the effectiveness of hypnotherapy for easier and less long-term treatment. The study involved seven adult patients with various types of phobias. Each participant received one 60-minute hypnotherapy session and was interviewed a week later. In the course of the study, the benefits of one session of hypnotherapy for people with specific phobias were investigated. All participants noted that a single session of hypnotherapy helped improve their phobic response, including changes in feelings, thoughts, and physical responses. From this standpoint, a single session of hypnotherapy can be the path to successful treatment, reducing time, emotional impact, and costs. Six participants who experienced phobic stimuli after the session experienced improvements in phobic response. A positive test result for the effect of a phobic stimulus indicates the effectiveness of hypnotherapy as a treatment method. Participants who had chronic anxiety and avoided treatment reported improvements in their phobia after hypnotherapy. Techniques such as special location, three-picture technique, and future rehearsal proved useful for some participants, but others found regression work less useful. Further research may help clarify how these methods can be adapted to the needs of clients and avoid the negative effects of failed previous attempts to treat phobias. Data from this study suggests that this more cost-effective approach may be beneficial for some clients. As a single session, it can offer a short and effective treatment, as a stand-alone intervention, or as a short supplement to other treatments. Further research is needed to confirm the validity of these results and assess the long-term therapeutic value of this approach, together with the use of specific techniques that were used in this study.

As for depression, researchers are also exploring the potential benefits of hypnotherapy as an adjunct treatment. It can help manage symptoms of depression and improve mood by working with the subconscious mind and mental processes. H. Khazraee et al. [298] examined *the effects of hypnotherapy on depression, self-focus, and psychological inflexibility* in women with severe depressive disorder in a randomised clinical trial. 34 participants were randomly assigned to hypnotherapy groups and a control group waiting for treatment.

The main group was given 8 personal, 60-minute weekly therapy sessions and audio cassettes with hypnosis elements for daily use. The study established that women who received hypnotherapy felt better compared to those who did not receive such treatment, both right after it and 2 months later. It also determined that women after hypnotherapy had fewer symptoms of depression, were better in control of their actions, and were less rigid in their perception of their situation. Thus, hypnotherapy is an effective method in treating depression and improving the well-being of people with severe depression.

Another study by the authors K. Fuhr et al. [299] sought to determine whether hypnotherapy is no less effective than CBT, the gold standard of psychotherapy, in terms of percentage reduction of depressive symptoms in mild to moderate major depressive disorder. The study involved 152 patients with major depressive disorder. Participants were randomly assigned to groups receiving CBT or GT. The primary outcome was the average percentage of improvement in depressive symptoms assessed using the Montgomery-Asberg depression scale before and after treatment. Thus, the study showed that both treatments are effective for depressive disorders. The results of follow-up observations 6 and 12 months after the end of treatment confirm the main results. The study has some limitations, including the absence of a control group without treatment. The main finding is that this is the first study to show that HT is no less effective than CBT in major depressive disorder, using strict methodological standards.

A. Haipt et al. [300] also considered *the effectiveness of two types of psychotherapy*: *CBT and hypnotherapy for patients with mild to moderate depression*. The authors used infrared spectroscopy to measure patients' brain activation during emotional tasks. Patients who received hypnotherapy showed a decrease in functional connectivity between certain parts of the brain responsible for emotional processing. This decrease was more noticeable in those who had a lower tendency to ruminate. Hypnotherapy activated emotional processes, especially in the right hemisphere of the brain, which is responsible for the perception of emotions. Therewith, CBT focused on changing negative thoughts and behaviours. Changes in activation in the right hemisphere of the brain during hypnotherapy depended on the patients' level of reflection. Those who ruminated less showed less activation, while those who ruminated more showed more activation. Hypnotherapy has been shown to have an effect on depression, especially in patients with a lower tendency to ruminate. Changes in activation in the right hemisphere of the brain during hypnotherapy were not associated with specific emotions. Limitations of the study included a small sample and limitations of the data processing methodology. Future research may consider other aspects of the effects of hypnotherapy on the brain and its interaction with rumination and the severity of depression symptoms.

Hypnotherapy can be a useful adjunct in the treatment of panic attacks, but it is not always the main method of therapy. Panic attacks are an intense state of anxiety accompanied by physiological symptoms such as rapid heartbeat, sweating, shortness of breath, etc. J. Cottraux [301] states that hypnotherapy can be useful for people with panic attacks in the following aspects: reducing anxiety, managing symptoms, identifying and understanding the causes of attacks, and developing coping strategies. During hypnotherapy, it is possible to explore the subconscious causes of panic attacks, which helps a person better understand their emotions and reactions. However, it is important to remember that hypnotherapy should only be performed by a qualified therapist who has experience in dealing with mental disorders. Hypnotherapy can be part of a comprehensive approach to treating panic attacks, which may include psychotherapy, pharmacotherapy, and other methods.

In the case of PTSD, hypnotherapy can be used to reduce symptoms such as recurring memories and nightmares. It helps restore emotional stability and helps victims cope with the consequences of a traumatic experience. C. B. J. Lesmana et al. [302] conducted a study that examined the effectiveness of Spiritual-Hypnosis-Assisted Therapy (SHAT) in the treatment of PTSD in adults who have suffered childhood traumatic events. PTSD is a serious disorder that can occur in response to traumatic events in childhood. This can include natural or humanitarian disasters, war, violence, rape, and other horrific events. Traumatic events in childhood have a strong impact on mental health and may be the main cause of psychopathological disorders in adulthood. The study evaluated the bio-beneficial effectiveness of SHAT in terms of changes in cortisol levels and PTSD symptoms. Cortisol is a hormone that has a strong effect on

HPA-visceral OTA and can be used to evaluate the biological effects of psychotherapy and pharmacotherapy. Participants were divided into a SHAT group and a control group receiving fluoxetine. SHAT is a culture-sensitive therapy that combines elements of psychodynamic hypnosis, CBT, and humanistic therapy. The study established that SHAT was substantially more effective than fluoxetine in reducing PTSD symptoms, while both methods had similar effects on cortisol modification. SHAT was identified to be effective in reducing PTSD symptoms and resulted in comparable pharmacological treatment modification of cortisol in the HPA-visceral axis. The study showed that SHAT is effective in treating PTSD, especially in patients with childhood trauma, and is important for understanding and treating PTSD. However, more research is needed to confirm these results and understand the mechanisms of SHAT's effects on mental health.

This study identified the effectiveness of culturally adapted clinical hypnotherapy in the treatment of PTSD. It showed that clinical hypnotherapy, adapted to cultural characteristics, substantially reduced the clinical symptoms of PTSD compared to pharmacological treatment. The mechanisms of action of such therapy are aimed at reframing traumatic memories and using a positive approach, considering cultural beliefs and practices. The study established that both clinical hypnotherapy and pharmacological treatment resulted in reduced cortisol levels. This result is consistent with data from previous studies but shows that clinical hypnotherapy has a similar effect on cortisol modification, which is a surprising discovery. The study had several limitations, such as the lack of experimental control over possible lifestyle cortisol modifiers and a short follow-up time. It is proposed to include the interaction of the immuno-neuroendocrine system as a biological marker to fully understand the therapeutic potential of clinical hypnotherapy. It is also recommended to use more accurate and advanced measurement tools, such as functional magnetic resonance imaging and positron emission tomography, to observe changes that occur in the central nervous system. SHAT has a substantial effect in reducing the clinical symptoms of PTSD and has a relatively similar effect on cortisol modification compared to pharmacological treatment. Understanding the neuroendocrine system response in participants with PTSD is important for further development of more effective interventions.

PTSD is a serious problem in both adults and children who have witnessed or directly experienced military operations or other disasters. Hypnotherapy, along with other techniques, is useful in the treatment of this pathology. Research shows that hypnotherapy can help PTSD patients reduce anxiety levels, calm the nervous system, and ease stress symptoms. In children, especially, hypnotherapy can be an effective method, as they may be more prone to visual and imaginative techniques. However, it is important to consider the individual characteristics of each patient and their readiness for this type of therapy. In addition to hypnotherapy, other techniques such as CBT, emotional regulation, and exposure therapy are also used to treat PTSD. A combination of different techniques may be most effective on a case-by-case basis, helping patients with PTSD return to normal life and overcome the effects of war trauma. It is important that the therapist has the appropriate qualifications and experience in working with PTSD patients and providing support and understanding of the difficulties that the patient faces during treatment.

There is currently some uncertainty in the scientific community regarding the diagnosis of PTSD in children, as no consensus has yet been reached on the level and severity of symptoms that are clinically substantial or require therapeutic intervention. Many experts believe that hypnotherapy targeting PTSD symptoms can be helpful. Thus, researchers C. Y. Xian-Yu et al. [303] believe that psychotherapy for PTSD should be aimed primarily at restoring mental activity disrupted as a result of trauma. The appearance of PTSD in a child indicates that the impact of a traumatic event substantially exceeded the resources that the child could use at that time. As a result of trauma, the main protective layers of the psyche were disrupted, and the protective mechanisms could not work adequately. Therefore, much of the work on trauma recovery is aimed at finding and strengthening the child's resources and strengthening adaptive defence mechanisms.

The main goal of psychotherapy is to create a new cognitive model in which a traumatic event occupies a certain place among other life events. Thus, a traumatic event ceases to play a "special, exceptional role" and becomes part of a life experience. New emotional experiences are not interpreted as a reminder of trauma. The central point of psychotherapy for a person with PTSD is to integrate the alien, unacceptable, terrible, and incomprehensible that happened to them into their self-image. Four therapy strategies can achieve this goal: maintaining adaptive self-skills, developing a positive attitude to symptoms, reducing avoidance, and changing meaning attribution. The goals of psychotherapy include reducing anxiety and restoring a sense of personal integrity and control over life.

Working with people suffering from PTSD has three stages: stabilisation, trauma resolution, and integration. The first stage involves establishing therapeutic relationships with the child and their parents to create an atmosphere of safety and trust, collecting a trauma-specific medical history, and stabilising the condition. Establishing trusting relationships is important because the interpersonal components of traumatic experiences tend to manifest themselves when interacting with a therapist. High rates of rejection of therapy after a serious trauma can be caused by fear of repeated trauma. In the second stage, the work is conducted to reproduce memories and experiences. It is important to avoid excessive haste or traumatic material, as this can disrupt the recovery process. The third stage is to integrate into everyday life and ensure a gradual transition from alertness to a sense of security. The fundamental rule of PTSD therapy is to accept the pace of work and self-discovery that the client offers. Dealing with PTSD can also include informing family members about the need to deal with traumatic experiences, as home support can help avoid stress.

Overall, new research continues to demonstrate the prospects of hypnotherapy as an effective and integral component of psychotherapy treatment for various psychiatric disorders. Recent research in the field of hypnosis in the context of pain management has discovered substantial potential for this method in reducing pain and improving the quality of life of patients. Some studies show that hypnotherapy can be effectively used to reduce pain in a variety of medical situations, including surgical procedures, childbirth, chronic pain, migraines, and other conditions. One such study was conducted by T. Thompson et al. [132], who investigated *the effectiveness of hypnosis in reducing pain* and determined factors affecting its effectiveness in their meta-analysis. A systematic analysis was conducted that included reviewing 6 large databases to identify studies comparing the effects of hypnotic inductions with control conditions for pain assessment, threshold, and pain tolerance using experimentally induced pain patterns in healthy participants. The meta-analysis identified 85 studies with 3,632 participants that demonstrated the analgesic effects of hypnosis for all pain indicators. In particular, it was established that the effectiveness of hypnosis substantially depends on the hypnotic propensity of participants, with the greatest reduction in pain observed in high- and medium-slope individuals. The conclusion is that hypnotic intervention can provide a substantial reduction in pain for most people and, therefore, can be an effective alternative to pharmaceutical intervention.

However, more high-quality studies are needed to be accepted in the chronic pain population. Thus, this meta-analysis highlights the substantial potential of hypnosis in pain management and points to the need for further research to understand the mechanisms of its action and its use in clinical practice. The uniqueness of this study lies in several key aspects. Firstly, it focuses on determining the effectiveness of hypnotherapy as a method of pain management, which is a relevant direction in modern medical science. Secondly, the study uses meta-analysis, which allows evaluating the results of not one, but several separate papers that have similar methods and indicators. This allows for more objective results and more general conclusions about the effectiveness of hypnotherapy in pain management. Thirdly, the study draws attention to the effect of participants' hypnotic propensity on the effectiveness of hypnotherapy. This expands the understanding of how and for whom this method can be most useful. In addition, the study points to the practical importance of the results and the potential of hypnosis as an alternative method of pain management, which is important for clinical practice and improving the quality of medical care for patients with chronic pain. Therefore, this study is unique in its approach, scale, and importance in the context of developing new pain treatments.

A study conducted by T. Gardner et al. [304] considers *the use of clinical hypnosis in the treatment of chronic pain*, identifying a few key points. Primarily, the study confirms that clinical hypnosis can effectively improve outcomes in patients with chronic pain. However, there are several obstacles to implementing this technique, including misunderstandings about the essence of hypnosis, low confidence in performing hypnotherapy, and anxiety about integrating hypnosis with current therapeutic approaches. In contrast, the study also identified factors that contribute to the introduction of clinical hypnosis, in particular, changes in knowledge and perceptions after training and openness to learning techniques and skills. In addition, the online training score was positive, especially because of its structure and reliability. Summarising, the successful implementation of clinical hypnosis requires the development of training programmes that consider existing misunderstandings, allow learning knowledge and skills, and adapt to the context of the intervention. This study has important implications for physicians practising in the field of pain management. It highlights the need to support healthcare professionals' training in the use of clinical hypnosis for patients with chronic pain to successfully integrate this technique into clinical practice.

The study focuses on analysing the obstacles and factors that contribute to the introduction of clinical hypnosis into the practice of chronic pain management and developing an online training programme for clinics using clinical hypnosis. Obstacles include uncertainty and expectations of clinics about the place of clinical hypnosis in current practice, safety issues, risk management, and setting limits on the scope of interdisciplinary clinics' practice. Proponents embrace clinics' open approach to learning and implementing clinical hypnosis and changing their perceptions after training. The study also highlights the need to develop training programmes that address existing misunderstandings about clinical hypnosis, allow learning knowledge and skills, and adapt to the context of the intervention. The structure and prospects of the online training programme were positively received, which contributed to changes in the attitude of clinics to clinical hypnosis. In conclusion, the introduction of clinical hypnosis may be a substantial therapy option for managing chronic pain, but its successful implementation requires carefully designed training programmes and further research to maintain effectiveness and successful implementation in clinical practice.

The latest research in the industry on the *treatment of phantom pains with hypnosis* shows encouraging results in reducing pain intensity and improving patients' quality of life. Phantom pain is felt in a lost or amputated part of the body. Hypnotherapy is used to change the perception of pain by activating various mechanisms in the brain and mental processes. Some studies have shown that hypnotherapy can help patients with phantom pain reduce the intensity of pain and reduce the stress levels associated with the condition. Hypnosis helps redirect the patient's attention from pain to feelings of relaxation and comfort. The study by V. L. Moura et al. [305] examines the problem of phantom limb pain (PLP) and the role of various mind-body therapies in its treatment. According to the results of the study, pain in the phantom limb is a substantial source of chronic pain in most amputees during their clinical journey. Pharmacological treatments for this condition often have only moderate efficacy and can lead to undesirable side effects. There is growing evidence for the therapeutic effectiveness of mind-body therapy in reducing chronic pain. Therefore, it is advisable to investigate their role in reducing amputation-related pain. A literature review identified research on hypnosis, mental exercises, and biofeedback, including visual mirror biofeedback. Most studies reflect substantial variability in the use of mind-body therapies, which makes definitive conclusions impractical. However, there are results that suggest that the mind–body approach to PLP treatment is promising, and specific methods may offer temporary or long-term relief, either alone or in combination with traditional methods. The authors recommend conducting additional and better studies investigating the effectiveness and mechanisms of action of mind-body therapies in the treatment of PLP. Therefore, the study highlights the importance of examining mind-body therapies in the treatment of PLP and the need for further work to understand their effectiveness and mechanisms of action.

Recent research in the treatment of phantom pain using hypnotherapy draws attention to an important aspect of collaboration with the patient in understanding and transforming their perception of pain. This approach includes the use of visualisation techniques, positive enhancers, and relaxation techniques. A study by D. A. Oakley et al. [306] examines the theoretical background, reviews the available literature, and offers a new perspective on the treatment of pain in the phantom limb using hypnotic images. The paper's methodology presents 2 new clinical cases in which hypnotic imaging procedures were used to reduce pain in the phantom limb and analyses 10 previous clinical studies with a similar approach.

The results of these studies distinguish 2 main treatment strategies: cognitive/visual approaches and motor/visual approaches. A common discovery is the need to view the phantom limb as a real part of the body, accept its existence as a real mental representation, and avoid overcoming the sensation of pain in the phantom limb only through the amputated limb. The conclusion of the study highlights that hypnotic treatments seem to be a useful addition to established strategies for treating pain in the phantom limb but require further, more systematic research. When analysing this paper, it is important to consider that it offers a new perspective on the treatment of phantom pain, emphasising the use of hypnotic images. The results of the study show that this approach can be effective, especially when the approach of the phantom limb is taken as a real part of the body and not just as a missing part.

In the context of clinical practice, it is important to consider the need for further research on this approach and its implementation in real practice. Given the potential benefits of hypnotic images in treating phantom pain, researchers and clinicians can develop more effective and innovative treatment strategies for patients facing this complex condition. Although further research is still needed to fully understand the effectiveness of hypnotherapy in treating phantom pain, the available data show the prospects of this method as one of the possible components of a comprehensive approach to managing this complex condition. Hypnotherapy helps reduce the perception of pain by changing the psychological response to it. It can affect pain perception by activating certain mechanisms in the brain, such as the production of endorphins, which are natural analgesics, and reducing the activity of areas of the brain responsible for pain perception.

Some studies have also shown that hypnotherapy can have a long-term effect in reducing pain after completing sessions, making it an attractive method for treating chronic pain and other long-term conditions. Overall, recent research on hypnosis confirms its importance and effectiveness as a tool for managing pain in a variety of medical settings. Hypnosis is a powerful tool that can be used to improve various aspects of life, including increased concentration, improved memory, weight loss, addiction relief, and improved athletic skills. In the study conducted by T. G. Wolf et al. [6], the results of examinations of functional and/or metabolic changes in brain activity using hypnosis, using various imaging techniques, were reviewed.

The use of various imaging techniques, such as EEG, MRI, and PET, allowed the researchers to investigate changes in brain activity during hypnosis. Many studies have shown positive changes in brain activity during hypnosis, including increased θ and α activity. The results point to the role of certain brain regions in the effects of hypnosis on pain and stress perception. However, to better compare the results and standardise research methods, it is recommended to implement a homogeneous approach. Overall, studies support the possibility of using hypnosis to reduce pain and stress, but more research is needed to better understand the mechanisms of action and optimise its use.

Recent findings in the field of hypnotherapy for psychosomatic diseases point to the potential benefits of this approach in treatment. One of the research areas is investigating the effectiveness of hypnotherapy in treating pain symptoms such as headache, back pain, arthritis, and other chronic pain. Some studies suggest that hypnotherapy can help reduce pain intensity and improve the quality of life in patients with various psychosomatic illnesses. Research also shows that hypnotherapy can be helpful in reducing the symptoms associated with stress, anxiety, and depression that often accompany psychosomatic illnesses. Studies are being conducted using hypnotherapy to reduce stress and anxiety levels in patients with various medical conditions. In addition, some studies have shown that hypnotherapy can affect the autonomic nervous system, which is responsible for the body's physiological responses to stressful situations. This can have a positive effect on the functioning of the body's organs and systems. Contrary to these prospects, more research is needed to understand the mechanisms of action of hypnotherapy in the context of psychosomatic diseases and to establish its optimal methods and protocols for use. Some studies also examine individual factors that may influence a patient's response to hypnotherapy.

The effectiveness of hypnotherapy in drug addiction can vary depending on many factors, according to researcher C. Connelly [307], including the type of drug, duration of addiction, psychological and physical characteristics of a person, and other factors. Some studies and clinical experience suggest that hypnotherapy can be useful as an adjunct to other forms of addiction treatment but may not always be sufficient on its own. Each person's response to hypnotherapy may be different. For some people, hypnosis can be an effective tool for changing psychological attitudes and reducing the desire to use drugs, while for others, it can have limited or minor benefits. Hypnotherapy is often most effective when used in conjunction with other treatments, such as psychotherapy, medication therapy, support programmes, etc. The success of hypnotherapy and other forms of treatment often depends on the person's motivation and willingness to change their habits and desire to support the recovery process. It is important that the hypnotherapist has the appropriate qualifications and experience in working with people suffering from drug addiction. In general, hypnotherapy can be a useful element of a comprehensive approach to drug addiction treatment, but it is important to consider the individual characteristics and needs of each client when choosing treatment methods.

T. O'keefe [308] examines several key aspects of clinical hypnotherapy. The essence of clinical hypnotherapy is its powerful effect on mental and behavioural changes aimed at helping clients who have *problems with alcohol, drugs, or addiction* to quickly move to sobriety and recovery, contributing to deep unconscious changes. The importance of training and experience of a hypnotherapist determines the success of hypnotherapy, which emphasises the importance of training and professional development in this area. A step-by-step approach to addiction treatment includes methods for recognising addiction, developing step-by-step treatment plans, and rapid therapeutic changes during discontinuation of various types of prescription drugs, amphetamines, and opioids, and helps clients build resilience for long-term sobriety. The text uses keywords related to hypnosis, hypnotherapy, substance abuse, addiction, addiction rehabilitation, and addiction treatment. The general conclusion is that clinical hypnotherapy can be an effective tool in treating alcohol and drug addiction, and the success of this therapy depends on the training, experience, and skills of the hypnotherapist.

Hypnotherapy can be a useful method in the treatment of obsessive-compulsive disorders (OCD), which are characterised by the presence of non-intentional thoughts, ideas, or images (obsessions) that cause anxiety, compulsive actions or rituals (compulsions) that are performed to reduce anxiety. Changing psychological perceptions can help change negative psychological perceptions and beliefs associated with obsessions, promote relaxation, and reduce anxiety. Developing anxiety management strategies can teach the client relaxation and control over anxiety, helping to reduce the need for compulsive activities. Developing adaptive responses to obsessions can help the client develop adaptive strategies for responding to obsessions that do not involve compulsive actions. Improving self-control helps improve the ability to think rationally in OCD-related situations. Relaxation and stress relief techniques can be taught to the client, which can help reduce anxiety levels and improve mental health. Stimulating positive changes in mental feelings can encourage positive changes in mental feelings and beliefs, which helps reduce obsessions and compulsions. It is important that hypnotherapy be performed by a qualified therapist with experience in dealing with OCD and other psychiatric disorders. Hypnotherapy can be effective, especially when combined with other forms of psychotherapy and support.

Hypnotherapy can be a useful method *in the treatment of gaming addiction* or pathological gambling. D. Pickering et al. [309] believed that this method consists of the hypnotherapist using hypnosis to penetrate the client's subconscious and direct it to positive changes in the mental state and behaviour. Some ways that hypnotherapy can be useful for treating gambling addiction are as follows: changing psychological perceptions can help change psychological perceptions and associations associated with gambling, reduce the desire to play, and change attitudes toward gambling. Managing stress and anxiety through hypnosis can promote relaxation and reduce anxiety, which is important for those dealing with gambling addiction and helps reduce the desire to gamble. Increased motivation for change can also be a consequence of hypnotherapy, helping to strengthen the client's desire to improve their health and life. Developing strategies to prevent a return to gambling can also be effective through hypnotherapy, which helps develop strategies to manage temptation and distortions of thoughts that may arise in the future. It is important to remember that hypnotherapy should be performed by a qualified therapist with experience in dealing with mental disorders, including gambling addiction. Hypnotherapy can also be part of a comprehensive approach to treating gambling addiction, which may include psychotherapy, group support, medication therapy, and other methods.

Hypnotherapy can be a useful method *in the treatment of shopping addiction* or pathological shopping. This approach involves using hypnosis to influence the client's subconscious mind to change psychological perceptions and behaviour. Some ways that hypnotherapy can help treat shopping addiction: changing stereotypes and associations can help change the psychological patterns associated with shopping. By influencing the subconscious mind, the perception of shopping as a way to solve problems or fill emotional voids can be influenced. Managing one's emotional state and stress can help reduce the levels of anxiety and stress that often lead to compulsive shopping. Relaxing and reducing emotional stress can help clients maintain control over their behaviour. Developing new behavioural strategies can help develop new strategies for managing restraint and controlling consumer needs. Clients can learn to distinguish between a real need and an impulsive desire. Changing intrinsic motivation and self-determination can promote positive self-esteem and support customers' intrinsic motivation to change their attitude toward shopping. In a study by O. Vasiliu [310], the pathology under consideration, which includes elements of bipolar affective disorder (BPAD), OCD, and impulse control disorders, draws attention to the heterogeneity of the populations included in the review reports. It is worth noting that some of them show proximity to BS, some show greater impulsivity, and others are associated with elements of the OCD spectrum. Different combinations of these parameters in the same patient are also theoretically possible, which is confirmed by this population's high level of comorbidities.

According to a review of 24 sources, including case reports (n = 8), clinical trials (n = 12), and systematic reviews/meta-analyses (n = 4), psychotherapy, especially group CBT, appears to be effective for patients with this pathology. Psychotherapy is recommended for patients with BPAD and OCD. Combined pharmacotherapy and psychotherapy may also be effective, although the evidence for this strategy is less substantial. As for pharmacological preparations, the use of serotonergic antidepressants and topiramate is recommended, especially in patients with concomitant mood disorders. The strengths of this review are shown in the inclusion of both primary and secondary sources obtained through a systematic literature search and in the formulation of evidence-based recommendations with potential clinical benefits.

However, the limitations associated with including data from studies with many comorbidities that may affect the results are notable. In addition, not all researchers conducted a detailed examination of the presence of concomitant psychiatric diseases at the initial stage, which makes it difficult to interpret the data obtained. Further studies are expected to evaluate larger patient groups and more homogeneous participants and identify the long-term effects of therapy in patients with this pathology. It is important to remember that hypnotherapy should be performed by a qualified therapist with experience in dealing with mental disorders, including shopping addiction. This method can be an important component of a comprehensive approach to the treatment of shopping addiction, which may include psychotherapy, group support, drug therapy, and other methods.

Hypnotherapy can be a useful method in treating addiction to junk food, such as overeating, gluttony, or addiction to junk foods. It aims to change the psychological factors underlying this problem. Self-awareness and mental reprogramming are key aspects of hypnotherapy, where the client understands the reasons for their addiction and reprograms their ideas about food. F. Delestre et al. [311] argue that obesity is a serious problem in Western countries, and its management is becoming a subject of discussion. Some clinical studies have examined hypnosis, but it is quite difficult to draw definitive conclusions about its effectiveness. It has been investigated how hypnosis and autosuggestion affect nutritional control in obese adults with high levels of impulsivity. Adults with a BMI of 30 to 40 kg/m^2 and high levels of impulsivity were examined for a year in the study. Patients received training on proper nutrition at seminars. In the hypnosis group, participants underwent 8 hypnosis sessions, and the level of impulsivity, changes in anthropometric indicators, and other parameters were measured. It was identified that the group where hypnotherapy was performed showed a lower level of impulsivity and weight loss compared to the control group. This suggests the possible effectiveness of hypnosis in controlling nutrition and weight loss in obese adults with high levels of impulsivity.

Additional benefits of hypnotherapy include managing stress and emotions, creating positive associations with healthy eating, and developing healthy eating strategies. Increasing motivation and self-control is another important aspect that helps customers better manage their food preferences and make more informed choices. Hypnotherapy can be an effective tool in combating addiction to junk food, but it is important to note that it should be used in combination with other treatment methods and strategies.

Hypnotherapy is one of the methods *of treatment of enuresis, especially in children,* according to one of the researchers, N. Tomic [312]. Enuresis, or night-time urinary incontinence, is a common disorder in children, especially at an early age. Hypnotherapy in this context can be an effective tool for changing behavioural habits and positively influencing the psychological state of a child. Hypnotherapy can help children create a positive mental image of themselves and their ability to control their urination during sleep. Self-confidence can have a positive effect on their psychological state and reduce anxiety associated with enuresis. Hypnotherapy can also use relaxation and stress relief techniques to help children fall asleep calmly and continuously. A calm and relaxed state can reduce the likelihood of nighttime urinary incontinence. In addition, hypnotherapy can help children change their psychological patterns and associations associated with accidental urination during sleep and serve as a form of psychological support. This method allows the child to feel in control of their body and the sleep process, which contributes to positive self-determination and psychological well-being. It is important to remember that hypnotherapy should be performed by a qualified specialist who has experience working with children and understanding the specific features of their psyche.

Hypnotherapy and sports achievements may have an important correlation, as hypnotherapy can help athletes improve their focus, self-control, self-confidence, and overall psychological resilience, which affects their performance and achievements in sports. For example, there are effective ways to explain the correlation between hypnotherapy and athletic performance. The researchers noted increased focus and concentration, or vice versa, deep relaxation and concentration, which helps athletes focus on their performance, game strategy, or training, minimising the impact of external distracting factors. Increase self-confidence and positive thinking. Hypnotherapy can help athletes overcome internal obstacles, such as doubts about their own abilities or fears of failure. Through positive subconscious programming, hypnotherapy promotes confidence and supports the mental success factor. Managing stress and reducing anxiety. Athletes often face stress and anxiety, especially before important competitions or while recovering from an injury. Hypnotherapy can teach them relaxation and stress relief strategies that help improve their psychological resilience, increase motivation, and purposefulness. Hypnotherapy can help athletes set and achieve specific goals, increase their motivation and self-confidence, speed up recovery, and relieve pain. Hypnotherapy can be an effective tool for speeding up recovery from injuries and relieving pain, helping athletes return to training and competition faster. Therefore, hypnotherapy can play an important role in improving athletic performance by improving the mental processes and emotional well-being of athletes.

The study by Z. Li and S. X. Li [313] identified several key points regarding *the psychological training of athletes and the use of hypnosis in sports.* Psychological training for athletes is considered an integral part of training, as it helps to maintain calmness, reduce anxiety, and manage stress before competitions. Hypnosis can be useful for athletes in increasing self-esteem, reducing anxiety, overcoming fears, improving technique, and restoring energy after training and during competitions. However, the text emphasises some limitations of the use of hypnosis in sports, such as the lack of methodologically sound research and the lack of empirical data on the mechanisms of its action. Despite the fact that hypnosis is a simple and effective way to regulate the physical and psychological state without special requirements for conditions and equipment. Such observations highlight the importance of research in this area and the potential of hypnosis as a tool for the psychological training of athletes. The general conclusion is that hypnosis may have some potential to improve athletes' performance and maintain their psychological well-being, but more scientific research and empirical evidence are needed to confirm its effectiveness. Despite this benefit, it is worth considering the possible risks of hypnosis for the brain.

Hypnosis, as a method of psychotherapy, is quite safe for most people, but in some cases, it can cause unexpected effects that require attention and understanding. These effects include deterioration of the mental state, panic attacks, and hyperstimulation. The first thing to consider is an individual predisposition to reactions to hypnosis. While most people may feel calm and relaxed during a hypnosis session, some may react with a worsening mental state. For example, in people with a history of mental disorders or anxiety, hypnosis can cause negative emotional reactions or worsen symptoms. In addition, panic attacks can occur during hypnosis sessions in people with an increased tendency to anxiety disorders. Deep immersion in a hypnotic state can trigger unexpected reactions in people who already have a tendency to panic attacks or agoraphobia. The possibility of hyperstimulation, when a deep state of hypnosis can lead to excessive brain stimulation or cause the impression of loss of control, is also notable. This can be particularly problematic for people with hyperactivity or increased vulnerability to feelings of loss of control. All these unintended effects of hypnosis highlight the importance of competent and professional use of this method of psychotherapy. Before conducting a hypnosis session, it is necessary to consider individual characteristics and potential risks and have the skills to regulate the effects of hypnotic influence on the client's mental state.

Hypnosis, as a phenomenon of a mental state, can cause people to lose control of their behaviour and reactions. This loss of control can be partial or complete, depending on the individual characteristics of the person, the depth of the hypnotic state, the ability of the hypnotist, and the nature of the hypnosis session. Loss of control during hypnosis can have many negative consequences. A person may perform actions or make decisions that they would not normally have performed or taken. This can lead to unintended consequences and create risks to the safety and well-being of the individual and the environment.

For example, during a hypnotic session, a person may lose consciousness of the environment and perform dangerous actions, such as driving a car or using equipment, without the necessary control and attention. In addition, loss of control can lead to psychological discomfort and feelings of loss of autonomy and self-control, which can affect a person's emotional state. It is important that the hypnotherapist has appropriate training and experience in conducting sessions and adheres to ethical standards and safety standards to minimise the risk of loss of control during hypnosis. In addition, a person who is going to go to a hypnosis session should be aware of the possibility of losing control and carefully relate to the chosen hypnotherapist and the conditions of the session.

During hypnotherapy, there may be a situation where unplanned effects on the subconscious mind can have unpredictable consequences. This is reflected in the fact that under the influence of hypnosis, a person becomes more likely to accept instructions and subconscious suggestions that can affect their thinking, emotional state, and behaviour. It is precisely because of the instability of the subconscious response to hypnotherapy influences that it is important to ensure professional ethics and careful selection of hypnotherapy methods and strategies. Experienced hypnotherapists should have a deep understanding of psychological processes and be prepared to adapt their approach to the individual characteristics of each client. Avoiding unexpected influences during hypnosis requires careful analysis and the development of individual approaches to each case. It is important to consider the context, goals and needs of each client, provide appropriate support and monitor the possible negative effects of hypnosis on the client's mental health and emotional state. In general, hypnosis is a powerful tool that can be useful for many people in different areas of life, but before using it, it is necessary to consider the risks and conduct it only under the qualified supervision of a specialist.

5.2 Discussion of Practical Aspects of Using Knowledge about the Brain, Hypnosis, and Neural Networks in Various Fields, Such As Medicine, Psychology, Technology, and Training

Hypnotherapy, as a treatment method, can have various applications in solving various disorders. Patients with depression can change negative thought patterns that are typical of depressive states by using hypnotherapy, which promotes relaxation and cognitive restructuring. This could entail teaching the brain to process things more optimistically and reinterpreting unpleasant experiences. Through guided imagery and deep breathing, hypnotherapy helps people with anxiety disorders control their stress and regulate their own behaviour. Hypnosis is also used in the treatment of chronic pain since it can alter how pain is perceived and encourage relaxation. It has also shown promise in addiction treatment, assisting patients in reframing their connection to addictive behaviours and creating more constructive coping strategies.

Notwithstanding its advantages, hypnotherapy has drawbacks. Because it could make symptoms worse, it might not be appropriate for people with severe mental diseases or psychotic disorders. Additionally, each person's vulnerability and level of commitment to the treatment process determine how effective it is. While some people see notable changes, others might only see minor advantages.

The use of cochlear implants for people who are profoundly deaf is a prime example of how the brain may adapt to outside influences. The brain's amazing ability to reorganise itself is demonstrated by the auditory cortex's substantial neuroplastic alterations that occur after implantation to make room for new electrical inputs. Research shows that stronger auditory pathways and better speech perception result from early implantation [314]. This story emphasises how useful it is to comprehend neuroplasticity in order to create successful therapeutic approaches in a variety of professions.

The theory and practice of hypnosis distinguish between several types that have their own unique characteristics and applications (Fig. 24). First of all, there is *a reproduction-activation type* associated with memory stimulation. This type of hypnosis promotes the activation of memories and the reproduction of information under the influence of a hypnotherapist, especially in hypnotrance [315]. The second type, known as *stopping hypnosis,* is aimed at relieving stress and psychotrauma, in particular, after criminal events. This method is aimed at the psychological recovery of people who have experienced traumatic situations by influencing the subconscious. The third type of hypnosis is the method of *juxtaposition*, which allows perceiving events from a different angle to get new information and understand the situation from different standpoints. This approach can help people change their beliefs and reactions to events happening around them, improving their understanding of themselves and the world around them through deep self-reflection under the influence of a hypnotherapist. This method can be dangerous, as it can cause mental trauma and distort the perception of the event.

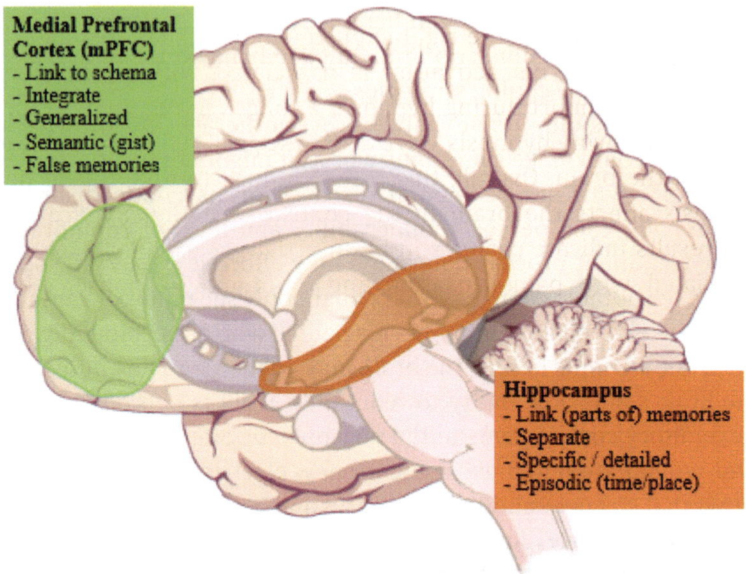

Figure 24 The role of the hippocampus and medial prefrontal cortex in memory processes and information integration.

Studies on the duration of hypnotherapy for various disorders may vary depending on the specific type of disorder, individual patient characteristics, and research methodology. However, some general conclusions can be drawn based on available research. Some studies have shown that hypnotherapy can be effective in treating depression [316]. This can lead to substantial relief from depressive symptoms, and the duration of this relief may vary for each individual. Some studies have shown an effect lasting up to several months, but additional sessions may be required to maintain the results. For stress and anxiety, hypnotherapy can also be effective as it can give a person stress management skills and positively affect their emotional state. The effect may be noticeable even after several sessions. Hypnotherapy is effective in treating phobias and PTSD. The duration of the effect may vary, but many patients experience an improvement in their condition after a course of hypnotherapy. However, some may require additional support or therapy to maintain results.

In the treatment of addictions, hypnotherapy can be used to change negative habits and influence psychological factors associated with addiction. The duration of the effect may depend on the specific type of addiction and individual characteristics of the patient. The general trend is that the effectiveness of hypnotherapy can be substantial, but achieving sustainable results may require a certain number of sessions and support or counselling after the end of therapy. In addition, individual outcomes may vary depending on various factors, including the patient's motivation and perception of therapy.

Hypnosis, as a method of therapy, can take many forms and types depending on its use and purpose. For example, clinical hypnosis is used for medical and psychotherapeutic purposes to treat physical and mental illnesses and to relieve stress, anxiety, phobias, and other psychological problems. *Clinical hypnosis* is a psychotherapeutic approach that uses the hypnotic state to achieve mental and physical goals. This method is based on the idea that a deep, relaxed state, accompanied by concentration and altered consciousness, can help a person change their thoughts, feelings, or behaviour. In clinical hypnosis, the therapist uses the hypnotic state to promote the patient's mental well-being [317]. Hypnotherapy in the form of clinical hypnosis can be effective for relieving pain and symptoms of illnesses such as migraines and increasing pain tolerance during procedures such as dental interventions or surgeries. Clinical hypnosis can be used to treat various mental problems, such as anxiety, depression, PTSD, and phobias. Hypnotherapy can be an effective method for changing negative habits or developing new healthy habits, such as managing stress, changing eating habits, quitting smoking, etc. Hypnosis can help boost self-confidence, overcome negative beliefs, and develop a positive attitude towards oneself. Clinical hypnosis is often used as part of a comprehensive treatment, but it can also be an effective independent method of psychotherapy. The main idea is that a deep state of relaxation and concentration can help unlock the potential for self-control and change in a person.

The use of knowledge about the brain and hypnosis in therapy represents various practical aspects covering clinical hypnosis and its various applications. Clinical hypnosis, used for medical and psychotherapeutic purposes, can relieve stress, anxiety, phobias, and other psychological problems. This method is based on the idea that a deep, relaxed state, accompanied by concentration and altered consciousness, contributes to changes in a person's thoughts, feelings, and behaviour. The use of hypnotherapy as part of clinical hypnosis can have substantial effects on the treatment of physical and mental conditions, including pain relief, relief of migraine symptoms, anxiety, depression, PTSD, and other mental health problems. In addition, hypnotherapy can be an effective method for changing negative habits and developing useful ones, such as managing stress or quitting smoking. An important part of clinical hypnosis is to increase self-confidence, overcome negative beliefs, and develop a positive attitude towards oneself. This method is often used as part of a comprehensive treatment and can also act independently as an effective method of psychotherapy. The conclusion is that the deep state of relaxation and concentration characteristic of hypnosis can open up the potential for self-management and personality changes, contributing to the achievement of mental well-being and improving the quality of life of patients.

Self-hypnosis technique is when a person independently triggers a state of hypnosis to achieve certain goals, such as relaxation, improving well-being, or achieving certain goals. This technique can be used to relax, improve well-being, relieve stress, increase motivation, and overcome certain negative habits or viruses. The main element of self-hypnosis is to focus on a specific object, thought, or sensation [318]. This can be visualising a calm place, concentrating on your breathing, or repeating certain positive affirmations. Through concentration and the use of visualisation techniques, a person can trigger a state of hypnosis characterised by deep relaxation and increased openness to subconscious influences. During a hypnotic state, a person can formulate positive goals or positive programmes for their subconscious mind, such as maintaining self-confidence, relieving stress, or improving well-being. After achieving certain goals or completing a self-hypnosis session, the person gradually returns to a full state of cheerfulness and consciousness. Self-hypnosis can be a useful tool for self-development and self-affirmation, but it is important to use it with an understanding and consideration of the individual characteristics of each person. In addition, for maximum effectiveness, it may be helpful to get instructions and practice under the guidance of an experienced hypnotherapy specialist.

Using knowledge about the brain and hypnosis, the self-hypnosis technique stands out as a powerful tool for achieving a variety of goals, from relaxation to increased motivation. The main element of self-hypnosis is focusing on a specific object, thought, or sensation, which promotes deep relaxation and increases openness to subconscious influences. During a hypnotic state, positive goals or programmes for the subconscious mind can be formulated, such as maintaining self-confidence or relieving stress. It is important to consider the individual characteristics of each person and use self-hypnosis with understanding. It is also useful to get instructions and practice under the guidance of an experienced hypnotherapy specialist for maximum effectiveness. Self-hypnosis can be not only a useful tool for self-development and self-affirmation but also help in overcoming various negative habits or viruses, providing a person with the opportunity to actively influence their mental and physical health.

Examples of disorders in which self-hypnosis can be used may include conditions such as anxiety disorders, phobias, chronic pain, insomnia, and some aspects of depression and stress. Self-hypnosis can be helpful in managing stress, reducing anxiety, relieving pain, and improving sleep quality. However, there are risks associated with the use of self-hypnosis, in particular, the possibility of unpleasant or undesirable effects if self-hypnosis techniques are not used properly or if the patient is not sufficiently trained. For example, in an uncontrolled environment, self-hypnosis can deepen the state of anxiety or cause negative emotional reactions. It is also important to keep in mind that self-hypnosis is not a panacea for all disorders and cannot replace professional medical or psychotherapy care.

The effectiveness of self-hypnosis can be high, especially with proper training and careful practice. The benefit is that the patient can learn to independently control their emotions and responses to stressful situations, reduce anxiety, relieve pain, and improve sleep quality without the need for constant medical support. However, it is important that self-hypnosis be used responsibly and under the supervision of a qualified professional or psychotherapist.

One specific example where self-hypnosis can be used is the treatment of chronic pain, such as in the case of migraines. A person suffering from migraines can learn self-hypnosis techniques to reduce the intensity of pain and the duration of seizures. This may include muscle relaxation, visualising positive images, breathing and meditation techniques. However, there are risks associated with self-hypnosis in the treatment of migraines. For example, incorrect use of techniques can lead to minor relief or even exacerbation of the condition. In addition, some people may not be susceptible enough to self-hypnosis, and therefore, this technique may be less effective for them.

In the case of migraines, the effectiveness of self-hypnosis can be high if the person knows how to apply the techniques correctly and regularly uses them in their practice. This can help reduce drug addiction and improve the quality of life of a person suffering from migraines. In migraines, self-hypnosis can be a useful tool for reducing the intensity of pain and improving the quality of life. One of the methods of self-hypnosis that can be used for migraines is the relaxation technique.

During migraines, a person can use relaxation self-hypnosis to reduce the muscle tension and stress that often accompany the condition. For example, the patient may start by concentrating on the breath, gradually reducing the frequency and depth of breathing. After that, they can visualise themselves in a calm and pleasant place, such as a beach or forest, where they feel relaxed and pain-free. During relaxation self-hypnosis, it is important to distract yourself from pain and negative thoughts, focus on pleasant sensations, and visualise relaxation. This can help reduce the degree of tension that can often make migraine pain worse.

However, it is important to remember that self-hypnosis can have different efficacy for different people, and it is not a panacea for all cases of migraines. In addition, before using self-hypnosis to treat migraines, it is important to consult a doctor to determine the appropriate treatment and rule out possible contraindications.

Developed by E. Erickson, a well-known American psychiatrist and psychotherapist, *Erickson's hypnosis* is used in therapy to stimulate subconscious changes and solve problems through metaphors and symbols [319]. This approach is used in therapy to stimulate subconscious changes and solve problems through metaphors and symbols. Erickson believed that the subconscious mind responds better to metaphors, stories, and symbols than to straightforward directives. In hypnotherapy, he used metaphorical stories to encourage patients to reflect and change themselves.

Erickson believed that each person is unique, so he treated each patient individually, considering their unique needs, perception style, and personality traits. Erickson used hypnosis as a tool to promote subconscious changes, enhance reflection, and stimulate creative thinking. An important part of Erickson's approach was understanding and working with patient resistance. Erickson believed that resistance could indicate deeper problems that needed attention. Erickson saw every change, regardless of its scale, as a positive step forward. He emphasised the importance of confidence in success and supported patients in their development. Erickson's hypnosis has become a popular approach in psychotherapy and hypnotherapy and is used by specialists to solve various psychological problems, including anxiety, depression, phobias, etc. It allows working with subconscious mechanisms and positively influencing the behaviour and emotional state of patients. The hypnotherapist can use metaphorical stories and symbols to help the client understand the root of their anxiety experiences and find ways to overcome them. Hypnotherapy can be effective in reducing pain caused by chronic illnesses or after an injury. With the help of metaphors, the client can learn new ways to feel pain and respond to it. Hypnotherapy can be used to change negative beliefs and habits, such as experiencing unworthiness or bad eating habits. Through the use of metaphors and positive reinforcement, a person can develop new, useful strategies for behaviour and thoughts. Hypnotherapy can help a person understand their strengths, values, and goals and unlock their inner potential. With the help of metaphors and visualisation techniques, the client can develop positive aspects of their personality. Hypnotherapy can help a person improve communication skills, understanding, and empathy. By working with metaphors and visualisations, the client can develop more understanding and empathy for other people. These examples show how Erickson's hypnosis can be an effective tool for achieving psychological well-being and personal development. An important component of successful hypnotherapy is cooperation between the client and the therapist and openness to experimentation and change.

By looking at a specific disorder, namely PTSD and its treatment with Erickson's hypnotherapy, both the benefits and risks of this method in terms of effectiveness can be determined. As for the benefits, it's worth noting that Erickson's hypnotherapy can help reduce the intensity of PTSD symptoms, including repetitive memories and affective responses. In addition, it can improve the quality of life by helping to reduce anxiety and improve the patient's emotional state. However, there are certain risks. Some patients may experience discomfort or increased anxiety during hypnotherapy, especially when recalling traumatic events. In addition, if hypnotherapy is not performed professionally or the individual needs of the patient are not considered, it can lead to negative effects, including increased anxiety or feelings of depression. In general, the effectiveness of Erickson's hypnotherapy in the treatment of PTSD depends on the individual characteristics of the patient, the quality of hypnotherapy sessions, and the interaction between the patient and the therapist. Understanding these aspects can help ensure effective and safe treatment of PTSD through hypnotherapy.

Hypnotherapy can be one of the components of the complex treatment of depressive disorders and suicidal thoughts. However, the effectiveness of hypnotherapy in these cases depends on many factors, including the individual characteristics of the patient, the severity of symptoms, motivation for change, and the professionalism of the therapist. Some studies suggest that hypnotherapy may be helpful in reducing symptoms of depression and stress and improving mood and overall mental health. It can help patients focus on the positive aspects of their lives, change negative thoughts and beliefs, and find strategies to deal with stress. However, it is important to understand that hypnotherapy is not a panacea for depression or suicidal thoughts. It should be used as part of a comprehensive treatment approach that may include psychotherapy, pharmacotherapy, social support, and other methods. If a person has depressive symptoms or suicidal thoughts, it is important to see a qualified professional for evaluation and treatment. Psychotherapy, including hypnotherapy, can be part of an effective treatment plan, but it should be prescribed and conducted under the supervision of an experienced and licensed psychotherapist.

Another interesting and effective approach to hypnotherapy is *neurohypnosis*, which uses the principles of brain neuroplasticity to change mental states and improve mental health. The basic idea of neurohypnosis is that the brain can change and adapt over the course of life, and hypnosis tools can promote this neuroplasticity to change mental states and improve mental health [320]. Neurohypnosis uses hypnosis techniques to stimulate certain areas of the brain and influence mental processes. These techniques can include deep relaxation, visualisations, metaphors, and other activities that help change the mind and subconscious mind. Neurohypnosis is based on the principle of neuroplasticity, which means the brain's ability to reprogram its structure and functions according to new experiences and influences. The use of hypnotic techniques can help create new neural connections and change brain activity. Neurohypnosis can be used to reduce stress, anxiety, depression, PTSD, and other mental health problems. It can help people focus on the positive aspects of their lives, change negative thoughts and beliefs, and find ways to deal with stress. Neurohypnosis can be a useful tool in psychotherapy, but it is important to consider that it should be used under the supervision of an experienced hypnotherapy professional. As with any form of psychotherapy, the effectiveness of neurohypnosis can vary depending on the individual characteristics of each person and the nature of their problems.

Neurohypnosis can be used in various areas of psychotherapy and personal development. A hypnotherapist can use deep relaxation hypnosis to reduce stress and anxiety levels. Using positive metaphors and visualisations can help reduce the level of arousal in the nervous system and increase the sense of calmness. Neurohypnosis can be helpful in changing negative beliefs and switching attention to the positive aspects of life. Using metaphors and symbols can help the client rethink their worldview and consider alternative ways of thinking. Neurohypnosis can promote positive self-perception and increase self-confidence. Using hypnotic techniques, such as positive reinforcement and success visualisation, can help a person feel stronger and more confident. Neurohypnosis can be an effective tool for reducing the intensity of pain and improving tolerance to it. The use of hypnotic techniques can help change the perception of pain and increase comfort levels. Neurohypnosis can help a person improve their sleep quality and reduce insomnia. A hypnotherapist can use special hypnosis techniques to promote relaxation and create favourable conditions for deep sleep.

These examples demonstrate the diversity of applications of neurohypnosis in psychotherapy and personal development. An important component of the successful use of this approach is cooperation between the client and the therapist and openness to new experiences and changes.

In conclusion, neurohypnosis is an interesting and effective approach in hypnotherapy that uses the principles of brain neuroplasticity to improve mental health and change mental states. This approach is based on the fact that the brain can change and adapt over the course of life, and hypnosis techniques promote this neuroplasticity by encouraging the brain to create new neural connections and change activity. Neurohypnosis can be used to reduce stress, anxiety, depression, and other mental health problems and to improve self-perception, self-confidence, and the ability to cope with stress. In practice, neurohypnosis allows for a variety of psychotherapeutic effects, including reducing stress levels, changing negative beliefs, improving sleep quality, and reducing pain intensity. An important condition for the successful use of neurohypnosis is cooperation between the client and the therapist and openness to new experiences and changes. Thus, neurohypnosis can be a useful tool in various areas of psychotherapy and personal development, but it is important to use it under the supervision of an experienced specialist in hypnotherapy since the effectiveness can vary depending on the individual characteristics of each person and the nature of their problems.

Neurohypnosis, as a method of influencing the psyche and behaviour, is becoming increasingly popular. It is especially useful in the treatment of depression, anxiety disorders, and PTSD. One specific example of the use of neurohypnosis is the use of deep relaxation and meditation techniques to reduce the symptoms of anxiety disorders. During neurohypnosis sessions, special sounds or visualisations can be used to help reduce stress and anxiety levels in the body. As for depression, neurohypnosis can be used to stimulate the production of neurotransmitters such as serotonin, which is responsible for mood and emotional state. Such techniques may include sound therapy or visualisation of positive images.

As for PTSD, neurohypnosis can help in processing traumatic memories and reduce the hypervision response. This can be achieved through the use of special audio therapy programmes that promote relaxation and mental health. However, as with any treatment, there are risks and limitations. For example, improperly used neurohypnosis can worsen a state of anxiety or depression, especially if the therapist does not have sufficient experience in dealing with mental disorders. In addition, some people may find an insufficient response to such methods due to the individual characteristics of their nervous system. In general, the effectiveness of neurohypnosis may vary depending on each specific situation and individual characteristics of the patient. It is important that neurohypnosis is conducted under the guidance of a qualified specialist with medical education and experience in this field.

A hypnotherapy technique that involves immersion in the patient's subconscious mind to examine previous memories or events that may affect their psychological state is *regressive hypnosis*. This method involves immersion in the subconscious mind to explore previous memories or events that may affect a person's psychological state [14]. The basic idea is to examine and understand the causes of certain mental problems, identify and resolve traumatic events or conflicts that may be hidden by consciousness. Regressive hypnosis can help the client discover and understand traumatic events from their past that may underlie psychological problems in modern life. This allows for identifying and resolving unresolved conflicts or injuries that may be important for further treatment. Through regressive hypnosis, the therapist can help the client understand the roots of their psychological problems, which may include early childhood traumas, negative experiences or relationships, and other events that have shaped their worldview and behaviour. Regressive hypnosis can be an effective tool for treating mental health problems such as depression, anxiety, PTSD, etc. By discovering and understanding the roots of problems, the client can go through the process of improving their mental state and restoring their emotional well-being. The experience of regressive hypnosis can help the client better understand themselves, their internal mechanisms, and their reactions to events. It can promote education and self-knowledge, which is important for further development and self-realisation. Regressive hypnosis requires great care and experience on the part of a hypnotherapist, as there may be a risk of causing negative emotions or identifying painful subjects. It is important that regressive hypnosis is performed by professionals with appropriate training and experience in this field.

The use of regressive hypnosis in hypnotherapy is a powerful tool for investigating and solving psychological problems, especially those related to traumatic events or conflicts from the past. This technique allows diving into the patient's subconscious to identify and understand the factors that may underlie their psychological problems. Regressive hypnosis is aimed at determining and understanding traumatic events that may be hidden in the subconscious. This allows the client to understand the roots of their problems and start the process of solving them. By identifying unresolved conflicts or injuries, the client can gain important insights and develop strategies for further treatment and self-improvement. This approach can be particularly effective for treating a variety of mental health problems, such as depression, anxiety, and PTSD. Understanding the roots of these problems can help the client experience emotional recovery and improve their emotional well-being. However, it is important to remember that regressive hypnosis requires caution and experience on the part of a hypnotherapist. There is a risk of causing negative emotions or discovering painful subjects, so it is important that this process is conducted under the guidance of professionals with appropriate training and experience. Accordingly, regressive hypnosis can promote education and self-knowledge, which is important for the further development and self-realisation of the individual.

These different types of hypnosis may have different purposes, methods, and effectiveness depending on the context of their use. It is important to use hypnosis with professionalism and ethical standards in mind. Types and methods of hypnosis may differ in characteristics, such as technicality, approach to therapy, goals, and methods of influencing the patient. Different types of hypnosis can use different techniques and measures to achieve a hypnotic state.

The choice of hypnotherapy method for various disorders depends on several factors, including the disorder itself, the individual characteristics of the patient, their condition, and previous experience. It is important to consider the nature and characteristics of the disorder. For example, exposure hypnotherapy can be an effective method for treating phobias, which allows the patient to gradually get used to insurance situations in a safe environment. Each person is unique, so it is important to consider their personality, perception style, and previous experience with hypnotherapy. For example, for a person with a high level of anxiety, it may be important to choose milder methods of hypnosis. Sometimes, the patient's condition and physical and emotional readiness for hypnotherapy can influence the choice of method. For example, for people with serious mental disorders, it may be important to choose methods that allow the patient to remain more in control. If the patient has previous experience with hypnotherapy and knows which methods are more suitable for them, it is important to consider this experience when choosing a treatment method. It is necessary to consider the individual needs and preferences of the patient. Some people may be more open to certain hypnotherapy techniques, while others may feel uncomfortable with other techniques.

In general, the choice of hypnotherapy method should be individualised and consider all of the above factors to ensure the most effective and safe treatment. For example, classical hypnotherapy can use deep relaxation and imagination, while Erickson's hypnosis can use milder and more metaphorical techniques. Different methods of hypnosis may have different approaches to therapy and interaction with the patient. For example, clinical hypnotherapy may focus on treating specific mental health problems, while self-hypnosis may provide the patient with tools to manage their condition independently. Different methods of hypnosis can be aimed at achieving different goals and used in different fields. For example, neurohypnosis can be aimed at improving mental health and developing brain capacity, while regressive hypnosis can be used to assess previous memories and traumas. Different methods of hypnosis may require different levels of openness and receptivity on the part of the patient. Some techniques may be more direct and intense, while others may be gentler and more empathetic. Each method of hypnosis has its own advantages and limitations, and the choice of a particular method may depend on the needs of the patient, the style of the therapist, the specifics of the situation, and the goals of therapy. It is important that the therapist has deep knowledge and experience in using the chosen hypnosis method and appropriate skills to work with different patients.

Knowledge about the brain and hypnosis opens up a wide range of possibilities for practical application in clinical practice. Different types of hypnosis have different goals, methods, and effectiveness, depending on the context of their use. Professionalism and compliance with ethical standards for performing hypnotherapy are of paramount importance. Methods of hypnosis can vary in technique, approaches to therapy, and methods of influencing the patient. For example, classical hypnotherapy relies on deep relaxation and imagination, while Erickson's hypnosis uses soft and metaphorical techniques. Different methods of hypnosis may also have different approaches to therapy and interaction with the patient. Different methods of hypnosis can be aimed at achieving different goals and used in different fields. For example, neurohypnosis can be aimed at improving mental health and developing brain capacity, while regressive hypnosis can be used to assess previous memories and traumas. Each method of hypnosis has its own advantages and limitations, and the choice of a particular method depends on the patient's needs, the therapist's style, and the purpose of therapy. The therapist must have deep knowledge and experience in using the chosen hypnosis method and appropriate skills to work with different patients. Therefore, knowledge of the brain and hypnosis opens up wide opportunities for application in clinical practice, but it is important to adhere to high standards of professionalism and ethics in the use of hypnotherapy.

Currently, hypnotherapy takes its place among other methods of treating psychological disorders as an additional or combined approach. This means that hypnotherapy is used as part of a comprehensive approach to treatment, along with other forms of psychotherapy, pharmacotherapy, social support, and other methods. Hypnotherapy is often used as an adjunct to other treatments, such as CBT, psychodynamic therapy, interpersonal therapy, etc. Hypnotherapy can be integrated with various psychotherapy techniques, depending on the needs of the individual patient and the characteristics of their disorder. Some studies and clinical observations suggest that hypnotherapy may be effective for certain psychiatric disorders, such as anxiety, phobias, PTSD, and chronic pain. Hypnotherapy also finds its application in medical practice to treat physical problems such as chronic pain and immune system depression. It is important to keep in mind that the effectiveness of hypnotherapy may vary depending on the individual characteristics of the patient, the nature of their disorder and the specifics of the therapeutic process. Therefore, while hypnotherapy is not always the primary treatment for psychological disorders, it can be a useful and effective component of a comprehensive approach to patients' mental well-being.

Hypnotherapy techniques can be classified according to various criteria, such as approaches to hypnosis, goals, and methods of exposure. For example, among the approaches to hypnosis, *directive hypnosis* is notable when the therapist tells the patient what to do during hypnosis, that is, gives certain instructions or performs visualisation exercises. Directive hypnosis is a form of hypnotherapy in which the therapist uses certain directive techniques to control the hypnotic process and influence the patient [321]. The basic idea is for the therapist to actively manage the hypnosis session by giving the patient specific instructions or performing specific exercises. The therapist gives the patient specific instructions on how to respond during a hypnotic state. This may include visualisation exercises, behavioural experiments, or special relaxation techniques. The therapist acts as the leader of the hypnosis session, actively directing the process and determining the direction of therapy. It can adjust the strategy depending on the patient's response and the choice of methods that are most appropriate for the specific situation.

A directive approach allows the therapist to effectively achieve the goals of a hypnotherapy session. It can help the patient understand their problem, work on specific aspects of behaviour, or contribute to achieving specific goals and changes. Directive hypnosis can be especially useful in situations where there is a need to monitor the process and ensure its effectiveness. This approach can help make sessions more systematic and results-based. While directive hypnosis can be a useful tool in psychotherapy, it is important to remember that each person responds to hypnosis individually, and the therapist's approach should consider the patient's needs and character. The use of a directive approach can only be effective if the therapist has a deep understanding and ability to interact with the patient. Directive hypnosis can be useful in the treatment of various pathologies and psychological problems, but its effectiveness may vary depending on the specific situation and individual characteristics of the patient. Directive hypnosis can help patients with phobias or anxiety disorders reduce their anxiety levels, develop coping strategies, and reduce their fear of specific situations or objects. Managing chronic pain can be challenging, but directive hypnosis can help patients change their perception of pain, reduce its intensity, and improve their quality of life.

Hypnotherapy can be useful in treating depressive disorders, helping patients change their negative thoughts and emotional state, boost their self-esteem, and find new ways to deal with life's challenges. Directive hypnosis can be a useful tool in treating people with PTSD, helping them relax, regain control of their emotions, and reduce the impact of traumatic memories. For children with learning or cognitive problems, such as dyslexia or concentration disorders, directive hypnosis can be a useful tool for improving attention, memory, and learning skills. Notably, the effectiveness of hypnotherapy, including directive hypnosis, can be individual and depend on many factors, including the depth of mental problems, patient motivation, and collaboration with the therapist.

The approach of nondirective hypnosis allows the patient to freely explore their thoughts, emotions, and memories during hypnosis without being actively influenced by the therapist. Nondirective hypnosis is a form of hypnotherapy in which the therapist creates conditions for the patient to freely explore their thoughts, emotions, and memories without actively influencing them. In this approach, the therapist acts as an intermediary or supportive factor, creating a safe and supportive environment for the patient's self-reflection and self-expression. During nondirective hypnosis, the patient has the opportunity to independently explore their inner world, emotions, memories, and beliefs without active guidance or intervention from the therapist. The therapist strives to create an atmosphere of openness and empathy where the patient feels supported and understood. It encourages the patient to reflect on their inner experiences and express them without restrictions.

Nondirective hypnosis aims to help the patient focus on their own thoughts, emotions, and experiences. This can help develop a deep understanding of oneself and one's own needs. During nondirective hypnosis, the patient can discover new ideas, beliefs, and opportunities that promote their self-knowledge and self-development. The therapist adheres to the principle of maintaining patient autonomy, allowing them to independently determine the direction and depth of their research during hypnotherapy. Nondirective hypnosis can be especially useful for those patients who feel more comfortable opening up and exploring their inner world on their own, without the active influence of the therapist. This approach allows the patient to focus on their own needs and development, promoting deep self-awareness and personal growth. Nondirectional hypnosis consists of the method of influence of the therapist on the patient and his process of self-expression during hypnotherapy.

It is necessary to understand some of the most important differences between these two approaches. In directive hypnosis, the therapist acts as an active participant, providing the patient with specific instructions and tasks during hypnotherapy. In nondirective hypnosis, the therapist acts more as a facilitator or supportive factor, creating a supportive environment for the patient's self-reflection and self-expression without actively influencing their processes. In directive hypnosis, the patient may be more focused on performing specific tasks and instructions given by the therapist, whereas in nondirective hypnosis, the patient has more freedom to explore their thoughts, emotions, and memories without feeling much pressure or influence from the therapist. In directive hypnosis, the focus of therapy may be more on achieving specific goals set by the therapist, while in nondirective hypnosis, the focus may be more extended, allowing the patient to identify and solve their own problems and questions. In directive hypnosis, the therapist is more responsible for conducting the session and controlling the treatment process, while in nondirective hypnosis, more emphasis is placed on the patient's responsibility for their own process of self-expression and change. Both approaches should consider each patient's needs and characteristics and the specific circumstances of the therapy session. Each approach can have its own advantages and applications depending on the context and needs of the patient.

Among the goals of hypnotherapy, there are two most important ones: *treatment of physical or mental health problems* (treating illnesses, relieving stress, anxiety, phobias, depression, etc.) or *self-development and self-knowledge* (hypnotherapy can be used to increase self-awareness, develop personality, and inner potential) [322]. Hypnotherapy is a complex psychotherapeutic method that is used to treat various physical and mental problems and for self-development and self-knowledge. The goals of hypnotherapy can range from relieving stress, anxiety, phobias, and depression to reducing symptoms of physical illnesses and improving the patient's psychoemotional state. In hypnotherapy, the therapist can act as an active participant in the session, providing the patient with specific instructions and tasks or as a facilitator, creating a supportive environment for self-reflection and self-expression. The patient may also be assigned specific tasks or have more freedom to explore their thoughts and emotions. Hypnotherapy is used to increase self-awareness and develop the individual and the inner potential of the individual. It helps the client better understand their inner motivations, experiences, and thoughts, which contributes to deep self-knowledge and self-improvement. Hypnotherapy goes beyond conventional treatment, allowing people to tap into their inner resources to gain a deeper understanding of themselves and the world around them. It is a complex method that combines scientific approaches with psychological and physiological principles, which allows achieving substantial results in the treatment and self-development of the individual.

Hypnotherapy uses a variety of methods of influence, which can be divided into three categories. The first category is *visualisation and imagination*, which consists of creating positive mental images and situations, contributing to improving well-being and developing a positive attitude to life. The second category is *relaxation and breathing exercises*, which are based on the principles of deep relaxation and breath control. These techniques are aimed at relieving stress, calming the nervous system, and improving overall well-being. The third category is *regressive therapy*, which covers immersion in past events or memories. This method helps the client understand and solve problems that arise from certain life situations or events and develop new approaches to solving them. Each of these methods has its unique characteristics and can be effective in different situations, depending on each client's needs and individual characteristics. These classifications help therapists choose appropriate methods and approaches based on the needs and characteristics of each patient. The combination of different methods and approaches often provides the most effective result of hypnotherapy. Hypnotherapy can be useful for many psychiatric disorders, but its effectiveness may vary depending on the specific condition and characteristics of the patient.

There are mental disorders in which hypnotherapy may be less effective or even not recommended as the main treatment method. In severe cases of schizophrenia, when reality disorders and other symptoms predominate, hypnotherapy may be less effective because the patient may have a limited connection to reality and self-awareness. In cases of serious feeding disorders where there is a strong internal struggle with food and self-esteem, hypnotherapy may be less effective. In severe cases of depression and other mood disorders with a high risk of suicide or serious complications, hypnotherapy may be less safe and is not used as the primary treatment. In cases where PTSD is accompanied by serious symptoms, such as intense flashbacks, affective disorders, and acute anxiety, hypnotherapy may require specialised approaches and may not always be effective immediately. In general, it is important to remember that hypnotherapy is not a universal method of treatment, and its use should be based on the individual needs, capabilities, and characteristics of each patient and on the recommendations of a qualified specialist.

The use of brain knowledge has extensive practical applications of various hypnotherapy approaches in various fields, including medicine, psychology, education, technology, and business. Knowledge of the brain is an important factor in the development of new methods for the diagnosis and treatment of various neurological and psychiatric disorders, including stroke, depression, schizophrenia, and other diseases. A deep examination of the brain allows understanding its complex structure, functioning, and pathologies underlying these diseases. Research on the nervous system, including the brain, opens up new perspectives in understanding the mechanisms of disorder occurrence, identifying new targets for therapy, and developing innovative treatments. The use of hypnosis and knowledge about the brain in medicine and neuroscience opens up wide opportunities for improving the diagnosis, treatment, and rehabilitation of various diseases and disorders. The practical use of hypnosis in medicine is evident in a wide range of its applications to improve the physical and mental state of patients.

The main areas of use of hypnosis in medicine include pain, stress, and anxiety control, treatment of phobias and mental disorders, support for rehabilitation and recovery from injuries and surgeries, and management of symptoms of chronic diseases. Using hypnosis to control pain helps patients reduce the feeling of pain and stress associated with medical procedures. It can also be effective in reducing anxiety and stress levels in patients with psychiatric disorders. Treating phobias and disorders helps improve patients' quality of life, and supporting rehabilitation helps them recover faster from injuries and surgeries. *In medicine,* hypnotherapy has proven useful in treating chronic pain, phobias, anxiety disorders, depression, and many other mental and physical conditions. With hypnosis, one can reduce the feeling of pain, relieve stress, improve sleep quality, and increase self-awareness. It can also be used to support patients during treatment, reduce unwanted side effects of medications, and improve the effectiveness of rehabilitation measures.

Psychosomatic disorders are diseases that occur or worsen due to the influence of mental factors on a person's physical condition. These disorders can include diseases such as headache, constipation, asthma, arthritis, ulcers, and others that are often associated with stress, anxiety, or depression. Hypnotherapy, which uses hypnosis to change a patient's conscious state, has proven to be very effective in treating psychosomatic disorders. Hypnotherapy is a powerful and effective method of treating psychosomatic disorders. By influencing mental factors and activating the patient's internal resources, hypnosis can help relieve stress, relieve symptoms, and improve overall health. Understanding the relationship between the psyche and physical health opens up new possibilities for using hypnotherapy in the treatment of psychosomatic diseases.

In neuroscience, knowledge of the brain allows understanding the basic mechanisms of brain functioning and identifying connections between different brain structures and their impact on the mental and physical state of a person. Using modern neuroimaging techniques, such as functional magnetic resonance imaging and electroencephalography, the brain mechanisms of attention, memory, speech, emotions, and other cognitive functions are examined [14]. This helps to develop new methods for the diagnosis and treatment of neurological and mental illnesses and optimise rehabilitation programmes for patients with brain disorders. Practical applications of hypnosis and brain knowledge in medicine and neuroscience are important for the development of modern medical practice and for achieving more effective results in the treatment and rehabilitation of patients. They open up new opportunities to improve the quality of life of people facing various diseases and disorders and contribute to the development of scientific and clinical experience in these areas.

Based on neuroscience data, scientists examine various aspects of the brain, including its morphology, structure, functional organisation, and molecular and cellular processes. Understanding these aspects allows identifying specific mechanisms underlying various diseases and determining new targets for developing treatment approaches. One of the key areas of research is the examination of neuroplasticity—the ability of the brain to change and adapt to changes in the environment. Understanding the mechanisms of neuroplasticity opens up opportunities for developing new methods of rehabilitation and restoration of brain function after injuries, strokes, and other lesions. In addition, brain research provides an opportunity to develop new diagnostic technologies, such as functional magnetic resonance imaging, electroencephalography, and positron emission tomography. These methods allow not only to detect disorders in the brain but also to study its activity and relationships between different areas.

Therefore, brain knowledge plays an important role in the development of medicine, contributing to the improvement of methods for the diagnosis and treatment of various neurological and mental diseases, which is a big step in improving the quality of life of patients.

Understanding brain function plays an important role in the modern world, where learning and education are considered key elements of the successful development of society. Knowing the brain allows for a better understanding of the processes of sensation, perception, learning, and memory that underpin people's thinking, behaviour, and reactions to the world around them. The practical aspects of using hypnosis and brain knowledge in psychology and pedagogy can be quite diverse and cover a wide range of applications. The practical use of hypnosis in psychology is diverse and includes various aspects that are aimed at achieving different goals and developing the mental well-being of the individual. The use of hypnosis in teaching opens up several interesting prospects. First of all, it can help increase students' concentration and focus by reducing stress and anxiety levels, which will contribute to a high-quality understanding of the material. The second aspect is to improve memorisation and reproduction of information, as hypnotherapy can help stimulate memory and improve learning efficiency. In addition, hypnosis can help students overcome their fear of exams and performances, providing calmness and self-confidence in critical moments. Finally, it can promote positive thinking and self-confidence, helping overcome obstacles and achieve high academic performance.

Hypnotherapy can be considered a potential *method for improving learning and memorisation processes,* but its effectiveness depends on various factors, including the psychological readiness of the individual, the depth of the hypnotic state, types of information, individuality, and trust in the process [323]. People respond differently to hypnotherapy because of their psychological structure, and some may reach a deep hypnotic state that can be useful for learning and remembering. Hypnotherapy can help improve memorisation and learning, especially those types of information that require a lot of attention and concentration. Each person has their own unique learning and memorisation methods, and it is important to find the method that works for a particular person. Trust in the hypnotherapy process and the therapist also affects its success. Announcing this, it should be noted that hypnotherapy is not a universal solution, and its effectiveness may vary depending on the individual conditions and characteristics of each person. One specific example of the use of hypnotherapy in the field of learning and memorisation may include a situation where a student faces substantial difficulties in memorising material before difficult exams. Let us imagine that a student regularly attends hypnotherapy sessions before starting their studies. During these sessions, the hypnotherapist can promote a state of deep relaxation and concentration that helps the student better absorb and remember information. This can be especially useful in situations where the student is experiencing stress or psychological pressure before exams. After several sessions of hypnotherapy, a student may notice improvements in their ability to concentrate, improve memory, and reduce anxiety levels during training. This can contribute to better exam results and a generally positive learning experience. This example demonstrates how hypnotherapy can be an effective tool for improving the learning process and memorising information in students.

In pedagogy, knowledge about the brain helps to understand how children learn, how they perceive information, and how best to adapt the learning process to their needs. Using techniques based on the principles of brain development, such as active learning, using a variety of teaching methods, and using critical thinking techniques, can improve learning and promote children's cognitive development. In addition, the use of hypnosis in learning can help students improve concentration and memorisation, which helps improve learning performance and academic performance. In pedagogy, understanding brain processes can also help teachers create a supportive learning environment that promotes better material assimilation and develops students' creative thinking.

Therefore, brain knowledge and the use of hypnosis can be important tools in psychology and pedagogy that contribute to improving mental health and learning. The use of these methods in practice can greatly facilitate the solution of problems and the achievement of goals in these areas. The study of the brain contributes to the development of effective learning methods that optimise the processes of mastering knowledge and skills. Through scientific research on the brain, teachers and educators can develop innovative pedagogical strategies aimed at improving student learning efficiency and development. Understanding brain processes allows for better adaptation of curricula and methods to the needs of different categories of students, providing them with the most comfortable conditions for mastering knowledge.

Knowledge of the brain also opens up new horizons in understanding the individual characteristics of each person, helping create personalised approaches to learning and developing each individual. This allows considering the diversity of individual needs, abilities and learning styles, contributing to a more successful and deep awareness of the educational material. Ultimately, brain knowledge is an important foundation for further development of education and learning. It not only helps people better understand themselves and the world around them but also opens up new opportunities for creating more effective and innovative pedagogical strategies aimed at improving the quality of education and developing society in general.

However, it is worth remembering that the use of hypnosis for educational purposes should meet ethical standards, be conducted under the guidance of a qualified specialist in hypnotherapy, and consider the individual characteristics of each student and their approaches to examination. First, hypnotherapy is used to treat a variety of mental disorders, including anxiety, depression, phobias, and PTSD. This approach allows promoting changes in consciousness and working with the subconscious causes of these disorders. Secondly, hypnosis can be used to control pain, especially in medicine during surgery, childbirth, and the treatment of chronic pain. Thirdly, hypnotherapy helps improve psychological well-being, reduce stress, increase self-esteem, and positively rebuild relationships. Fourthly, the use of hypnosis promotes personality development, introspection, concentration, and self-regulation. Finally, hypnosis can be helpful in weight management and managing habits such as smoking or overeating. It is important to remember that to achieve successful results in hypnotherapy, qualified psychotherapists who have specialised training in hypnotherapy and understand the ethical aspects of using this method must be attracted. Understanding hypnosis as a psychological tool helps professionals work effectively with clients and contribute to their psychological well-being.

In psychology, hypnosis is often used as a therapeutic tool to treat various mental disorders, such as phobias, anxiety, depression, PTSD, and many others. The deep relaxation state characterised by hypnosis allows a person to immerse themselves in consciousness and interact with subconscious processes, which can help solve mental problems. An important component of successful hypnosis use is the training and qualification of medical professionals who can effectively use this method. However, there are problems, such as the lack of standardised approaches to performing hypnotherapy and ambiguity in determining its effectiveness. The latest research in the field of hypnosis is aimed at improving the methods of applying this technique, understanding the mechanisms of its action at the brain level, and developing standardised protocols for use in medical practice. This contributes to the development of more effective and safe treatments and helps ensure a better outcome for patients.

Hypnotherapy can serve as a treatment for alcohol dependence; however, its effectiveness and correlation depend on numerous factors, such as the degree of addiction, the patient's psychological state, and motivation to give up alcohol. This refers to the variety of hypnotherapy techniques that can be used. For example, positive programming includes the use of positive instructions and programming to change psychological attitudes towards alcohol, including positive visualisation and maintaining motivation for a healthy lifestyle. In addition, hypnotherapy can promote the development of self-control and stress management skills, which are important for avoiding alcohol consumption in stressful situations. In addition, it promotes the development of new adaptive habits and the replacement of bad habits of alcohol consumption. Hypnotherapy also helps in understanding and dealing with the subconscious causes that lead to alcohol addiction, such as emotional trauma, fears, and low self-esteem.

Notably, the effectiveness of hypnotherapy may vary on a case-by-case basis and is not guaranteed. It is often used as part of a comprehensive approach to alcohol addiction treatment that includes medication therapy, psychotherapy, community support, and other methods. It is important that the patient receives an individually selected and comprehensive approach to the treatment of addiction. Let us assume that a patient with an alcohol addiction has started hypnotherapy as part of their comprehensive treatment plan. The therapist uses the method of hypnosis to work with the patient's subconscious and change his attitude towards alcohol. During a hypnotherapy session, the patient is in a deep state of relaxation, which allows them to better focus on positive instructions and programming. The therapist introduces the patient into a state of hypnosis, after which he gives positive instructions aimed at changing his attitudes towards alcohol. During the session, the patients portray situations in which they feel a desire to consume alcohol and then reflect on situations in which they make positive substitutes for this desire. The therapist uses hypnotherapy techniques to enhance positive visualisations and develop new, healthy habits. After several sessions of hypnotherapy, the patients notice that their desire to consume alcohol has decreased, and healthy habits become more natural for them. Although hypnotherapy may not be an exclusive solution for treating alcohol addiction, in this particular case, it helps the patient in a comprehensive approach to their problem.

During hypnotherapy aimed at improving memory or treating addiction, certain changes in brain activity occur. Hypnosis helps to change the psychophysiological state, causing a state of concentration and increased attention. During hypnotherapy, various processes occur in the brain that should be considered. Firstly, under the influence of hypnosis, the activity of the anterior part of the brain, known as the prefrontal cortex, decreases. This can increase vulnerability to subconscious influences and ideas. Secondly, hypnosis can lead to activation of the limbic system responsible for emotions and memory. This activation can improve concentration and memorisation. Third, hypnotherapy can enhance cooperation between different parts of the brain, which helps improve information processing and memorisation. In addition, in cases of addiction treatment, hypnosis can promote the development of new coping strategies and self-regulation by altering the mental mechanisms underlying addiction. Finally, hypnosis can increase a person's openness to positive psychotherapeutic effects, which contributes to the effectiveness of treatment. An example is the use of hypnotherapy to improve a student's memory before an exam. During hypnosis, positive instructions and visualisation can be used to increase concentration and make learning material easier to remember. This can help improve exam results and change the psychophysiological state of the brain to achieve the desired effect.

Treatment of nightmares and fears in children by hypnotherapy methods can be effective but requires a careful approach and cooperation with a qualified psychotherapist who specialises in hypnotherapy for children. There are several methods that can be used in hypnotherapy to treat nightmares and fears in children [324]. Visualisation and positive imagery during hypnotherapy can help children learn to use visualisation to create positive images and scenarios that will help them calm down and relieve the stress associated with nightmares. Relaxation and breathing exercises—the child can learn various relaxation techniques and breathing exercises that will help them calm down before going to bed and during nighttime awakenings. Working with the subconscious mind—a hypnotherapist can work with a child's subconscious mind to help understand the roots of their fears and nightmares. This may include dealing with emotional trauma, stress, or experiences that affect the child. Creating positive associations—hypnotherapy can help a child create positive associations with restful and safe sleep, which will help reduce the negative associations associated with nightmares. Parental support—it is important that parents are included in the treatment process and that they support their children in their efforts to overcome their fears. A hypnotherapist can offer tips and strategies for parents on how to support their child at home.

It is important to remember that hypnotherapy for children should be performed by a qualified specialist with experience working with children and knowledge of their characteristics. It is also important to consider the individual needs and characteristics of each child when choosing hypnotherapy methods. An example of a hypnotherapy method for treating nightmares and fears in children is the scenario of using visualisation and positive imagery. During a hypnotherapy session, the child can imagine calm, safe images and scenes that help them calm down and overcome fear. Various processes occur in the child's brain during such hypnotherapy. When a child goes into a state of hypnosis, their brain becomes more vulnerable to the influence of positive instructions and images provided by the therapist. Visualising calm moments helps activate parts of the brain responsible for emotional regulation, such as the amygdala and hypothalamus. These processes can help reduce the activity of the amygdala, which is responsible for fear sensations, and help increase the production of positive neurotransmitters, such as serotonin, which provides a sense of calmness and bliss. In addition, working with the subconscious mind can help the child understand their fears and emotions, which can lead to a decrease in the intensity of nightmares and fears in the future. Such sessions can also change your child's perception of the night, making it more conducive to restful sleep and rest.

Hypnotherapy is a powerful tool for dealing with childhood trauma in adulthood. Childhood trauma can involve emotional, physical, or psychological experiences that leave traces in a person's subconscious mind throughout their life and can substantially affect their mental and emotional well-being. Hypnotherapy allows accessing the subconscious layers of the brain, where memories and emotional impressions of traumatic events are stored. By using special hypnosis techniques, the therapist can help the client discover and process these traumatic memories, allowing them to review and reevaluate their perceptions and reactions to these events. Hypnotherapy can also help the client find ways to get rid of the negative effects of childhood trauma by developing the ability to self-regulate, increase self-awareness, and perceive their own emotions. Thus, hypnotherapy can be an effective tool for dealing with traumatic experiences learned in childhood and helping to improve a person's mental and emotional state in adulthood.

Knowledge of the brain, its functions, and reactions is key to implement the *investigation of crimes and understanding the psychological mechanisms behind criminal actions.* As the basis for understanding human behaviour, the brain is the object of deep research aimed at identifying links between its physiological processes and criminal actions [325]. Analysis of brain activity in the context of crime allows identifying certain patterns and psychological properties that may be associated with the risk of committing crimes or with certain types of crime. The use of hypnosis and brain knowledge has important practical applications in criminology and law enforcement. Hypnotherapy can be used to facilitate the memories of witnesses or victims, which helps in the investigation of crimes. The use of hypnosis can help restore memory and trigger memories of events that may be useful for law enforcement.

Knowledge of the brain helps criminologists and law enforcement officers understand the psychological mechanisms that may underlie the commission of crimes. Understanding the psychological aspects allows creating profiles of criminals and developing strategies for their detection and prevention. Knowledge of the brain helps forensic experts conduct more objective and accurate examinations. For example, brain activity analysis can be used to identify mental disorders that may be related to the commission of crimes. Hypnotherapy can be an effective tool in crime prevention and criminal rehabilitation. It helps to understand and change the mental processes that may be associated with the commission of crimes and contributes to the restoration of social adaptation. Hypnotherapy can be used to psychologically influence suspects or witnesses to obtain important information about the commission of crimes or identify the circumstances of a case.

Hypnosis is a method of influence that can have some practical applications in the context of standard cases of forgetting, amnesia, and stressful situations, in particular, traumatic events. The examination of hypnosis as a forensic tool for investigating crimes is necessary, as it can open up new opportunities for identifying and eliminating the effects of hypnotic exposure on a person. All of these practical aspects of using hypnosis and brain knowledge in criminology and law enforcement highlight the importance of investigating brain processes and their impact on human behaviour to effectively counter crime and ensure justice.

Approaches to using brain knowledge in the fight against crime cover not only theoretical models but also practical methods. For example, the examination of the brain's response to stressful situations, the use of neuropsychological tests to identify features of cognitive function depending on the type of crime, or the use of neuromarketing methods to analyse the impact of advertising materials on the perception and behaviour of individuals with certain risk factors. Considering the variety of methods and approaches based on the concepts of brain activity, it is possible to develop more effective strategies for crime prevention and counteraction. Specialists should be warned about the obligation to preserve the confidentiality of the investigation and the inadmissibility of interference in the depths of a person's consciousness that do not relate to the circumstances that need to be established, since violation of this requirement means a serious violation of the constitutional rights and freedoms of a person.

Now, the problem of using the possibilities of hypnosis in the fight against crime is at the stage of formation. There is every reason to believe that in the future, a new scientific branch will be formed in criminalistics—hypnology, which will have full recognition as a method fixed by science and proven by practice, which will contribute to obtaining information about the circumstances of the event in question.

Brain knowledge is also used to create prevention and rehabilitation programmes aimed at reducing the risk of crime among various social groups, including adolescents, veterans, and people with mental health disorders. Understanding the mechanisms of brain activity helps identify vulnerability to criminal behaviour and develop customised intervention strategies to prevent crime and improve the quality of life of people at risk. This evidence-based approach to crime prevention contributes to the creation of a safer and more sustainable society. Discussing these aspects allows for a better understanding of how brain knowledge affects the development of various spheres of life and how it can be used to achieve certain goals in scientific, medical, technological, and social fields.

The application of hypnotherapy methods in shopping malls is a complex but interesting idea that requires careful consideration. Using the principles of hypnotherapy to stimulate sales and improve customer engagement includes sound and visual stimuli such as music, nature sounds, and calm voices, which can create an atmosphere of relaxation in the retail space, helping to reduce the level of stress in customers and increase the length of their stay. The use of colours, lighting, and design aimed at creating comfort and calm can positively affect the emotional state of customers and increase their interest in products. Using scents that are associated with pleasant sensations can stimulate appetite and increase customer interest in products. The use of psychological techniques, such as subconscious exposure through text messages or special signs and symbols, can increase customers' attention to certain products or promotions. Recreation areas or trial tastings can create an atmosphere where customers can take a break from the hustle and bustle of the store, which positively affects their perception and desire to shop.

It is necessary to consider the ethical aspects of using hypnotherapy in sales halls and avoid methods that may cause discomfort to customers. The effectiveness of such methods also depends on the individual characteristics of customers and their response to incentives. All these methods of hypnotherapy used in supermarket sales halls have their own scientific explanations and mechanisms of influence on humans. Sound and visual stimuli can affect a person's emotional state by activating neural networks in the brain responsible for emotional sensations. Music and nature sounds can promote relaxation and reduce stress levels, making shoppers more open to the store's influence and offerings. Colours, lighting, and store design can activate certain parts of the brain, evoking positive associations and feelings of comfort. This can increase the overall mood of customers and increase their interest in products. Scents can activate neural pathways associated with the sense of smell, which in turn can affect the mood, appetite, and emotional state of customers.

The use of psychological methods, such as subconscious influence through text messages or special characters, can activate subconscious mechanisms in customers, which contributes to their attention to certain products or promotions [326]. Recreation areas are created to reduce overall stress levels and provide a positive customer experience. They can help increase the length of time customers stay in the store and make them more likely to shop.

In general, all of these methods use science-based principles of psychology and neuroscience to change a person's emotional state and behaviour in a shopping environment. However, it is important to consider the ethical aspects of using such methods and their impact on individual rights and freedoms.

Hypnotherapy can act as a useful tool for overcoming self-doubt when performing on stage and conducting business training. Hypnotherapy techniques for this purpose include positive imagery and visualisation, working with the subconscious mind, stress management and relaxation, improving self-esteem and self-respect, and breathing and relaxation techniques. During hypnotherapy, it is possible to work on creating positive images and scenarios of successful performances, identifying and changing subconscious beliefs and attitudes, learning relaxation and stress management techniques, increasing self-esteem and faith in your abilities and learning breathing and relaxation techniques. It is important to remember that the effectiveness of hypnotherapy may vary depending on each person's characteristics and needs. In addition, hypnotherapy can only be effective in combination with other psychological and behavioural techniques and practical exercises to develop speaking and training skills. This works through the effect of hypnotherapy on various aspects of the human psyche, which may be associated with self-doubt when performing on stage or conducting training sessions.

For example, positive imagery and visualisation can activate certain parts of the brain responsible for generating positive emotions and confidence. During hypnotherapy, a person can reproduce scenarios of successful performances in their imagination, which helps reduce anxiety and increase self-confidence. Working with the subconscious mind can help identify and change negative beliefs or attitudes that may hinder one's personality during speeches or training sessions. A hypnotherapist helps reprogram these beliefs into positive and supportive ones, which helps increase self-confidence. Relaxation and stress management techniques can reduce the level of anxiety and tension before performing, which makes the person calmer and more confident. Improving self-esteem and self-respect can also be achieved through hypnotherapy, as it helps a person view themselves and their abilities in a more positive light. Learning breathing and relaxation techniques helps to control one's physical responses to stressful situations, which provides greater stability and self-confidence during performances. Consequently, these hypnotherapy techniques help turn negative emotions and beliefs into positive ones, stimulating self-improvement and self-development.

Knowledge about the brain and hypnosis is also important for the *application in advertising and marketing*. Understanding the psychological mechanisms of consumers, their perception of advertising messages, and the impact of advertising on their decisions allows them to develop more effective marketing strategies. This approach helps to create ads that better meet the needs of the target audience. This opens up opportunities for developing more insightful and targeted marketing campaigns that promote more successful customer engagement and increase the effectiveness of advertising efforts.

Understanding how the brain works and what mechanisms influence consumer behaviour allows marketers and advertisers to develop more effective strategies for communicating with their audience. Scientific discoveries in the field of neuromarketing allow an understanding of those factors that encourage consumers to make decisions about buying goods or services. The use of hypnosis and brain knowledge in business and marketing opens up wide opportunities for influencing consumers, improving productivity, and achieving success in various aspects of entrepreneurship. Using brain knowledge is crucial for developing effective marketing, advertising, and sales strategies. These strategies are aimed at attracting consumer attention, increasing sales, and creating successful business strategies in a competitive environment. Practical aspects of using hypnosis in business include using hypnotherapy techniques to increase employee productivity and motivation. For example, hypnotherapy sessions can help reduce stress levels, improve concentration and focus, and improve workflow efficiency. In marketing, brain knowledge allows understanding the psychological mechanisms of consumer behaviour and using them to create successful strategies for promoting products and services.

For example, by learning about the brain's responses to stimuli, marketers can develop more effective advertising campaigns using colours, words, and images perceived by the brain to be more appealing to the target audience. In addition, hypnosis techniques can be used in sales to increase the impact on customers and convince them of the need to purchase goods or services. Brain knowledge helps understand the mechanisms that influence consumer decisions and adapt marketing strategies to suit their needs and preferences. Thus, the use of hypnosis and brain knowledge in business and marketing can substantially improve the efficiency of an enterprise, provide competitive advantages in the market, and contribute to success in today's business environment.

The use of neuroscience research allows analysing the brain's responses to various types of advertising, product design, and pricing strategies. For example, examining brain activity during the perception of advertising content can help determine the effectiveness of advertising campaigns and improve them to maximise the impact on the target audience. Business strategies built on the basis of scientific data on the brain can include such aspects as optimising the store space to encourage purchases, developing products that consider the psychological characteristics of consumers and developing personalised advertising strategies aimed at individual needs and wishes of customers. In general, the use of brain knowledge in marketing, advertising, and sales is a key factor for achieving success in today's business environment, where competition is constantly growing and consumers are becoming more demanding and informed.

Knowledge about the brain and its functioning plays an important role in various areas of technology, including the *development of artificial intelligence systems, neural interfaces, robotics,* and other advanced areas. Understanding how the brain works allows engineers and researchers to create more efficient and productive artificial intelligence systems that can model the complex cognitive functions of the human brain. The use of hypnosis and brain knowledge in the field of technology and innovation has a wide range of practical applications that contribute to the development and improvement of both human health and technological advances [327].

The practical use of hypnosis in technology opens up interesting possibilities, although it is not a typical application. Some *applications and programmes can use hypnotherapy* to provide users with relaxation and stress management tools. Sounds, music, voice prompts, and visual effects can help relieve stress and improve the user's well-being [312]. Some brain training apps may contain elements of hypnosis to improve focus and concentration. This can be useful for students, athletes, or anyone looking to improve their performance and efficiency. Mobile apps and online platforms can offer audio recordings of hypnotherapy for self-development, improving self-confidence, reducing stress, improving sleep, and other aspects of personal growth. Some virtual games or applications may contain elements of hypnosis to create the impression of a deeper immersion of the user in the game or to create a relaxing or meditative effect.

Some apps and programmes use hypnotherapy to promote relaxation, relieve stress, and improve mental health. Some of the examples include:

- Calm: The Calm app offers hypnotherapy audio recordings to calm down, improve sleep, and relieve anxiety. It contains a variety of meditation exercises that can be useful for reducing stress in everyday life.
- Headspace: Headspace offers hypnotherapy sessions focused on various aspects of mental health, including stress, anxiety, and insecurity. This app also includes meditation techniques and exercises to improve emotional well-being.
- Hypnosis and Guided Meditation is an app that provides access to a wide range of hypnotherapy sessions and meditation exercises. It focuses on various aspects of life, such as improving sleep, relieving stress, increasing self-esteem, and improving concentration.
- Insight Timer: This app offers a wide range of hypnotherapy sessions and meditation exercises that cover various aspects of mental health, including relaxation, mental health development, and working with emotions.

These apps and programmes can be useful for those looking for support in relieving stress, improving sleep, and overall mental well-being. However, it is important to consider individual needs and beliefs when choosing a hypnotherapy method and to consult a specialist in case of serious mental health problems. However, it is important to remember that the use of hypnosis in technology must be ethical and consider the potential risks to users. Before using any programme or app that contains elements of hypnotherapy, it is important to consult a qualified health professional or psychologist.

In addition, it is necessary to consider the individual characteristics of users and take a careful approach to the use of hypnosis in technology.

During hypnosis, feelings of happiness may occur, but these feelings may be temporary. Changes in brain activity play a key role in sensations during hypnosis. Studies using functional magnetic resonance imaging show that during hypnotherapy, there may be changes in the activity of various parts of the brain associated with emotions and self-awareness. Hypnosis can also affect the levels of neurotransmitters such as endorphins and serotonin, which are responsible for feelings of happiness and pleasure. During hypnosis, parts of the brain associated with positive emotions can be activated, leading to temporary feelings of increased mood and emotional well-being. However, this happy feeling can quickly disappear after hypnotherapy due to stressful situations and everyday problems. The effect of hypnosis may be temporary, and a lasting result may require support, training, and additional practices such as psychotherapy or self-development. Therefore, while hypnotherapy can create temporary improvements in the emotional state, it often takes more time, effort, and support to achieve lasting results.

In terms of hormones, the effects of hypnotherapy may be related to various neurotransmitters and hormones that affect emotional state and mental well-being. During hypnosis, feelings of happiness may occur or vice versa due to various biochemical and neurophysiological processes. Primarily, hypnotherapy can stimulate the release of endorphins—natural opiates in the brain that are associated with reduced pain and increased mood. This can cause feelings of euphoria and joy. In addition, hypnosis can affect the level of serotonin, the happiness hormone that regulates mood and emotional state. Increasing serotonin levels can promote feelings of pleasure and boost people's mood. In addition, hypnosis can increase levels of dopamine, the reward and pleasure hormone, which promotes feelings of joy and satisfaction. Finally, during hypnosis, cortisol, the stress hormone, may decrease. Reducing cortisol levels helps reduce stress and anxiety, which can improve the emotional state. Thus, various biochemical changes that occur in the body during hypnosis can lead to both feelings of happiness or pleasure and a change in mood in the opposite direction. In general, hypnotherapy can affect the levels of various hormones and neurotransmitters, which can lead to improved emotional state and mental well-being. However, it is important to understand that the response to hypnotherapy can be individual and depends on various factors, such as personality, context, and type of therapy.

In the field of innovation, understanding the brain's response to various stimuli allows developing new technologies for VR, AR, and brain-computer interfaces that open up new opportunities in entertainment, learning, and medicine. Using hypnosis together with these technologies can help improve therapy outcomes, learning, and productivity. Neural interfaces, which are based on knowledge about brain activity, allow interacting with computer systems without direct physical intervention, which has great potential for the development of future Interfaces and medical applications, such as prosthetic management and rehabilitation of patients with motor disorders.

In robotics, knowledge of brain mechanisms helps develop biometric robots that can perform tasks efficiently by mimicking complex aspects of human brain function and motor skills. This can include robots that reproduce certain aspects of interacting with the environment, learning, and adapting to changing conditions, allowing them to adapt to different tasks and situations. Thus, the use of brain knowledge in technology is key to creating new intelligent systems and devices that can substantially improve the quality of life of people and the development of society in general.

Hypnotherapy and AR can be used to treat people with phobias and phantom pain through several mechanisms. Exposure therapy through AR is an innovative approach to treating phobias and fears. This method uses a virtual environment created using AR technology to expose patients to situations or objects that cause their phobias [178].

Hypnotherapy, in this context, acts as an important support element, helping patients achieve the level of relaxation and emotion management needed for successful exposure. Changing perception through hypnosis is another aspect of treatment where hypnotherapy helps retrain the patient's brain to perceive pain and stress. During a hypnosis session, a person may redirect their attention or overestimate their perception of pain or stress, which helps reduce the symptoms of phobias or phantom pain.

Hypnotherapy and AR can be used together to improve the effectiveness of the therapeutic process. Their combination opens up wide opportunities in the treatment of various disorders and improving mental well-being. For example, in the treatment of phobias and anxiety disorders, hypnotherapy can be combined with AR to create a virtual environment where the patient can be exposed to their phobias or anxiety situations in a controlled and safe environment.

In addition, the use of AR during hypnotherapy can help patients with PTSD process traumatic events in a controlled environment, helping to reduce symptoms and improve well-being. In addition, this combination can be useful in medical rehabilitation, where patients can recover through virtual simulations of movement and exercise and in improving overall mental health and emotional well-being.

These examples demonstrate how hypnotherapy combined with AR can be used for various treatment purposes and to improve patients' quality of life. The combination of hypnotherapy and AR can improve the effectiveness of treatment, as patients can examine their responses to phobias or pain in a controlled environment. Hypnotherapy, in this case, helps support patients in this process, ensuring their concentration and emotional comfort. Improving self-knowledge and self-regulation is another vital component of hypnotherapy in the context of treating phobias and phantom pains. Hypnotherapy helps patients understand the root causes of their fears and pain and teaches them self-regulation techniques that help them better manage symptoms and improve their mental state. These methods can be effective in the complex treatment of phobias and phantom pains, but it is important that they are used under the supervision of a qualified specialist who has experience treating these problems.

Scientific research in the brain, hypnosis, and neural networks is crucial for understanding the complexity and functioning of human neural metabolism. These studies provide insights into how various aspects of the brain work, including the mechanisms of perception, representation, memory, and emotion regulation. A deep understanding of brain physiology and pathology opens up new horizons in the treatment of mental and neurological disorders. Investigating the effect of hypnosis on brain activity allows for revealing its potential to influence the mental state of a person and control various brain functions. An important aspect of research is the use of neural networks to model complex brain processes and understand the relationships between different areas and structures. This allows creating more accurate and effective methods for diagnosing and treating various disorders of neural activity. Ultimately, the integration of data from brain research, hypnosis, and neural networks opens up promising opportunities for developing innovative approaches to the treatment and rehabilitation of patients with various mental and neurological diseases.

The use of hypnotherapy and neural networks for practical medical, psychological, and psychiatric purposes is a promising area of research and application that can revolutionise the treatment of various diseases. These applications cover a wide range of pathologies, including, but not limited to, depression, anxiety, chronic pain, and other mental and physical conditions. The introduction of such innovative methods involves the creation of new approaches to diagnosis and treatment based on a thorough analysis of psychophysiological mechanisms and the interaction of neuronal networks in the context of pathological processes.

The results of research in this area can open up new prospects for improving the effectiveness and accessibility of treatment for various diseases, contributing to improving the quality of life of patients and strengthening their psychoemotional state. The need for further research on hypnosis and its effects on brain function and the development of new intervention technologies is marked by substantial opportunities to expand our understanding of psychophysiological mechanisms. Advanced research can help clarify the relationship between hypnotherapy and brain function, identify the basis for the effectiveness of hypnosis in the treatment of various mental and physical diseases, and develop new techniques and approaches for introducing hypnotherapy into clinical practice.

Continuing research in these areas can contribute to improving therapeutic strategies and improving the effectiveness of hypnotherapy as a tool for influencing a person's mental and physical health. When using hypnotherapy and neural networks, it is important to carefully consider ethical aspects, as this ensures the safety, confidentiality and respect for the rights of patients. In the field of medical practice where these methods are used, it is necessary to adhere to standards of professional conduct and ethical standards, in particular, to ensure informed consent of patients and compliance with the principles of confidentiality. In addition, it is important to consider ethical principles when collecting, processing, and storing patient data, ensuring their privacy and protection from unauthorised access.

Ethics in the use of hypnotherapy and neural networks also implies avoiding any form of influence or manipulation on the patient and instead encourages the development of mutual understanding and cooperation between the patient and the therapist based on mutual trust and respect. This approach reflects the importance of ethical principles in ensuring high-quality and effective medical care. Overall, further research on the brain, hypnosis, and neural networks has great potential to develop new technologies and treatments that can improve the quality of life of people with various mental and neurological disorders.

Conclusions

By analysing the history and modernity of research on the brain, hypnosis, and neural networks, key points that demonstrate their relationship and impact on understanding the human psyche and behaviour can be identified. Historically, hypnosis has been considered a method of treatment and consciousness exploration, first applied in the 18th century by Franz Anton Mesmer. He believed he could correct "magnetic imbalance" in a patient's body, which became the first known attempt at using hypnosis. Although his theories were debunked, interest in hypnosis as a psychotherapeutic tool has persisted to the present day. In modern research, hypnosis is used to analyse brain function, particularly its response to altered states of consciousness. The use of methods such as functional MRI and EEG allows scientists to visualise brain activity during hypnosis, exploring how hypnotic states affect various brain processes. On the other hand, the development of neural networks and artificial intelligence has opened new avenues for modelling and understanding the cognitive functions of the brain. Artificial neural networks, which mimic the structure and functions of biological networks, have become an important tool in examining cognitive processes and developing new methods for treating mental disorders.

The integration of knowledge and methods from different scientific disciplines, such as psychology, neurobiology, and computer science, opens up new ways for a deep understanding of the brain and its functioning. This approach allows researchers to combine knowledge of brain structure, its biochemical processes, and behavioural aspects with cutting-edge modelling and data processing technologies, significantly expanding the boundaries of the current understanding of mental processes and disorders. From a psychological perspective, the integration of knowledge helps

DOI: 10.1201/9781003716976-7

to uncover mechanisms of cognitive functions, such as attention, memory, thinking, and emotional reactions, by analysing behavioural manifestations and internal mental states. The neurobiological perspective focuses on examining the anatomical structure of the brain, neurons, synapses, and the role of neurotransmitters in regulating brain activity and signal transmission. Computer science, in turn, contributes to the development of algorithms for artificial intelligence and machine learning, including artificial neural networks that mimic the way biological neural networks process information in the brain. This not only allows for modelling mental processes at a new level of complexity but also for creating innovative diagnostic tools and methods for treating mental disorders. An interdisciplinary approach to analysing the brain contributes to the discovery and interpretation of its complex relationships with mental phenomena, using data from experiments, clinical practice, and theoretical models. This includes examining brain plasticity, mechanisms of learning and memory, the impact of emotions on cognitive processes, and the role of hypnosis and other altered states of consciousness. Such integration not only contributes to the development of theoretical knowledge but also has practical applications in medicine, education, and psychotherapy, enabling the development of more effective methods of diagnosis, treatment, and rehabilitation for a wide range of neurological and mental disorders. This approach also has significant implications for the development of neuroethics and understanding moral and philosophical questions related to consciousness, will, and identity.

Research in neuroscience, hypnosis, and neural networks reveals potential for practical applications. These can fundamentally change approaches to diagnosis, treatment, and rehabilitation in medicine and contribute to the development of new technological solutions in artificial intelligence and automation. These studies open up new horizons for understanding complex processes in the human brain and provide opportunities to develop effective methods of treating a range of mental and neurological disorders. With an understanding of the mechanisms of hypnosis and its impact on brain neural networks, doctors and psychotherapists gain a powerful tool for treating conditions such as chronic pain, depression, anxiety disorders, post-traumatic stress disorder, as well as addictions. Hypnotherapy can be used as an alternative or complement to conventional treatment methods, offering patients a non-pharmacological way to manage symptoms.

The development and use of artificial neural networks, which mimic the properties of biological brain neural networks, pave the way for creating intelligent computer systems capable of learning, adapting, and self-improving. These technologies find applications in speech recognition, natural language processing, robotics, automated control systems, medical diagnosis, and many other fields, significantly expanding the horizons of artificial intelligence possibilities. Understanding of the processes of learning, memory, and cognitive development that occur in the brain can be applied to develop more effective educational programmes and teaching methods. The use of neural networks allows for the creation of adaptive learning systems that can cater to individual students' needs, thereby optimising the learning process and ensuring better assimilation of material.

Utilising knowledge about the interaction between hypnosis, neural networks, and brain activity can enhance psychotherapy methodologies, enabling therapists to understand and manage patients' mental processes more deeply. This opens up new possibilities for treating mental disorders, considering the individual characteristics of consciousness and subconsciousness. The importance of the practical application of brain, hypnosis, and neural network research cannot be overstated. Integrating this knowledge into medicine, technology, education, and psychology promotes the creation of innovative solutions that improve people's quality of life and stimulate further research in these exciting fields of science.

The continuous process of learning in the fields of science and psychology is a fundamental principle that underscores the inexhaustible nature of knowledge acquisition and the unlimited possibilities for knowledge development. In this context, neuroscience is at the forefront of exploring the complex processes that occur in the human brain, influencing people's thoughts, behaviour, emotions, and mental health. Each discovery in this field not only adds a piece to the overall understanding of the brain but also opens up new areas for further research. For example, the discovery of new neurotransmitters or neural pathways may lead to the development of innovative methods for treating neurological and mental disorders. Similarly, understanding the mechanisms of learning and memory in the brain opens up prospects for the development of more effective pedagogical strategies and teaching technologies.

In the field of psychology, new knowledge about cognitive processes and emotional states can contribute to a better understanding of human behaviour and the development of new approaches in psychotherapy. This is particularly important in light of the growing number of psychological challenges people face in the modern world.

Technological progress, such as the development of functional magnetic resonance imaging, electroencephalography, and other visualisation methods, opens up new opportunities for deeper analysis of the brain in action. These technologies allow researchers to literally "look into" the brain and observe its activity in real time, providing unprecedented access to studying its functions. Furthermore, the continuous growth of data volumes and the development of computational technologies open up new avenues for the analysis and interpretation of brain information. Artificial intelligence and machine learning are becoming indispensable tools in neuroscience, enabling the processing of large datasets and the identification of complex patterns that were previously inaccessible. In this context, continuous learning not only encourages further scientific development but also reminds us of the need to be open to new ideas and approaches. The world of science is dynamic and constantly changing, and achieving a deeper understanding of the complex processes that occur in the brain is only possible through continuous learning and research.

In the history of science, there are often moments when discoveries that initially seemed counterintuitive or radical later became important steps in the development of knowledge. Examples from physics (such as quantum mechanics) and biology (the discovery of the structure of DNA) underscore the importance of openness to new ideas. This openness is particularly relevant in the analysis of the brain and consciousness, where the complexity of the system and the interactions between its components present significant challenges for researchers. Modern neuroscience is on the cusp of significant discoveries due to the development of technologies such as functional magnetic resonance imaging, which allow for a more detailed visualisation of brain activity than ever before. Such visualisation not only helps better understand the functioning of the brain in normal conditions but also allows examining pathological conditions such as depression, schizophrenia, and autism. However, without openness to new ideas and a willingness to reconsider existing theories and hypotheses, many potential discoveries may go unnoticed.

The examination of hypnosis as a state of altered consciousness also requires researchers to be open to new ideas. Conventionally seen as an art or even charlatanism, hypnosis is now being investigated as a valuable window into understanding consciousness, which can provide unique information about brain function. The development of artificial neural networks, inspired by the structure and functions of biological neural networks, is another example of how being open to new ideas leads to significant breakthroughs. Artificial networks enable the modelling of complex cognitive processes, opening new avenues for understanding the mechanisms of thinking and learning. They also form the basis for the development of cutting-edge technologies in fields ranging from medicine to artificial intelligence.

Scientific openness extends beyond merely accepting new ideas; it also entails a willingness to critically reconsider one's own beliefs and hypotheses. This creates a healthy scientific environment where continuous development is possible and where each discovery is seen as a step toward a deeper understanding of the complexities of the brain and consciousness. Thus, openness to the new is a fundamental principle in the analysis of the brain, hypnosis, and neural networks, prompting continuous development, discoveries, and innovations in this exciting field of science. Research into the interaction between the brain, hypnosis, and neural networks opens up new horizons for understanding the human psyche and advancing science. They remind us of the unlimited potential of the human mind and inspire the further pursuit of knowledge and discoveries.

This book illuminates the complex interrelationships between the brain, hypnosis, and neural networks, revealing progress in understanding the mechanisms behind hypnotic states and their impact on mental processes. Oriented towards historical development and contemporary approaches to the analysis of hypnosis, the study analyses neuronal mechanisms, paving the way for innovative treatment methods and a deeper understanding of brain activity. Establishing goals and objectives serves as navigation for research, guiding the collection and analysis of information, which contributes to the rational allocation of resources and determination of expected results.

The value of the study is underpinned by the potential to uncover new aspects of brain activity, particularly through the analysis of hypnosis and its influence on neural networks. By examining the studies of eminent scientists of the past, such as J.-M. Charcot and H. Bernheim, a historical context is provided for the development of hypnosis research, highlighting the progressive expansion of scientific understanding of this phenomenon. Modern research methods, including functional MRI and EEG, allow for a deeper dive into brain activity, particularly during hypnosis, opening up new perspectives for medicine and psychotherapy. The purpose of the study is to elucidate the interaction between hypnosis, brain activity, and neural network function, which is significant for the scientific community and has practical applications in the treatment of mental disorders. Understanding these relationships can lead to the development of new hypnotherapy-based treatment methods and improve individuals' psychological well-being. Thus, this study plays a key role in the advancement of psychology, neurobiology, and medicine, offering new approaches to understanding the human brain and its functioning. It promotes the integration of scientific knowledge and practical application, paving the way for innovations in treatment and the understanding of human consciousness.

A comprehensive understanding of the importance of examining the impact of hypnosis on the brain and neural networks opens up limitless prospects for the scientific world. Hypnosis remains one of the most interesting and mysterious areas of psychology and neuroscience. Scientists are attracted to it because it has the potential to reveal new depths of consciousness, imagination, and neurophysiology. The book extensively examines how hypnosis affects brain function and neural networks, exploring the possible mechanisms of this influence and current research in this area. Understanding these aspects can be crucial for clinical research, the development of new hypnotherapy methods, and unlocking the potential of neural networks in the context of understanding and altering consciousness. This subject becomes even more relevant as it helps uncover the mysteries of human cognition and opens up new possibilities for applying this knowledge in various fields, from medicine to psychology and even future technologies.

Therefore, the analysis of information highlights the importance of exploring the anatomy and physiology of the brain to understand its functions and capabilities. The brain is responsible for a wide range of activities, including perception, processing, and storage of information, motor control, regulation of internal functions, and others. Research into the anatomy of the brain reveals how its various parts collaborate to perform these functions, and neuroplasticity and neurogenesis demonstrate that the brain can actively change throughout a person's life, which is important for learning and adaptation to change.

The relevance of developing advanced research methods lies in their ability to deepen the understanding of neuroplasticity and neurogenesis, which in turn is crucial for developing treatment and rehabilitation methods for neurological disorders and injuries. The purpose of the study is also to improve methods for integrating the brain with computers and artificial intelligence technologies to create more effective interfaces. Further research into brain anatomy promises to uncover more mysteries of this neurobiological wonder and find applications for this knowledge in medicine, technology, and psychology. With the development of these areas, significant progress can be expected in understanding and treating various brain disorders and improving people's quality of life.

The current use of hypnosis in research and practical medicine demonstrates the wide range of possibilities of this phenomenon. It is used as a tool for exploring various aspects of the mind and neurophysiology, as well as for treating various medical and psychological problems. Research on the use of hypnosis is focused on understanding its mechanisms of action on the human brain and psyche. Studies in this area revealed a number of interesting facts, such as the impact of hypnosis on neuronal activity, the mechanisms of hypnotic state formation, and their relationship with various mental functions. In addition, the possibilities of using hypnosis in the treatment of mental disorders, phobias, chronic pain, and other medical problems are being explored.

The future development of hypnosis promises to further reveal its potential as a tool for improving mental and physical health. It may become an additional therapeutic method in psychiatry and psychotherapy, as well as find applications in various fields of medicine, including anaesthesiology, rehabilitation, and chronic pain management. The advancement of new technologies, such as functional magnetic resonance imaging, can help uncover more insights into the mechanisms of hypnosis on the brain and nervous system. Overall, hypnosis remains an intriguing and promising area of research that can significantly contribute to the advancement of medical science and the enhancement of quality of life.

Scientific and technological breakthroughs, which have laid the foundation for pioneering research, unveil neurophysiological and psychophysiological aspects of brain functions and their response to hypnosis. Modern brain imaging methods, including functional magnetic resonance imaging and electroencephalography, allow for detailed examination of activity in various brain regions. These methods open new horizons for understanding mental processes and their dynamics. The use of transcranial magnetic stimulation and brain–computer interfaces paves the way for innovative research and influence on brain activity. These technologies are not only valuable tools for studying the brain but also hold great promise for modifying brain functions.

In the context of hypnosis, interest in neurophysiological approaches opens up new perspectives. Research focuses on uncovering the brain mechanisms underlying hypnotic states and their impact on mental health. This section highlights studies as key tools for understanding and influencing brain functions and investigating the phenomenon of hypnosis within the context of contemporary neuroscientific discoveries. The study of the brain and hypnosis is identified as a crucial area of scientific inquiry today, as their interconnections help reveal and comprehend the deepest aspects of human mind and psyche functions. These studies offer opportunities to enrich scientific knowledge about brain activity and consciousness mechanisms, as well as practical applications in medicine, psychotherapy, and other fields aimed at improving the quality of life.

The continuous development and deeper understanding of brain and hypnosis phenomena form the foundation for further research in these vital areas. It is noted that understanding the impact of hypnosis on brain activity not only enriches the theoretical foundation of neuroscience but also opens broad prospects for practical use in various spheres of life. Activity in this field can lead to the development of new treatment methods and approaches to psychotherapy, enhancing the overall understanding of human consciousness.

Brain research can be classified into several key areas, uncovering different aspects of this complex organ's functioning. Structural neuroscience employs advanced technologies such as three-dimensional magnetic resonance imaging and virtual reality for a detailed study of brain anatomy. This enables researchers to understand how different structural elements interact and influence brain functions. Functional neuroscience investigates the activity of various brain regions during different tasks or in response to various stimuli. The application of methods such as functional magnetic resonance imaging allows for the acquisition of crucial real-time data about brain function. Psychophysiology and psychology focus on understanding the relationship between brain activity and mental processes. They employ innovative techniques such as deep learning and fMRI analysis to uncover complex interactions within the brain during various states and activities.

Brain stimulation and interfaces are advancing to enable the modulation of brain activity or interaction with the brain using technologies such as transcranial magnetic stimulation and brain-computer interfaces. Brain simulation and artificial intelligence are used to develop models that mimic brain processes, improving AI and solving complex tasks by replicating natural principles of information processing in the brain.

The potential practical applications of hypnosis research include the development of individualised hypnotherapy techniques tailored to specific issues or conditions for each patient, which can enhance outcomes and provide better clinical effects. In addition, the use of hypnosis in medical practice for the treatment of pain, stress, and psychosomatic disorders can become an effective tool for improving patients' mental and physical well-being. Research highlights the potential of hypnosis to enhance memory and comprehension abilities, opening new prospects for its use in educational processes. New trends in hypnosis research underscore the importance of incorporating hypnotherapy into rehabilitation programmes to facilitate adaptation and recovery of both physical and mental health. Scientific discoveries in the field of neuroscience and neuroimaging provide an opportunity to better understand the role of the brain in shaping the psycho-emotional state of patients before medical procedures. In addition, the development of technological tools for monitoring hypnotic states can significantly ease the objective measurement and analysis of the depth and nature of hypnosis in patients, opening new possibilities for accurately determining the efficacy of hypnotherapy and understanding its impact on brain activity.

Contemporary methods of hypnosis research integrate advanced technologies and innovative scientific approaches, encompassing neurophysiological studies, cognitive approaches, clinical applications, and transdisciplinary methodologies. Neurophysiological studies employ functional magnetic resonance imaging and electroencephalography to examine brain activity during hypnosis, allowing the understanding of how hypnosis influences perception, attention, and other cognitive processes. Cognitive approaches focus on investigating mechanisms of attention, memory, and perception within the context of hypnotic phenomena, exploring how hypnosis can modify or enhance these processes. The clinical applications of hypnosis are viewed through the lens of therapeutic intervention, with particular emphasis on treating chronic pain, anxiety disorders, addictions, and other mental conditions, highlighting the efficacy of hypnotic techniques in clinical practice. Transdisciplinary research combines approaches from various fields, such as psychology, neurology, cognitive science, and computer technology, to develop cutting-edge methods for analysing and applying hypnosis, including virtual reality and artificial intelligence. These innovative research methods expand the understanding of hypnosis as a complex phenomenon, opening new possibilities for its application in medicine, psychotherapy, and even educational processes, thereby enhancing treatment effectiveness and the development of cognitive skills.

Discovering the mechanisms by which hypnosis interacts with brain activity unveils extensive prospects for developing innovative psychotherapeutic techniques aimed at treating various mental disorders. A profound understanding of the connections between hypnosis and brain activity is fundamental for refining psychotherapeutic methods and improving their overall efficacy. An important area is the development of individualised psychotherapeutic programmes based on detailed studies of patients' unique brain activity characteristics and their responses to hypnotherapy, which is crucial for effectively treating mental disorders, given the uniqueness of each individual. Hypnotherapy, in particular, shows significant potential in treating psychosomatic disorders, where an individual's emotional state directly impacts physical health. Innovative methods aimed at adjusting the brain mechanisms underlying these disorders can offer new approaches to their treatment. Specifically, understanding the impact of hypnosis on brain regions responsible for processing stress and trauma can contribute to the development of effective psychotherapeutic strategies for overcoming post-traumatic stress disorders.

The use of hypnosis can also be effective in correcting sleep disorders and restoring emotional balance, where innovative approaches can include hypnotherapy sessions to improve sleep quality and regulate emotions. Further expansion of the possibilities of treating mental disorders can be achieved through the use of virtual reality and other modern technologies that allow integrating hypnosis with other methods of psychotherapy while ensuring the ethical and safe use of the latest approaches. Personalisation of treatment based on a detailed analysis of the functional activity of the patient's brain can substantially increase the effectiveness of hypnotherapy interventions, and the use of neurofeedback and real-time cognitive interfaces opens up new opportunities for individualisation and optimisation of psychotherapy sessions. These advanced techniques can substantially improve the availability and effectiveness of hypnotherapy for a wide range of mental disorders, opening up new horizons in understanding and treating mental conditions.

Research on the interaction between hypnosis and brain activity paves the way for new insights into the scientific understanding and clinical use of this phenomenon despite the existing challenges. Progress in this area encourages further expansion of knowledge about the mechanisms of interaction between the brain and hypnotic states using modern technologies and innovative methodologies. The development of the latest neural interfaces that enable reading and interaction with brain activity promises substantial progress in understanding how hypnosis affects various aspects of neural activity, opening up prospects for more precise control of hypnotic states and the development of targeted hypnotherapy strategies. Advanced big data processing algorithms derived from neuroimaging research can uncover complex behavioural patterns and correlations between hypnosis and brain activity, helping to identify key mechanisms underlying hypnotic exposure. This, in turn, will allow developing personalised hypnotherapy programmes that consider the unique neurophysiological characteristics of each person, increasing the effectiveness of treating various conditions, from stress to chronic pain.

The integration of virtual reality with hypnotherapy opens up new possibilities for creating controlled research environments, allowing for the examination of the effect of hypnosis on the brain's response to various stimuli in detail. In addition, the use of real-time hypnosis monitoring technologies can revolutionise approaches to examining brain activity, enabling the direct observation of the dynamics of changes in the state of consciousness during hypnotic exposure. Such areas of research not only expand the scientific horizon but also offer specific methods for clinical practice, opening up new opportunities for treating and understanding the human psyche. These areas of research are only part of the potential that needs to be realised in the future when further scientific advances and a deeper understanding of brain processes will open up even more opportunities to explore the mysteries of the interaction between the brain and hypnosis.

Analysing the current state of hypnosis research opens up important horizons for the scientific community that cover a wide range of aspects, from neurophysiological foundations to clinical applications in the treatment of both mental and physical diseases. The examination of the mechanisms by which hypnosis affects the human psyche and behaviour, and the analysis of biological and sociocultural factors that modulate its effectiveness, contribute to a deep understanding of this phenomenon. Modern scientific efforts are also aimed at developing new techniques and methods of hypnosis, considering the ethical aspects of its use and forming an appropriate legal framework. These areas of research are aimed at improving mental health support methods and expanding treatment options.

The use of hypnosis in medical practice shows substantial potential in reducing pain and stress, improving sleep quality, and treating a number of psychological and physical conditions, such as anxiety, depression, and chronic pain. Hypnotherapy can help improve overall health by changing attitudes about specific problems or habits. There are cases when hypnosis helps unlock the potential of memory and creativity, but its effectiveness can vary substantially depending on the individual characteristics of a person. However, along with the potential benefits, hypnosis can carry certain risks and limitations. For example, some people may experience unpredictable reactions during hypnosis, including panic attacks or loss of consciousness, especially those who suffer from mental disorders. Unskilled use of hypnosis or its use in vulnerable individuals can negatively affect mental health. In addition, ethical issues related to identity control and information confidentiality during hypnotherapy require careful consideration and the development of appropriate regulatory frameworks.

In the context of general scientific and technological trends, hypnosis research fits into a wider range of innovative practices, including the development of the latest medical treatments, the use of artificial intelligence for diagnostics and personalised treatment, and the use of nanotechnology and blockchain. An analysis of the current state of hypnosis research illustrates its substantial potential in both clinical practice and psychological research, highlighting the importance of further scientific research in this area to expand understanding of the nature of consciousness and develop effective treatments.

The development of research in the field of hypnosis can substantially enrich the treatment options for a wide range of diseases, thus contributing to improving the standard of living of patients and strengthening their psycho-emotional well-being. Active investigation of the effects of hypnosis on brain function and the development of innovative intervention technologies play a vital role in expanding the understanding of psychophysiological mechanisms. Efforts in this direction can contribute to identifying the mechanisms by which hypnotherapy affects brain activity, examining the basics of the effectiveness of hypnosis in the treatment of mental and somatic disorders and introducing new methods of hypnotherapy into clinical practice.

Expanding the research base in this area can substantially improve therapeutic strategies and increase the effectiveness of using hypnosis as a means of influencing mental and physical health. When using hypnotherapy and technologies based on neural networks, it is important to pay attention to ethical requirements, ensure patient safety, confidentiality, and respect for their rights. In medical practice, this implies compliance with professional and ethical standards, in particular, ensuring informed consent and privacy of patient data. The ethical aspects of the use of hypnotherapy and neural networks are of particular importance, as they emphasise the need to avoid any form of manipulation of patients' minds, contributing instead to the development of a trusting and respectful relationship between the patient and the therapist. This approach not only meets high ethical standards in medicine but also ensures the effectiveness and quality of treatment. As a result, research in the field of hypnosis, neural networks, and brain activity opens up new opportunities for scientists to develop advanced treatments that promise substantial improvements in the treatment of various neurological and mental disorders, thus contributing to an improvement in the quality of life of patients.

References

1. Davidson, R. J., McEwen, B. S. (2012). Social influences on neuroplasticity: Stress and interventions to promote well-being. *Nature Neuroscience,* 15(5), 689–695. https://doi.org/10.1038/nn.3093
2. Pepperell, R. (2018). Consciousness as a physical process caused by the organization of energy in the brain. *Frontiers in Psychology,* 9, 2091. https://doi.org/10.3389/fpsyg.2018.02091
3. Yen, C., Lin, C. L., Chiang, M. C. (2023). Exploring the frontiers of neuroimaging: A review of recent advances in understanding brain functioning and disorders. *Life,* 13(7), 1472. https://doi.org/10.3390/life13071472
4. Tooley, U. A., Bassett, D. S., Mackey, A. P. (2021). Environmental influences on the pace of brain development. *Nature Reviews Neuroscience,* 22, 372–384. https://doi.org/10.1038/s41583-021-00457-5
5. Faerman, A., Spiegel, D. (2021). Shared cognitive mechanisms of hypnotizability with executive functioning and information salience. *Scientific Reports,* 11, 5704. https://doi.org/10.1038/s41598-021-84954-8
6. Wolf, T. G., Faerber, K. A., Rummel, C., Halsband, U., Campus, G. (2022). Functional changes in brain activity using hypnosis: A systematic review. *Brain Sciences,* 12(1), 108. https://doi.org/10.3390/brainsci12010108
7. Mordeniz, C. (2020). Hypnosis and hypnotherapy: Emerging of science-based hypnosis. In: *Hypnotherapy and Hypnosis.* IntechOpen. https://doi.org/10.5772/intechopen.94089
8. Yang, G. R., Wang, X. J. (2020). Artificial neural networks for neuroscientists: A primer. *Neuron,* 107(6), 1048-1070. https://doi.org/10.1016/j.neuron.2020.09.005
9. Chen, Z. S., Kulkarni, P. P., Galatzer-Levy, I. R., Bigio, B., Nasca, C., Zhang, Y. (2022). Modern views of machine learning for precision

psychiatry. *Patterns,* 3(11), 100602. https://doi.org/10.1016/j.patter.2022.100602

10. Desmarteaux, C., Streff, A., Chen, J. I., Houzé, B., Piché, M., Rainville, P. (2021). Brain responses to hypnotic verbal suggestions predict pain modulation. *Frontiers in Pain Research*, 2, 757384. https://doi.org/10.3389/fpain.2021.757384
11. Apelian, C., De Vignemont, F., Terhune, D. B. (2023). Comparative effects of hypnotic suggestion and imagery instruction on bodily awareness. *Consciousness and Cognition,* 108, 103473. https://doi.org/10.1016/j.concog.2023.103473
12. Tuominen, J., Kallio, S., Kaasinen, V., Railo, H. (2021). Segregated brain state during hypnosis. *Neuroscience of Consciousness*, 2021(1), niab002. https://doi.org/10.1093/nc/niab002
13. Paoletti, P., Dotan Ben-Soussan, T., Glicksohn, J. (2020). Inner navigation and theta activity: From movement to cognition and hypnosis according to the sphere model of consciousness. In: *Hypnotherapy and Hypnosis*. IntechOpen. https://doi.org/10.5772/intechopen.92755
14. Hudu Garba, M., Mamman, M. (2020). Hypnosis and hypnotherapy: The role of traditional versus alternative approach. In: *Hypnotherapy and Hypnosis*. IntechOpen. https://doi.org/10.5772/intechopen.91619
15. Catani, M., Sandrone, S. (2015). *Brain Renaissance: From Vesalius to Modern Neuroscience*. New York: Oxford University Press. https://doi.org/10.1093/med/9780199383832.001.0001
16. Walusinski, O., Bogousslavsky, J. (2020). Charcot, Janet, and French models of psychopathology. *European Neurology*, 83(3), 333–340. https://doi.org/10.1159/000508267
17. Waraich, M., Shah, S. (2018). The life and work of Jean-Martin Charcot (1825–1893): "The Napoleon of Neuroses". *Journal of the Intensive Care Society*, 19(1), 48–49. doi:10.1177/1751143717709420.
18. Peter, B. (2005). Bernheim, Hippolyte Marie. In: Stumm, G., Pritz, A., Gumhalter, P., Nemeskeri, N., Voracek, M. (eds). *Personenlexikon der Psychotherapie* (pp. 44–46). Vienna: Springer. https://doi.org/10.1007/3-211-29396-5_25
19. Elkins, G. R., Barabasz, A. F., Council, J. R., Spiegel, D. (2015). Advancing research and practice: The revised APA division 30 definition of hypnosis. *The American Journal of Clinical Hypnosis*, 57(4), 378–385. https://doi.org/10.1080/00029157.2015.1011465
20. Lankton, S. (2020). What Milton Erickson said about being Ericksonian. *The American Journal of Clinical Hypnosis*, 63(1), 4–13, https://doi.org/10.1080/00029157.2020.1754068

21. Petersen, S. E., Posner, M. I. (2012). The attention system of the human brain: 20 years after. *Annual Review of Neuroscience*, 35, 73–89. https://doi.org/10.1146/annurev-neuro-062111-150525

22. Hill, R. (2021). Ernest Lawrence Rossi: Genius and joy. *American Journal of Clinical Hypnosis*, 64(2), 87–89. https://doi.org/10.1080/00029157.2021.1981051

23. Laureys, S., Maquet, P., Faymonville, M. E. (2004). Brain function during hypnosis. In: Otte, A., Audenaert, K., Peremans, K., van Heeringen, K., Dierckx, R. A. (eds). *Nuclear Medicine in Psychiatry* (pp. 507–519). Heidelberg: Springer. https://doi.org/10.1007/978-3-642-18773-5_30

24. Hight, D. F., Kaiser, H. A., Sleigh, J. W., Avidan, M. S. (2020). Continuing professional development module: An updated introduction to electroencephalogram-based brain monitoring during intended general anesthesia. *Canadian Journal of Anaesthesia*, 67(12), 1858–1878. https://doi.org/10.1007/s12630-020-01820-3

25. Miranda, A., Bertoglio, D., Stroobants, S., Staelens, S., Verhaeghe, J. (2021). Translation of preclinical PET imaging findings: Challenges and motion correction to overcome the confounding effect of anesthetics. *Frontiers in Medicine*, 8, 753977. https://doi.org/10.3389/fmed.2021.753977

26. Kim, J. A., Davis, K. D. (2021). Magnetoencephalography: Physics, techniques, and applications in the basic and clinical neurosciences. *Journal of Neurophysiology*, 125(3), 938–956. https://doi.org/10.1152/jn.00530.2020

27. Chen, A., Phillips, K. A., Schaefer, J. E., Sonner, P. M. (2023). Community-derived core concepts for neuroscience higher education. *CBE Life Sciences Education*, 22(2), ar18. https://doi.org/10.1187/cbe.22-02-0018

28. Sheffler, Z. M., Reddy, V., Pillarisetty, L. S. (2023). Physiology, Neurotransmitters. In: *StatPearls*. Treasure Island: StatPearls Publishing. https://www.ncbi.nlm.nih.gov/books/NBK539894/

29. Javed, K., Reddy, V., Lui, F. (2023). Neuroanatomy, cerebral cortex. In: *StatPearls*. Treasure Island: StatPearls Publishing. https://www.ncbi.nlm.nih.gov/books/NBK537247/

30. Aggleton, J. P. (2024). *Memory and the Brain: Using, Losing, and Improving*. London: Routledge. https://doi.org/10.4324/9781003537649

31. Bem Junior, L. S., Lemos, N. B., de Lima, L. F. G., Dias, A. J. A., Neto, O. D. C. F., de Lira, C. C. S., Diniz, A. M. S., Rabelo, N. N., Barroso, L. K. V., Valença,

M. M., de Azevedo Filho, H. R. C. (2021). The anatomy of the brain—Learned over the centuries. *Surgical Neurology International*, 12, 319. https://doi.org/10.25259/SNI_200_2021

32. Wickens, A. P. (2014). *A History of the Brain: From Stone Age Surgery to Modern Neuroscience* (1st ed.). London: Psychology Press. https://doi.org/10.4324/9781315794549
33. Binder, J. R. (2015). The Wernicke area: Modern evidence and a reinterpretation. *Neurology*, 85(24), 2170–2175. https://doi.org/10.1212/WNL.0000000000002219
34. Ackerman, S. (1992). *Discovering the Brain.* Washington: National Academies Press. https://www.ncbi.nlm.nih.gov/books/NBK234146/
35. Tierney, A. L., Nelson, C. A. (2009). Brain development and the role of experience in the early years. *Zero to Three,* 30(2), 9–13. http://www.ncbi.nlm.nih.gov/pmc/articles/pmc3722610/
36. Bickle, J. (2016). Revolutions in neuroscience: Tool development. *Frontiers in Systems Neuroscience*, 10, 24. https://doi.org/10.3389/fnsys.2016.00024
37. Borghi, J. A., Van Gulick, A. E. (2018). Data management and sharing in neuroimaging: Practices and perceptions of MRI researchers. *PLoS One*, 13(7), e0200562. https://doi.org/10.1371/journal.pone.0200562
38. Brainard, J. (2018). Rethinking retractions. *Science*, 362(6413), 390–393. https://doi.org/10.1126/science.362.6413.390
39. Misra, M. K., Damotte, V., Hollenbach, J. A. (2018). The immunogenetics of neurological disease. *Immunology*, 153(4), 399–414. https://doi.org/10.1111/imm.12869
40. Brain basics: Know your brain. https://www.ninds.nih.gov/health-information/public-education/brain-basics/brain-basics-know-your-brain
41. Agnati, L. F., Guidolin, D., Cervetto, C., Maura, G., Marcoli, M. (2023). Brain structure and function: Insights from chemical neuroanatomy. *Life*, 13(4), 940. https://doi.org/10.3390/life13040940
42. Finnerup, N. B., Kuner, R., Jensen, T. S. (2021). Neuropathic pain: From mechanisms to treatment. *Physiological Reviews*, 101(1), 259–301. https://doi.org/10.1152/physrev.00045.2019
43. Gazzaniga, M. S. (2000). Cerebral specialization and interhemispheric communication: Does the corpus callosum enable the human condition? *Brain*, 123(7), 1293–1326. https://doi.org/10.1093/brain/123.7.1293

44. Yang, M., Yang, Z., Wang, P., Sun, Z. (2021). Current application and future directions of photobiomodulation in central nervous diseases. *Neural Regeneration Research*, 16(6), 1177–1185. https://doi.org/10.4103/1673-5374.300486

45. Guidolin, D., Marcoli, M., Tortorella, C., Maura, G., Agnati, L. F. (2019). Receptor-receptor interactions as a widespread phenomenon: Novel targets for drug development? *Frontiers in Endocrinology*, 10, 53. https://doi.org/10.3389/fendo.2019.00053

46. Slosky, L. M., Caron, M. G., Barak, L. S. (2021). Biased allosteric modulators: New frontiers in GPCR drug discovery. *Trends in Pharmacological Sciences*, 42(4), 283–299. https://doi.org/10.1016/j.tips.2020.12.005

47. Hökfelt, T., Broberger, C., Xu, Z. Q., Sergeyev, V., Ubink, R., Diez, M. (2000). Neuropeptides: An overview. *Neuropharmacology*, 39(8), 1337–1356. https://doi.org/10.1016/s0028-3908(00)00010-1

48. Cifuentes, F., Morales, M. A. (2021). Functional implications of neurotransmitter segregation. *Frontiers in Neural Circuits*, 2021, 15, 738516. https://doi.org/10.3389/fncir.2021.738516

49. Kandel, E. R., Siegelbaum, S. A., Mack, S. H., Koester, J. D. (2021). *Principles of Neuroscience, Sixth Edition*. New York: McGraw-Hill Education.

50. Kim, S., Sabatini, B. L. (2023). Analytical approaches to examine gamma-aminobutyric acid and glutamate vesicular co-packaging. *Frontiers in Synaptic Neuroscience*, 14, 1076616. https://doi.org/10.3389/fnsyn.2022.1076616

51. Guidolin, D., Tortorella, C., Marcoli, M., Cervetto, C., Maura, G., Agnati, L. F. (2021). Receptor-receptor interactions and glial cell functions with a special focus on G protein-coupled receptors. *International Journal of Molecular Sciences*, 22(16), 8656. https://doi.org/10.3390/ijms22168656

52. Chvátal, A., Syková, E. (2000). Glial influence on neuronal signaling. *Progress in Brain Research*, 125, 199–216. https://doi.org/10.1016/S0079-6123(00)25011-5

53. Kofuji, P., Araque, A. (2021). G-protein-coupled receptors in astrocyte-neuron communication. *Neuroscience*, 456, 71–84. https://doi.org/10.1016/j.neuroscience.2020.03.025

54. Guidolin, D., Marcoli, M., Tortorella, C., Maura, G., Agnati, L. F. (2020). Adenosine A_{2A}-dopamine D_2 receptor-receptor interaction in neurons and astrocytes: Evidence and perspectives. *Progress in Molecular

Biology and Translational Science, 169, 247–277. https://doi.org/10.1016/bs.pmbts.2019.11.004

55. Amato, S., Averna, M., Guidolin, D., Pedrazzi, M., Pelassa, S., Capraro, M., Passalacqua, M., Bozzo, M., Gatta, E., Anderlini, D., Maura, G., Agnati, L. F., Cervetto, C., Marcoli, M. (2022). Heterodimer of A_{2A} and oxytocin receptors regulating glutamate release in adult striatal astrocytes. *International Journal of Molecular Sciences*, 23(4), 2326. https://doi.org/10.3390/ijms23042326
56. Singer, W. (2021). Differences between natural and artificial cognitive systems. In: von Braun, J., S. Archer, M., Reichberg, G. M., Sánchez Sorondo, M. (eds). *Robotics, AI, and Humanity* (pp. 17–27). Cham: Springer. https://doi.org/10.1007/978-3-030-54173-6_2
57. Lee, D. (2020). *Birth of Intelligence: From RNA to Artificial Intelligence*. Oxford: Oxford University Press.
58. Agnati, L. F., Anderlini, D., Guidolin, D., Marcoli, M., Maura, G. (2022). Man is a "Rope" stretched between virosphere and humanoid robots: On the urgent need of an ethical code for ecosystem survival. *Foundations of Science*, 27(2), 311–325. https://doi.org/10.1007/s10699-021-09796-z
59. Agnati, L. F., Fuxe, K. (2000). Volume transmission as a key feature of information handling in the central nervous system: Possible new interpretative value of the Turing's B-type machine. *Progress in Brain Research*, 125, 3–19. https://doi.org/10.1016/S0079-6123(00)25003-6
60. Venturini, A., Passalacqua, M., Pelassa, S., Pastorino, F., Tedesco, M., Cortese, K., Gagliani, M. C., Leo, G., Maura, G., Guidolin, D., Agnati, L. F., Marcoli, M., Cervetto, C. (2019). Exosomes from astrocyte processes: Signaling to neurons. *Frontiers in Pharmacology*, 10, 1452. https://doi.org/10.3389/fphar.2019.01452
61. Woo, J., Choi, K., Kim, S. H., Han, K., Choi, M. (2021). Characterization of multiscale logic operations in the neural circuits. *Frontiers in Bioscience*, 26(10), 723–739. https://doi.org/10.52586/4983
62. George, S. R., Fan, T., Xie, Z., Tse, R., Tam, V., Varghese, G., O'Dowd, B. F. (2000). Oligomerization of mu- and delta-opioid receptors. Generation of novel functional properties. *The Journal of Biological Chemistry*, 275(34), 26128–26135. https://doi.org/10.1074/jbc.M000345200
63. Agnati, L. F., Guidolin, D., Cervetto, C., Borroto-Escuela, D. O., Fuxe, K. (2016). Role of iso-receptors in receptor-receptor interactions with a focus on dopamine iso-receptor complexes. *Reviews in the Neurosciences*, 27(1), 1–25. https://doi.org/10.1515/revneuro-2015-0024

64. Mores, K. L., Cassell, R. J., van Rijn, R. M. (2019). Arrestin recruitment and signaling by G protein-coupled receptor heteromers. *Neuropharmacology*, 152, 15–21. https://doi.org/10.1016/j.neuropharm.2018.11.010

65. Guo, S., Zhao, T., Yun, Y., Xie, X. (2022). Recent progress in assays for GPCR drug discovery. *American Journal of Physiology. Cell Physiology*, 323(2), C583-C594. https://doi.org/10.1152/ajpcell.00464.2021

66. Maggio, R., Fasciani, I., Carli, M., Petragnano, F., Marampon, F., Rossi, M., Scarselli, M. (2021). Integration and spatial organization of signaling by G protein-coupled receptor Homo- and Heterodimers. *Biomolecules*, 11(12), 1828. https://doi.org/10.3390/biom11121828

67. Valle-León, M., Callado, L. F., Aso, E., Cajiao-Manrique, M. M., Sahlholm, K., López-Cano, M., Soler, C., Altafaj, X., Watanabe, M., Ferré, S., Fernández-Dueñas, V., Menchón, J. M., Circuela, F. (2021). Decreased striatal adenosine A_{2A}-dopamine D_2 receptor heteromerization in schizophrenia. *Neuropsychopharmacology*, 46, 665–672. https://doi.org/10.1038/s41386-020-00872-9

68. Valle-León, M., Casajuana-Martin, N., Del Torrent, C. L., Argerich, J., Gómez-Acero, L., Sahlholm, K., Ferré, S., Pardo, L., Ciruela, F. (2023). Unique effect of clozapine on adenosine A_{2A}-dopamine D_2 receptor heteromerization. *Biomedicine & Pharmacotherapy*, 160, 114327. https://doi.org/10.1016/j.biopha.2023.114327

69. Fernández-Dueñas, V., Gómez-Soler, M., Valle-León, M., Watanabe, M., Ferrer, I., Ciruela, F. (2019). Revealing adenosine A_{2A}-dopamine D_2 receptor Heteromers in Parkinson's disease post-mortem brain through a new AlphaScreen-based assay. *International Journal of Molecular Sciences*, 20(14), 3600. https://doi.org/10.3390/ijms20143600

70. Marcoli, M., Agnati, L. F., Franco, R., Cortelli, P., Anderlini, D., Guidolin, D., Cervetto, C., Maura, G. (2023). Modulating brain integrative actions as a new perspective on pharmacological approaches to neuropsychiatric diseases. *Frontiers in Endocrinology*, 13, 1038874. https://doi.org/10.3389/fendo.2022.1038874

71. Li, K., Li, J., Zheng, J., Qin, S. (2019). Reactive astrocytes in neurodegenerative diseases. *Aging and Disease*, 10(3), 664–675. https://doi.org/10.14336/AD.2018.0720

72. Cragnolini, A. B., Lampitella, G., Virtuoso, A., Viscovo, I., Panetsos, F., Papa, M., Cirillo, G. (2020). Regional brain susceptibility to neurodegeneration: What is the role of glial cells? *Neural Regeneration Research*, 15(5), 838–842. https://doi.org/10.4103/1673-5374.268897

73. Fil, J., Dalchau, N., Chu, D. (2022). Programming molecular systems to emulate a learning spiking neuron. *ACS Synthetic Biology*, 11(6), 2055–2069. https://doi.org/10.1021/acssynbio.1c00625

74. Voss, P., Thomas, M. E., Cisneros-Franco, J. M., de Villers-Sidani, É. (2017). Dynamic brains and the changing rules of neuroplasticity: Implications for learning and recovery. *Frontiers in Psychology*, 8, 1657. https://doi.org/10.3389/fpsyg.2017.01657

75. Chamoun, M., Huppé-Gourgues, F., Legault, I., Rosa-Neto, P., Dumbrava, D., Faubert, J., Vaucher, E. (2017). Cholinergic potentiation improves perceptual-cognitive training of healthy young adults in three dimensional multiple object tracking. *Frontiers in Human Neuroscience*, 11, 128. https://doi.org/10.3389/fnhum.2017.00128

76. Teles, R. V. (2020). Phineas Gage's great legacy. *Dementia & Neuropsychologia,* 14(4), 419–421. https://doi.org/10.1590/1980-57642020dn14-040013

77. Singh, A., Abraham, W. C. (2017). Astrocytes and synaptic plasticity in health and disease. *Experimental Brain Research*, 235(6), 1645–1655. https://doi.org/10.1007/s00221-017-4928-1

78. Zenner, H. P., Delb, W., Kröner-Herwig, B., Jäger, B., Peroz, I., Hesse, G., Mazurek, B., Goebel, G., Gerloff, C., Trollmann, R., Biesinger, E., Seidler, H., Langguth, B. (2017). A multidisciplinary systematic review of the treatment for chronic idiopathic tinnitus. *European Archives of Oto-Rhino-Laryngology,* 274(5), 2079–2091. https://doi.org/10.1007/s00405-016-4401-y

79. Mateos-Aparicio, P., Rodríguez-Moreno, A. (2019). The impact of studying brain plasticity. *Frontiers in Cellular Neuroscience*, 13, 66. https://doi.org/10.3389/fncel.2019.00066

80. Allen, N. J., Lyons, D. A. (2018). Glia as architects of central nervous system formation and function. *Science*, 362(6411), 181–185. https://doi.org/10.1126/science.aat0473

81. Mansvelder, H. D., Verhoog, M. B., Goriounova, N. A. (2019). Synaptic plasticity in human cortical circuits: Cellular mechanisms of learning and memory in the human brain? *Current Opinion in Neurobiology*, 54, 186–193. https://doi.org/10.1016/j.conb.2018.06.013

82. Kumar, A., Pareek, V., Faiq, M. A., Ghosh, S. K., Kumari, C. (2019). Adult neurogenesis in humans: A review of basic concepts, history, current research, and clinical implications. *Innovation in Clinical Neuroscience*, 16(5–6), 30–37. https://www.ncbi.nlm.nih.gov/pmc/articles/PMC6659986/

83. Liu, P. Z., Nusslock, R. (2018). Exercise-mediated neurogenesis in the hippocampus via BDNF. *Frontiers in Neuroscience*, 12, 52. https://doi.org/10.3389/fnins.2018.00052

84. Morales-García, J. A., Kaneko, N., Herranz-Pérez, V. (2022). Editorial: New insights into adult neurogenesis and neurodegeneration: Challenges for brain repair. *Frontiers in Neuroscience*, 16, 868876. https://doi.org/10.3389/fnins.2022.868876

85. Abdissa, D., Hamba, N., Gerbi, A. (2020). Review article on adult neurogenesis in humans. *Translational Research in Anatomy*, 20, 100074. https://doi.org/10.1016/j.tria.2020.100074.

86. Moreno-Jiménez, E. P., Terreros-Roncal, J., Flor-García, M., Rábano, A., Llorens-Martín, M. (2021). Evidences for adult hippocampal neurogenesis in humans. *The Journal of Neuroscience*, 41(12), 2541–2553. https://doi.org/10.1523/JNEUROSCI.0675-20.2020

87. Kase, Y., Shimazaki, T., Okano, H. (2020). Current understanding of adult neurogenesis in the mammalian brain: How does adult neurogenesis decrease with age? *Inflammation and Regeneration*, 40, 10. https://doi.org/10.1186/s41232-020-00122-x

88. Owji, S., Shoja, M. M. (2020). The history of discovery of adult neurogenesis. *Clinical Anatomy*, 33(1), 41–55. https://doi.org/10.1002/ca.23447

89. Fares, J., Bou Diab, Z., Nabha, S., Fares, Y. (2019). Neurogenesis in the adult hippocampus: History, regulation, and prospective roles. *The International Journal of Neuroscience*, 129(6), 598–611. https://doi.org/10.1080/00207454.2018.1545771

90. Verma, V., Samanthapudi, K., Raviprakash, R. (2016). Classic studies on the potential of stem cell neuroregeneration. *Journal of the History of the Neurosciences*, 25(2), 123–141. https://doi.org/10.1080/0964704X.2015.1039904

91. Jeremic, R., Pekovic, S., Lavrnja, I., Bjelobaba, I., Djelic, M., Dacic, S., Brkic, P. (2023). Hyperbaric oxygenation prevents loss of immature neurons in the adult hippocampal dentate gyrus following brain injury. *International Journal of Molecular Sciences*, 24(5), 4261. https://doi.org/10.3390/ijms24054261

92. Triarhou, L. C., Manto, M. (2022). Postnatal neurogenesis beyond rodents: The groundbreaking research of Joseph Altman and Gopal Das. *Cerebellum*, 21(1), 1–8. https://doi.org/10.1007/s12311-021-01314-y

93. Schmitd, L. B., Perez-Pacheco, C., D'Silva, N. J. (2021). Nerve density in cancer: Less is better. *FASEB bioAdvances*, 3(10), 773–786. https://doi.org/10.1096/fba.2021-00046

94. Sambra, V., Echeverria, F., Valenzuela, A., Chouinard-Watkins, R., Valenzuela, R. (2021). Docosahexaenoic and arachidonic acids as neuroprotective nutrients throughout the life cycle. *Nutrients*, 13(3), 986. https://doi.org/10.3390/nu13030986

95. Hammond, D. C. (2013). A review of the history of hypnosis through the late 19th century. *The American Journal of Clinical Hypnosis*, 56(2), 174–191. https://doi.org/10.1080/00029157.2013.826172

96. Riskin, J. (2024). *Franz Anton Mesmer (1734–1815)*. Stanford: Stanford University. https://exhibits.stanford.edu/super-e/feature/franz-anton-mesmer-1734-1815

97. Yeates, L. B. (2016). Émile Coué and his method (I): The chemist of thought and human action. *Australian Journal of Clinical Hypnotherapy and Hypnosis*, 38(1), 3–27. https://psycnet.apa.org/record/2016-43121-002

98. Lynn, S. J., Laurence, J. R., Kirsch, I. (2015). Hypnosis, suggestion, and suggestibility: An integrative model. *The American Journal of Clinical Hypnosis*, 57(3), 314–329. https://doi.org/10.1080/00029157.2014.976783

99. Definition and description of hypnosis. (2014). https://www.apadivisions.org/division-30/about#:~:text=30%20Executive%20Committee%20prepared%20the,procedure%20designed%20to%20induce%20hypnosis

100. De Benedittis, G. (2021). Neural mechanisms of hypnosis and meditation-induced analgesia: A narrative review. *The International Journal of Clinical and Experimental Hypnosis*, 69(3), 363–382. https://doi.org/10.1080/00207144.2021.1917294

101. De Benedittis, G. (2015). Neural mechanisms of hypnosis and meditation. *Journal of Physiology*-Paris, 109(4–6), 152–164. https://doi.org/10.1016/j.jphysparis.2015.11.001

102. Jensen, M. P., Adachi, T., Tomé-Pires, C., Lee, J., Osman, Z. J., Miró, J. (2015). Mechanisms of hypnosis: Toward the development of a biopsychosocial model. *International Journal of Clinical and Experimental Hypnosis*, 63(1), 34–75. https://doi.org/10.1080/00207144.2014.961875

103. Jansen S. (2017). *Conversational hypnosis: Spotting unconscious moments: Scott Jansen hypnosis method; Introducing the ABSURD*

hypnosis formula. https://www.amazon.com/Conversational-Hypnosis-Spotting-Unconscious-introducing-ebook/dp/B074T8S1L2#detailBullets_feature_div

104. Vanhaudenhuyse, A., Laureys, S., Faymonville, M. E. (2014). Neurophysiology of hypnosis. *Clinical Neurophysiology*, 44(4), 343–353. https://doi.org/10.1016/j.neucli.2013.09.006
105. Halsband, U., Wolf, T. G. (2019). Functional changes in brain activity after hypnosis. Neurobiological mechanisms and application to patients with a specific phobia-limitations and future directions. *The International Journal of Clinical and Experimental* Hypnosis, 67(4), 449–474. https://doi.org/10.1080/00207144.2019.1650551
106. Landry, M., Lifshitz, M., Raz, A. (2017). Brain correlates of hypnosis: A systematic review and meta-analytic exploration. *Neuroscience and Biobehavioral* Reviews, 81(Pt A), 75–98. https://doi.org/10.1016/j.neubiorev.2017.02.020
107. Keppler, J. (2018). Shedding light on the fundamental mechanism underlying hypnotic analgesia. *Annals of Palliative Medicine*, 7(1), 170–176. https://doi.org/10.21037/apm.2017.04.03
108. Kekecs, Z., Bowers, J., Johnson, A., Kendrick, C., Elkins, G. (2016). The Elkins hypnotizability scale: Assessment of reliability and validity. *The International Journal of Clinical and Experimental Hypnosis*, 64(3), 285–304. https://doi.org/10.1080/00207144.2016.1171089
109. Elkins, G. R. (2016). *Handbook of Medical and Psychological Hypnosis*. New York: Springer Publishing Company.
110. Franquelo, M. A., Monteiro, D., Tomás-Marco, I., Capafons, A., Hernández-Mendo, A. (2022). Psychometric properties of the Valencia scale of attitudes and beliefs towards hypnosis (client version) (VSABTH-C) online version. *Current Psychology*, 41, 3205–3215. https://doi.org/10.1007/s12144-020-00779-0
111. Bastek, V., van Vliet, M. (2023). A whole new world of healing: Exploring medical hypnotherapy for pediatric patients: A review. *European Journal of Pediatrics*, 182(7), 3021–3032 (2023). https://doi.org/10.1007/s00431-023-04983-5
112. Werner, A., Uldbjerg, N., Zachariae, R., Wu, C. S., Nohr, E. A. (2013). Antenatal hypnosis training and childbirth experience: A randomized controlled trial. *Birth*, 40(4), 272–80. https://doi.org/10.1111/birt.12071
113. Sawni, A., Breuner, C. C. (2017). Clinical hypnosis, an effective mind-body modality for adolescents with behavioral and physical complaints. *Children*, 4(4), 19. https://doi.org/10.3390/children4040019

114. Kendrick, C., Sliwinski, J., Yu, Y., Johnson, A., Fisher, W., Kekecs, Z., Elkins, G. (2016). Hypnosis for acute procedural pain: A critical review. *The International Journal of Clinical and Experimental Hypnosis*, 64(1), 75–115. https://doi.org/10.1080/00207144.2015.1099405

115. Delivet, H., Dugue, S., Ferrari, A., Postone, S., Dahmani, S. (2018). Efficacy of self-hypnosis on quality of life for children with chronic pain syndrome. *The International Journal of Clinical and Experimental Hypnosis*, 66(1):43–55. https://doi.org/10.1080/00207144.2018.1396109

116. Rousseau-Salvador, C., Amouroux, R., Annequin, D., Salvador, A., Tourniaire, B., Rusinek, S. (2014). Anxiety, depression and school absenteeism in youth with chronic or episodic headache. *Pain Research &* Management, 19(5), 235–240. https://doi.org/10.1155/2014/541618

117. Franquelo, M. A., Pastrana-Brincones, J. L., Reigal, R. E., Morillo-Baro, J. P., Vázquez-Diz, J. A., Hernández-Mendo, A., Morales-Sánchez, V. (2021). Data mining for attitudinal and belief profiles determination towards hypnosis. *Sustainability*, 13, 7721. https://doi.org/10.3390/su13147721

118. Franquelo, M. A., Tomás, I., Capafons, A., Hernández-Mendo, A. (2021). Psychometric properties of the Valencia Scale of Attitudes and Beliefs Towards Hypnosis (therapist version) (VSABTH-T) online version. *Quality & Quantity*, 55, 203–219. https://doi.org/10.1007/s11135-020-01001-w

119. Moss, D., Willmarth, E. (2019). Hypnosis, anesthesia, pain management, and preparation for medical procedures. *Journal Annals of Palliative Medicine*, 8(4), 498–503. https://doi.org/10.21037/apm.2019.07.01

120. Willmarth, E., Davis, F., Fitzgerald, K. (2014). Biofeedback and integrative medicine in the pain clinic setting. *Biofeedback*, 42(3), 111–114. https://doi.org/10.5298/1081-5937-42.03.10

121. Werner, A., Uldbjerg, N., Zachariae, R., Nohr, E. A. (2013). Effect of self-hypnosis on duration of labor and maternal and neonatal outcomes: A randomized controlled trial. *Acta obstetricia et gynecologica Scandinavica*, 92(7), 816–823. https://doi.org/10.1111/aogs.12141

122. Kaiser, P., Kohen, D. P., Brown, M. L., Kajander, R. L., Barnes, A. J. (2018). Integrating pediatric hypnosis with complementary modalities: Clinical perspectives on personalized treatment. *Children*, 5(8), 108. https://doi.org/10.3390/children5080108

123. Kohen, D. P., Kaiser, P. (2014). Clinical hypnosis with children and adolescents—What? Why? How?: Origins, applications, and efficacy. *Children*, 1(2), 74–98. https://doi.org/10.3390/children1020074

124. Luo, Y., Keefer, L. (2022). The clinical value of brain-gut behavioral therapies for functional esophageal disorders and symptoms. *Neurogastroenterology and Motility*, 34(6), e14373. https://doi.org/10.1111/nmo.14373

125. Lowén, M. B., Mayer, E. A., Sjöberg, M., Tillisch, K., Naliboff, B., Labus, J., Lundberg, P., Ström, M., Engström, M., Walter, S. A. (2013). Effect of hypnotherapy and educational intervention on brain response to visceral stimulus in the irritable bowel syndrome. *Alimentary Pharmacology & Therapeutics*, 37(12), 1184–1197. https://doi.org/10.1111/apt.12319

126. Browne, P. D., den Hollander, B., Speksnijder, E. M., van Wering, H. M., Tjon A Ten, W., George, E. K., Groeneweg, M., Bevers, N., Wessels, M. M. S., van den Berg, M. M., Goede, J., Teklenburg-Roord, S. T. A., Frankenhuis, C., Benninga, M. A., Vlieger, A. M. (2019). Gut-directed hypnotherapy versus standard medical treatment for nausea in children with functional nausea or functional dyspepsia: Protocol of a multicentre randomised trial. *BMJ Open*, 9(4), e024903. https://doi.org/10.1136/bmjopen-2018-024903

127. Moser, G. (2014). The role of hypnotherapy for the treatment of inflammatory bowel diseases. *Expert Review of Gastroenterology & Hepatology*, 8(6), 601–606. https://doi.org/10.1586/17474124.2014.917955

128. Keefer, L., Taft, T. H., Kiebles, J. L., Martinovich, Z., Barrett, T. A., Palsson, O. S. (2013). Gut-directed hypnotherapy significantly augments clinical remission in quiescent ulcerative colitis. *Alimentary Pharmacology & Therapeutics*, 38(7), 761–771. https://doi.org/10.1111/apt.12449

129. Hoekman, D. R., Vlieger, A. M., Stokkers, P. C., Mahhmod, N., Rietdijk, S., de Boer, N. K., de Meij, T. G., Frankenhuis, C., D'Haens, G. R., Benninga, M. A. (2021). Hypnotherapy for irritable bowel syndrome-type symptoms in patients with quiescent inflammatory bowel disease: A randomized, controlled trial. *Journal of Crohn's & Colitis*, 15(7), 1106–1113. https://doi.org/10.1093/ecco-jcc/jjaa241

130. Ruddy, J., Taft, T., Keszthelyi, D. (2021). Gut-brain interactions in patients with inflammatory bowel disease and the role of hypnotherapy in managing symptoms. *Journal of Crohn's & Colitis*, 15(10), 1780–1781. https://doi.org/10.1093/ecco-jcc/jjab053

131. Becker, P. M. (2015). Hypnosis in the management of sleep disorders. *Sleep Medicine Clinics*, 10(1), 85–92. https://doi.org/10.1016/j.jsmc.2014.11.003

132. Thompson, T., Terhune, D. B., Oram, C., Sharangparni, J., Rouf, R., Solmi, M., Veronese, N., Stubbs, B. (2019). The effectiveness of hypnosis for pain relief: A systematic review and meta-analysis of 85 controlled experimental trials. *Neuroscience and Biobehavioral Reviews*, 99, 298–310. https://doi.org/10.1016/j.neubiorev.2019.02.013

133. Rosenbloom, B. N., Slepian, P. M., Azam, M. A., Aternali, A., Birnie, K. A., Curtis, K., Thaker, S., Ladak, S., Waisman, A., Clarke, H., Katz, J., Weinrib, A. Z. (2024). A randomized controlled trial of clinical hypnosis as an opioid-sparing adjunct treatment for pain relief in adults undergoing major oncologic surgery. *Journal of Pain Research*, 17, 45–59. https://doi.org/10.2147/JPR.S424639

134. Reid, D. B. (2017). Treating panic disorder hypnotically. *The American Journal of Clinical Hypnosis*, 60(2), 137–148. https://doi.org/10.1080/00029157.2017.1288608

135. Allison N. (2015). Hypnosis in modern dentistry: Challenging misconceptions. *Faculty Dental Journal*, 6(4), 172–175. https://doi.org/10.1308/rcsfdj.2015.172

136. Armfield, J. M., Heaton, L. J. (2013). Management of fear and anxiety in the dental clinic: A review. *Australian Dental Journal*, 58(4), 390–407. https://doi.org/10.1111/adj.12118

137. Facco, E., Zanette, G., Casiglia, E. (2014). The role of hypnotherapy in dentistry. *SAAD Digest*, 30, 3–6. https://pubmed.ncbi.nlm.nih.gov/24624516/#:~:text=It%20provides%20an%20effective%20sedation,with%20a%20severe%20gag%20reflex

138. Peretz, B., Bercovich, R., Blumer, S. (2013). Using elements of hypnosis prior to or during pediatric dental treatment. *Pediatric Dentistry*, 35(1), 33–36. https://pubmed.ncbi.nlm.nih.gov/23635894/

139. Meints, S. M., Edwards, R. R. (2018). Evaluating psychosocial contributions to chronic pain outcomes. *Progress in Neuro-Psychopharmacology & Biological Psychiatry*, 87(Pt B), 168–182. https://doi.org/10.1016/j.pnpbp.2018.01.017

140. Bastek, V. B., Groeneveld, E. M., van Vliet, M. J. (2022). Medical hypnotherapy; Even in a tertiary care setting a promising treatment. *International Journal of Contemporary Pediatrics*, 9(5), 418–424. https://doi.org/10.18203/2349-3291.ijcp20221065

141. Kuttner, L. (2020). Pediatric hypnosis: Treatment that adds and rarely subtracts. *The International Journal of Clinical and Experimental Hypnosis*, 68(1), 16–28. https://doi.org/10.1080/00207144.2020.1685329

142. Yeh, V. M., Schnur, J. B., Montgomery, G. H. (2014). Disseminating hypnosis to health care settings: Applying the RE-AIM framework. *Psychology of* Consciousness, 1(2), 213–228. https://doi.org/10.1037/cns0000012

143. Chester, S. J., Tyack, Z., De Young, A., Kipping, B., Griffin, B., Stockton, K., Ware, R. S., Zhang, X., Kimble, R. M. (2018). Efficacy of hypnosis on pain, wound-healing, anxiety, and stress in children with acute burn injuries: A randomized controlled trial. *Pain*, 159(9), 1790–1801. https://doi.org/10.1097/j.pain.0000000000001276

144. Flynn, N. (2019). Effect of an online hypnosis intervention in reducing migraine symptoms: A randomized controlled trial. *The International Journal of Clinical and Experimental Hypnosis*, 67(3), 313–335. https://doi.org/10.1080/00207144.2019.1612674

145. Rutten, J. M. T. M., Vlieger, A. M., Frankenhuis, C., George, E. K., Groeneweg, M., Norbruis, O. F., Tjon A Ten, W., van Wering, H. M., Dijkgraaf, M. G. W., Merkus, M. P., Benninga, M. A. (2017). Home-based hypnotherapy self-exercises vs individual hypnotherapy with a therapist for treatment of pediatric irritable bowel syndrome, functional abdominal pain, or functional abdominal pain syndrome: A randomized clinical trial. *JAMA Pediatrics*, 171(5), 470–477. https://doi.org/10.1001/jamapediatrics.2017.0091

146. Rexwinkel, R., Bovendeert, J. F. M., Rutten, J. M. T. M., Frankenhuis, C., Benninga, M. A., Vlieger, A. M. (2022). Long-term follow-up of individual therapist delivered and standardized hypnotherapy recordings in pediatric irritable bowel syndrome or functional abdominal pain. *Journal of Pediatric Gastroenterology and Nutrition*, 75(1), 24–29. https://doi.org/10.1097/MPG.0000000000003478

147. Noble, H., Hasan, S. S., Simpson, V., Whorwell, P. J., Vasant, D. H. (2022). Patient satisfaction after remotely delivered gut-directed hypnotherapy for irritable bowel syndrome during the COVID-19 era: Implications for future practice. *BMJ Open Gastroenterology*, 9(1), e001039. https://doi.org/10.1136/bmjgast-2022-001039

148. Rastogi, R. G., Arnold, T. L., Borrero-Mejias, C., Hastriter, E. V., Hickman, C., Karnik, K. T., Lewis, K. S., Little, R. D. (2021). Non-pharmacologic and mindful-based approaches for pediatric headache disorders: A review. *Current Pain and Headache Reports*, 25(12), 78. https://doi.org/10.1007/s11916-021-00993-w

149. Roullet, P., Pons, L.-L., Delmas, P., Weber, C., de Prigny, P. R. (2025). Esketamine facilitate psychotherapies for post-traumatic stress disorder: A retrospective case series of six patients. *European Journal*

of Trauma & Dissociation, 9(1), 100490. https://doi.org/10.1016/j.ejtd.2024.100490

150. Jong, M. C., Boers, I., van Wietmarschen, H. A., Tromp, E., Busari, J. O., Wennekes, R., Snoeck, I., Bekhof, J., Vlieger, A. M. (2019). Hypnotherapy or transcendental meditation versus progressive muscle relaxation exercises in the treatment of children with primary headaches: A multi-centre, pragmatic, randomised clinical study. *European Journal of Pediatrics*, 178(2), 147–154. https://doi.org/10.1007/s00431-018-3270-3
151. Carmody, T. P., Duncan, C., Simon, J. A., Solkowitz, S., Huggins, J., Lee, S., Delucchi, K. (2008). Hypnosis for smoking cessation: A randomized trial. *Nicotine & Tobacco Research*, 10(5), 811–818. https://doi.org/10.1080/14622200802023833
152. Coogle, J., Coogle, B., Quezada, J. (2021). Hypnosis in the treatment of pediatric functional neurological disorder: The magic glove technique. *Pediatric Neurology*, 125, 20–25. https://doi.org/10.1016/j.pediatrneurol.2021.08.011
153. Bayram-Weston, Z., Andrade-Sienz, M., Knight, J. (2022). Nervous system 1: Introduction to the nervous system. *Nursing Times*, 118, 3. https://www.nursingtimes.net/neurology/nervous-system-1-introduction-to-the-nervous-system-21-02-2022/
154. Curti, S., Davoine, F., Dapino, A. (2022). Function and plasticity of electrical synapses in the mammalian brain: Role of non-junctional mechanisms. *Biology*, 11(1), 81. https://doi.org/10.3390/biology11010081
155. Dendrou, C. A., Fugger, L., Friese, M. A. (2015). Immunopathology of multiple sclerosis. *Nature Reviews. Immunology*, 15(9), 545–558. https://doi.org/10.1038/nri3871
156. Ferreira-Vieira, T. H., Guimaraes, I. M., Silva, F. R., Ribeiro, F. M. (2016). Alzheimer's disease: Targeting the cholinergic system. *Current Neuropharmacology*, 14(1), 101–115. https://doi.org/10.2174/1570159x13666150716165726
157. Ginhoux, F., Greter, M., Leboeuf, M., Nandi, S., See, P., Gokhan, S., Mehler, M. F., Conway, S. J., Ng, L. G., Stanley, E. R., Samokhvalov, I. M., Merad, M. (2010). Fate mapping analysis reveals that adult microglia derive from primitive macrophages. *Science*, 330(6005), 841–845. https://doi.org/10.1126/science.1194637
158. Jäkel, S., Dimou, L. (2017). Glial cells and their function in the adult brain: A journey through the history of their ablation. *Frontiers in Cellular Neuroscience*; 11, 24. https://doi.org/10.3389/fncel.2017.00024

159. Javed, K., Reddy, V., Lui, F. (2023). Neuroanatomy, choroid plexus. In: *StatPearls*. Treasure Island: StatPearls Publishing. https://www.ncbi.nlm.nih.gov/books/NBK538156/

160. Sandoiu, A. (2018). The curious case of a blind woman who sees motion. https://www.medicalnewstoday.com/articles/322136

161. Kiernan, J. A., Rajakumar, N. (2013). *Barr's the Human Nervous System: An Anatomical Viewpoint*. Alphen aan den Rijn: Wolters Kluwer.

162. Zhang, J. (2019). Basic neural units of the brain: Neurons, synapses and action potential. https://doi.org/10.48550/arXiv.1906.01703

163. Knight, J., Nigam, Y., Cutter, J. (2020). *Understanding Anatomy and Physiology in Nursing*. London: Sage.

164. Marieb, E. N., Hoehn, K. N. (2018). *Human Anatomy and Physiology*. London: Pearson.

165. Mtui, E., Gruener, G., Dockery, P. (2015). *Fitzgerald's Clinical Neuroanatomy and Neuroscience*. Amsterdam: Elsevier.

166. Muzio, M. R., Fakoya, A. O., Cascella, M. (2022). Histology, Axon. In: *StatPearls*. Treasure Island: StatPearls Publishing.

167. Pardridge, W. M. (2012). Drug transport across the blood-brain barrier. *Journal of Cerebral Blood Flow and Metabolism*, 32(11), 1959–1972. https://doi.org/10.1038/jcbfm.2012.126

168. Patrikios, P., Stadelmann, C., Kutzelnigg, A., Rauschka, H., Schmidbauer, M., Laursen, H., Sorensen, P. S., Brück, W., Lucchinetti, C., Lassmann, H. (2006). Remyelination is extensive in a subset of multiple sclerosis patients. *Brain*, 129(Pt 12), 3165–3172. https://doi.org/10.1093/brain/awl217

169. Pawlina, W., Ross, M. H. (2015). *Histology: A Text and Atlas: With Correlated Cell and Molecular Biology.* Alphen aan den Rijn: Wolters Kluwer.

170. Yamins, D. L., Hong, H., Cadieu, C. F., Solomon, E. A., Seibert, D., DiCarlo, J. J. (2014). Performance-optimized hierarchical models predict neural responses in higher visual cortex. *PNAS Nexus*, 111(23), 8619–8624. https://doi.org/10.1073/pnas.1403112111

171. Johnson, M., Schuster, M., Le, Q. V., Krikun, M., Wu, Y., Chen, Z., Thorat, N., Viégas, F., Wattenberg, M., Corrado, G., Hughes, M., Dean, J. (2017). Google's Multilingual neural machine translation system: Enabling zero-shot translation. *Transactions of the Association for Computational Linguistics*, 5, 339–351. https://transacl.org/index.php/tacl/article/view/1081/248

172. Pereda, A. E. (2014). Electrical synapses and their functional interactions with chemical synapses. *Nature Reviews. Neuroscience*, 15(4), 250–263. https://doi.org/10.1038/nrn3708

173. Salzer, J. L., Zalc, B. (2016). Myelination. *Current Biology*, 26(20), R971-R975. https://doi.org/10.1016/j.cub.2016.07.074

174. Cramer, S. C., Sur, M., Dobkin, B. H., O'Brien, C., Sanger, T. D., Trojanowski, J. Q., Rumsey, J. M., Hicks, R., Cameron, J., Chen, D., Chen, W. G., Cohen, L. G., deCharms, C., Duffy, C. J., Eden, G. F., Fetz, E. E., Filart, R., Freund, M., Grant, S. J., Haber, S., Kalivas, P. W., Kolb, B., Kramer, A. F., Lynch, M., Mayberg, H. S., McQuillen, P. S., Nitkin, R., Pascual-Leone, A., Reuter-Lorenz, P., Schiff, N., Sharma, A., Shekim, L., Stryker, M., Sullivan, E. V., Vinogradov, S. (2011). Harnessing neuroplasticity for clinical applications. *Brain*, 134(6), 1591–1609. https://doi.org/10.1093/brain/awr039

175. Patton, K. T., Thibodeau, G. A. (2018). *Anthony's Textbook of Anatomy and Physiology.* Amsterdam: Elsevier.

176. Tortora, G. J., Derrickson, B. H. (2014). *Principles of Anatomy and Physiology*, 14th Edition. Hoboken: Wiley.

177. Van Hook, M. J., Nawy, S., Thoreson, W. B. (2019). Voltage- and calcium-gated ion channels of neurons in the vertebrate retina. *Progress in Retinal and Eye Research*, 72, 100760. https://doi.org/10.1016/j.preteyeres.2019.05.001

178. VanPutte, C., Regan, J., Russo, A., Seeley, R., Stephens, T., Tate, P. (2017). *Seeley's Anatomy and Physiology.* New York: McGraw-Hill.

179. Hiesinger, P. R. (2021). *The Self-Assembling Brain: How Neural Networks Grow Smarter.* Princeton: Princeton University Press.

180. Rolls, E. T. (2023). *Brain Computations and Connectivity.* Oxford: Oxford University Press.

181. Barrett, D. G., Morcos, A. S., Macke, J. H. (2019). Analyzing biological and artificial neural networks: Challenges with opportunities for synergy? *Current Opinion in Neurobiology*, 55, 55–64. https://doi.org/10.1016/j.conb.2019.01.007

182. Pircher, T., Pircher, B., Schlücker, E., Feigenspan, A. (2021). The structure dilemma in biological and artificial neural networks. *Scientific Reports*, 11, 5621. https://doi.org/10.1038/s41598-021-84813-6

183. Powell, H., Winkel, M., Hopp, A. V., Helmut, H. (2022). A hybrid biological neural network model for solving problems in cognitive planning. *Scientific Reports*, 12, 10628. https://doi.org/10.1038/s41598-022-11567-0

184. Suhaimi, A., Lim, A. W. H., Chia, X. W., Li, C., Makino, H. (2022). Representation learning in the artificial and biological neural networks underlying sensorimotor integration. *Science Advances*, 8(22), eabn0984. https://doi.org/10.1126/sciadv.abn0984

185. Smirnova, L., Caffo, B. S., Gracias, D. H., Huang, Q., Morales Pantoja, I. E., Tang, B., Zack, D. J., Berlinicke, C. A., Boyd, J. L., Harris, T. D., Johnson, E. C., Kagan, B. J., Kahn, J., Muotri, A. R., Paulhamus, B. L., Schwamborn, J. C., Plotkin, J., Szalay, A. S., Vogelstein, J. T., Worley, P. F., Hartung, T. (2023). Organoid intelligence (OI): The new frontier in biocomputing and intelligence-in-a-dish. *Frontiers in Science*, 1, 1017235. https://doi.org/10.3389/fsci.2023.1017235

186. Chen, Z., Liang, Q., Wei, Z., Chen, X., Shi, Q., Yu, Z., Sun, T. (2023). An overview of in vitro biological neural networks for robot intelligence. *Cyborg and Bionic Systems*, 4, 0001. https://doi.org/10.34133/cbsystems.0001

187. Seguin, C., Sporns, O., Zalesky, A. (2023). Brain network communication: Concepts, models and applications. *Nature Reviews Neuroscience*, 24, 557–574. https://doi.org/10.1038/s41583-023-00718-5

188. Mordvintsev, A., Olah, C., Tyka, M. (2015). Inceptionism: Going deeper into neural networks. https://research.google/blog/inceptionism-going-deeper-into-neural-networks/

189. Silver, D., Schrittwieser, J., Simonyan, K., Antonoglou, I., Huang, A., Guez, A., Hubert, T., Baker, L., Lai, M., Bolton, A., Chen, Y., Lillicrap, T., Hui, F., Sifre, L., van den Driessche, G., Graepel, T., Hassabis, D. (2017). Mastering the game of Go without human knowledge. *Nature*, 550, 354–359. https://doi.org/10.1038/nature24270

190. Vinyals, O., Babuschkin, I., Czarnecki, W. M., Mathieu, M., Dudzik, A., Chung, J., Choi, D. H., Powell, R., Ewalds, T., Georgiev, P., Oh, J., Horgan, D., Kroiss, M., Danihelka, I., Huang, A., Sifre, L., Cai, T., Agapiou, J. P., Jaderberg, M., Vezhnevets, A. S., Leblond, R., Pohlen, T., Dalibard, V., Budden, D., Sulsky, Y., Molloy, J., Paine, T. L., Gulcehre, C., Wang, Z., Pfaff, T., Wu, Y., Ring, R., Yogatama, D., Wünsch, D., McKinney, K., Smith, O., Schaul, T., Lillicrap, T., Kavukcuoglu, K., Hassabis, D., Apps, C., Silver, D. (2019). Grandmaster level in StarCraft II using multi-agent reinforcement learning. *Nature*, 575, 350–354. https://doi.org/10.1038/s41586-019-1724-z

191. Mei, X., Lee, H. C., Diao, K., Huang, M., Lin, B., Liu, C., Xie, Z., Ma, Y., Robson, P. M., Chung, M., Bernheim, A., Mani, V., Calcagno, C., Li, K., Li, S., Shan, H., Lv, J., Zhao, T., Xia, J., Long, O., Steinberger, S., Jacobi, A., Deyer, T., Luksza, M., Liu, F., Little, B. P., Fayad, Z. A., Yang, Y. (2020). Artificial

intelligence–enabled rapid diagnosis of patients with COVID-19. *Nature Medicine*, 26, 1224–1228. https://doi.org/10.1038/s41591-020-0931-3

192. Posner, M. I., Rothbart, M. K. (2011). Brain states and hypnosis research. *Consciousness and Cognition*, 20(2), 325–277. https://doi.org/10.1016/j.concog.2009.11.008

193. Carhart-Harris, R. L., Mayberg, H. S., Malizia, A. L., Nutt, D. (2008). Mourning and melancholia revisited: Correspondences between principles of Freudian metapsychology and empirical findings in neuropsychiatry. *Annals of General Psychiatry*, 7, 9. https://doi.org/10.1186/1744-859X-7-9

194. van Atteveldt, N., van Kesteren, M. T. R., Braams, B., Krabbendam, L. (2018). Neuroimaging of learning and development: Improving ecological validity. *Frontline Learning Research*, 6(3), 186–203. https://doi.org/10.14786/flr.v6i3.366

195. Naqvi, N. H., Morgenstern, J. (2015). Cognitive neuroscience approaches to understanding behavior change in alcohol use disorder treatments. *Alcohol Research*, 37(1), 29–38. https://www.ncbi.nlm.nih.gov/pmc/articles/PMC4476602/

196. Rainville, P., Hofbauer, R. K., Bushnell, M. C., Duncan, G. H., Price, D. D. (2002). Hypnosis modulates activity in brain structures involved in the regulation of consciousness. *Journal of Cognitive Neuroscience*, 14(6), 887–901. https://doi.org/10.1162/089892902760191117

197. National Institute of Neurological Disorders and Stroke. (2010). *Brain basics: Genes at work in the brain*. Bethesda: National Institutes of Health Publication. https://www.ninds.nih.gov/health-information/patient-caregiver-education/brain-basics-genes-work-brain

198. Kennedy, M. B. (2017). Biochemistry and neuroscience: The twain need to meet. *Current Opinion in Neurobiology*, 43, 79–86. https://doi.org/10.1016/j.conb.2017.01.004

199. Kamarajan, C., Porjesz, B. (2015). Advances in electrophysiological research. *Alcohol Research*, 37(1), 53–87. https://www.ncbi.nlm.nih.gov/pmc/articles/PMC4476604/

200. Maiseli, B., Abdalla, A. T., Massawe, L. V., Mbise, M., Mkocha, K., Nassor, N. A., Ismail, M., Michael, J., Kimambo, S. (2023). Brain-computer interface: Trend, challenges, and threats. *Brain Informatics*, 10(1), 20. https://doi.org/10.1186/s40708-023-00199-3

201. Jensen, M. P., Jamieson, G. A., Lutz, A., Mazzoni, G., McGeown, W. J., Santarcangelo, E. L., Demertzi, A., De Pascalis, V., Bányai, É. I., Rominger,

C., Vuilleumier, P., Faymonville, M. E., Terhune, D. B. (2017). New directions in hypnosis research: Strategies for advancing the cognitive and clinical neuroscience of hypnosis. *Neuroscience of Consciousness*, 3(1), nix004. https://doi.org/10.1093/nc/nix004

202. Barnes, J., McRobbie, H., Dong, C. Y., Walker, N., Hartmann-Boyce, J. (2019). Hypnotherapy for smoking cessation. *The Cochrane Database Systematic Reviews*, 6(6), CD001008. https://doi.org/10.1002/14651858.CD001008.pub3
203. Gao, W. J., Hu, Y., Ji, J. L., Liu, X. Q. (2023). Relationship between depression, smartphone addiction, and sleep among Chinese engineering students during the COVID-19 pandemic. *World Journal of Psychiatry*, 13(6), 361–375. https://doi.org/10.5498/wjp.v13.i6.361
204. Green, J. P. (1999). Hypnosis, context effects, and the recall of early autobiographical memories. *The International Journal of Clinical and Experimental Hypnosis*, 47(4), 284–300. https://doi.org/10.1080/00207149908410038
205. Williamson, A. (2019). What is hypnosis and how might it work? *Palliative Care*, 12, 1178224219826581. https://doi.org/10.1177/1178224219826581
206. Sekula, A. D., Downey, L., Puspanathan, P. (2022). Virtual reality as a moderator of psychedelic-assisted psychotherapy. *Frontiers in Psychology*, 13, 813746. https://doi.org/10.3389/fpsyg.2022.813746
207. de Bruijn, C. M. A., Hovy, S. W., Tromp, E., Benninga, M. A., Hall, K. T., Vlieger, A. M. (2023). Do polymorphisms predict hypnotherapy response in children with functional abdominal pain disorders: An explorative study. *Journal of Pediatric Gastroenterology and Nutrition*, 77(4), 486–490. https://doi.org/10.1097/MPG.0000000000003895
208. Brooker, E. (2020). Cognitive hypnotherapy. In: *Hypnotherapy and Hypnosis*. IntechOpen. http://dx.doi.org/10.5772/intechopen.91327
209. Cerritelli, F., Chiera, M., Abbro, M., Megale, V., Esteves, J., Gallace, A., Manzotti, A. (2021). The challenges and perspectives of the integration between virtual and augmented reality and manual therapies. *Frontiers in Neurology*, 12, 700211. https://doi.org/10.3389/fneur.2021.700211
210. Bernheim, H. (1891/1973). *Hypnotisme, suggestion, psychothérapie: Etudes nouvelles* (trans. New York, Aronson). Paris: Octave Douin.
211. Laurence, J. R., Perry, C. (1983). Hypnotically created memory among highly hypnotizable subjects. *Science*, 222(4623), 523–524. https://doi.org/10.1126/science.6623094

212. Laurence, J. R., Perry, C. (1988). *Hypnosis, Will, and Memory: A Psycho-legal History.* New York: Guilford Press.
213. Maguire, E. A., Gadian, D. G., Johnsrude, I. S., Good, C. D., Ashburner, J., Frackowiak, R. S., Frith, C. D. (2000). Navigation-related structural change in the hippocampi of taxi drivers. *PNAS Nexus*, 97(8), 4398–4403. https://doi.org/10.1073/pnas.070039597
214. Janet, P. (1889). *L'automatisme psychologique*. Paris: Félix Alcan.
215. Yapko, M. D. (1994). *Suggestions of Abuse: True and False Memories of Childhood Sexual Trauma.* New York: Simon & Schuster.
216. Poole, D. A., Lindsay, D. S., Memon, A., Bull, R. (1995). Psychotherapy and the recovery of memories of childhood sexual abuse: U. S. and British practitioners' opinions, practices, and experiences. *Journal of Consulting and Clinical Psychology,* 63, 426–437. https://doi.org/10.1037//0022-006x.63.3.426
217. Legault, E., Laurence, J. R. (2007). Recovered memories of childhood sexual abuse: Social worker, psychologist, and psychiatrist reports of beliefs, practices, and cases. *Australian Journal of Clinical & Experimental Hypnosis,* 35(2), 111–133. https://psycnet.apa.org/record/2007-19931-002
218. Ost, J., Wright, D. B., Easton, S., Hope, L., French, C. C. (2011). Recovered memories, satanic abuse, dissociative identity disorder and false memories in the UK: A survey of clinical psychologists and hypnotherapists. *Psychology, Crime & Law,* 19, 1–19. https://doi.org/10.1080/1068316X.2011.598157
219. Whitehouse, W. G., Orne, E. C., Orne, M. T., Dinges, D. F. (1991). Distinguishing the source of memories reported prior waking and hypnotic recall attempts. *Applied Cognitive Psychology*, 5(1), 51–59. https://psycnet.apa.org/doi/10.1002/acp.2350050105
220. Johnson, M. E., Hauck, C. (1999). Beliefs and opinions about hypnosis held by the general public: A systematic evaluation. *The American Journal of Clinical Hypnosis*, 42(1), 10–20. https://doi.org/10.1080/00029157.1999.10404241
221. Green, J. P., Page, R. A., Rasekhy, R., Johnson, L. K., Bernhardt, S. E. (2006). Cultural views and attitudes about hypnosis: A survey of college students across four countries. *The International Journal of Clinical and Experimental Hypnosis*, 54(3), 263–280. https://doi.org/10.1080/00207140600689439
222. Orne, M. T. (1951). The mechanisms of hypnotic age regression: An experimental study. *Journal of Abnormal and Social Psychology*, 46(2), 213–225. https://doi.org/10.1037/h0059971

223. Orne, M. T. (1979). The use and misuse of hypnosis in court. *The International Journal of Clinical and Experimental Hypnosis,* 27(4), 311–341. https://doi.org/10.1080/00207147908407571
224. Relinger, H. (1984). Hypnotic hypermnesia: A critical review. *The American Journal of Clinical Hypnosis*, 26(3), 212–225. https://doi.org/10.1080/00029157.1984.10404165
225. Erdelyi, M. H. (1994). The empty set of hypermnesia. *The International Journal of Clinical and Experimental Hypnosis*, 42(4), 379–390. https://doi.org/10.1080/00207149408409366
226. Erdelyi, M. H. (2006). The return of the repressed. *Behavioral and Brain Sciences*, 29(5), 535–551. https://psycnet.apa.org/doi/10.1017/S0140525X06479115
227. Malinoski, P., Lynn, S. J., Sivec, H. (1998). The assessment, validity, and determinants of early memory reports: A critical review. In: S. J. Lynn and K. M. McConkey (Eds.). *Truth in Memory* (pp. 109–136). New York: The Guilford Press.
228. Burgess, C. A., Kirsch, I. (1999). Expectancy information as a moderator of the effects of hypnosis on memory. *Contemporary Hypnosis*, 16(1), 22–31. https://doi.org/10.1002/ch.146
229. Barnier, A. J. (2002). Posthypnotic amnesia for autobiographical episodes: A laboratory model of functional amnesia? *Psychological Science,* 13(3), 232–237. https://doi.org/10.1111/1467-9280.00443
230. Barnier, A. J., Bryant, R. A., Briscoe, S. (2001). Posthypnotic amnesia for material learned before or during hypnosis: Explicit and implicit memory effects. *The International Journal of Clinical and Experimental Hypnosis*, 49(4), 286–304. https://doi.org/10.1080/00207140108410079
231. Barnier, A. J., Cox, R. E., O'Connor, A., Coltheart, M., Langdon, R., Breen, N., Turner, M. (2008). Developing hypnotic analogues of clinical delusions: Mirrored-self misidentification. *Cognitive Neuropsychiatry*, 13(5), 406–430. https://doi.org/10.1080/13546800802355666
232. Kihlstrom, J. F. (1997). Hypnosis, memory and amnesia. *Philosophical Transactions of the Royal Society of London. Series B, Biological Sciences*, 352(1362), 1727–1732. https://doi.org/10.1098/rstb.1997.0155
233. Kihlstrom, J. F. (1980). Posthypnotic amnesia for recently learned material: Interactions with 'episodic' and 'semantic' memory. *Cognitive Psychology*, 12(2), 227–251. https://doi.org/10.1016/0010-0285(80)90010-9
234. Kihlstrom, J. F. (1983). Instructed forgetting: Hypnotic and nonhypnotic. *Journal of Experimental Psychology. General*, 112(1), 73–79. https://doi.org/10.1037//0096-3445.112.1.73

235. Kihlstrom, J. F., Evans, F. J. (1971). Posthypnotic amnesia as disorganized retrieval. *Proceedings of the Annual Convention of the American Psychological Association*, 6 (Pt. 2), 775–776. https://psycnet.apa.org/record/1971-24153-001

236. Kihlstrom, J. F., Evans, F. J. (1976). Recovery of memory after posthypnotic amnesia. *Journal of Abnormal Psychology*, 85(6), 564–569. https://psycnet.apa.org/doi/10.1037/0021-843X.85.6.564

237. Kihlstrom, J. F., Wilson, L. (1984). Temporal organization of recall during posthypnotic amnesia. *Journal of Abnormal Psychology*, 93(2), 200–208. https://psycnet.apa.org/doi/10.1037/0021-843X.93.2.200

238. Cox, R., Barnier, R. (2003). Posthypnotic amnesia for a first romantic relationship: Forgetting the entire relationship versus forgetting selected events. *Memory*, 11(3), 307–318. https://doi.org/10.1080/09658210244000108

239. Cox, R., Barnier, R. (2010). Hypnotic illusions and clinical delusions: Hypnosis as a research method. *Cognitive Neuropsychiatry*, 15(1), 202–232. https://doi.org/10.1080/13546800903319884

240. Cox, R., Barnier, R. (2013). Shifting self, shifting memory: Testing the self-memory system model with hypnotic identity delusions. *The International Journal of Clinical and Experimental Hypnosis*, 61(4), 416–462. https://doi.org/10.1080/00207144.2013.810479

241. Cox, R., Bryant, R. (2008). Advances in hypnosis research: Methods, designs and contributions of intrinsic and instrumental hypnosis. In: M. R. Nash & A. J. Barnier (Eds.). *The Oxford Handbook of Hypnosis: Theory, Research, and Practice* (pp. 311–336). New York: Oxford University Press. https://doi.org/10.1093/oxfordhb/9780198570097.013.0012

242. Mendelsohn, A., Chalamish, Y., Solomonovich, A., Dudai, Y. (2008). Mesmerizing memories: Brain substrates of episodic memory suppression in posthypnotic amnesia. *Neuron*, 57(1), 159–170. https://doi.org/10.1016/j.neuron.2007.11.022

243. Brewin, C. R. (2001). Memory processes in post-traumatic stress disorder. *International Review of Psychiatry*, 13(3), 159–163. https://psycnet.apa.org/doi/10.1080/09540260120074019

244. Conway, M. A. (1996). Autobiographical knowledge and autobiographical knowledge. In: D. C. Rubin (Ed.). *Remembering our Past: Studies in Autobiographical Memory* (pp. 67–93). Cambridge: Cambridge University Press. https://psycnet.apa.org/doi/10.1017/CBO9780511527913.003

245. Conway, M. A., Pleydell-Pearce, C. W. (2000). The construction of autobiographical memories in the self-memory system. *Psychological Review,* 107(2), 261–288. https://doi.org/10.1037/0033-295x.107.2.261

246. MacCallum, F., McConkey, K. M., Bryant, R. A., Barnier, A. J. (2000). Specific autobiographical memory following hypnotically induced mood state. *The International Journal of Clinical and Experimental Hypnosis,* 48(4), 361–373. https://doi.org/10.1080/00207140008410366

247. Goddard, L., Dritschel, B., Burton, A. (1996). Role of autobiographical memory in social problem solving and depression. *Journal of Abnormal Psychology,* 105(4), 609–616. https://doi.org/10.1037//0021-843x.105.4.609

248. Basden, B. H., Basden, D. R., Coe, W. C., Decker, S., Crutcher, K. (1994). Retrieval inhibition in directed forgetting and posthypnotic amnesia. *The International Journal of Clinical and Experimental Hypnosis,* 42(3), 184–203. https://doi.org/10.1080/00207149408409351

249. Basden, B. H., Basden, D. R., Gargano, G. J. (1993). Directed forgetting in implicit and explicit memory tests: A comparison of methods. *Journal of Experimental Psychology: Learning, Memory, and Cognition,* 19(3), 603–616. https://psycnet.apa.org/doi/10.1037/0278-7393.19.3.603

250. Anderson, M. C., Green, C. (2001). Suppressing unwanted memories by executive control. *Nature,* 410(6826), 366–369 https://doi.org/10.1038/35066572

251. Bowers, K. S., Woody, E. Z. (1996). Hypnotic amnesia and the paradox of intentional forgetting. *Journal of Abnormal Psychology,* 105(3), 381–390. https://doi.org/10.1037//0021-843x.105.3.381

252. Oeberst, A., Blank, H. (2012). Undoing suggestive influence on memory: The reversibility of the eyewitness misinformation effect. *Cognition,* 125(2), 141–159. https://doi.org/10.1016/j.cognition.2012.07.009

253. Wagstaff, G. F., Cole, J., Wheatcroft, J., Anderton, A., Madden, H. (2008). Reducing and reversing pseudomemories with hypnosis. *Contemporary Hypnosis,* 25(3–4), 178–191. https://psycnet.apa.org/doi/10.1002/ch.366

254. Wagstaff, G. F., Wheatcroft, J. M., Burt, C. L., Pilkington, H. J., Wilkinson, K., Hoyle, J. D. (2011). Enhancing witness memory with focused meditation and eye-closure: Assessing the effects of misinformation. *Journal of Police and Criminal Psychology,* 26, 152–161. https://doi.org/10.1007/s11896-010-9082-7

255. Jensen, M. P., Patterson, D. R. (2014). Hypnotic approaches for chronic pain management: Clinical implications of recent research findings. *The American Psychologist*, 69(2), 167–177. https://doi.org/10.1037/a0035644

256. Jensen, M. P., Sherlin, L. H., Fregni, F., Gianas, A., Howe, J. D., Hakimian, S. (2014). Baseline brain activity predicts response to neuromodulatory pain treatment. *Pain Medicine*, 15(12), 2055–2063. https://doi.org/10.1111/pme.12546

257. Del Casale, A., Ferracuti, S., Rapinesi, C., De Rossi, P., Angeletti, G., Sani, G., Kotzalidis, G. D., Girardi, P. (2015). Hypnosis and pain perception: An Activation Likelihood Estimation (ALE) meta-analysis of functional neuroimaging studies. *Journal of Physiology, Paris*, 109(4–6), 165–172. https://doi.org/10.1016/j.jphysparis.2016.01.001

258. Larkin, M. (1999). Hypnosis makes headway in the clinic. *Lancet*, 353(9150), 386. https://doi.org/10.1016/S0140-6736(05)74964-3

259. Adachi, T., Fujino, H., Nakae, A., Mashimo, T., Sasaki, J. (2014). A meta-analysis of hypnosis for chronic pain problems: A comparison between hypnosis, standard care, and other psychological interventions. *The International Journal of Clinical and Experimental Hypnosis*, 62(1), 1–28. https://doi.org/10.1080/00207144.2013.841471

260. Yaqub, F. (2015). Pain in the USA: States of suffering. *Lancet*, 386(9996), 839. https://doi.org/10.1016/S0140-6736(15)00041-0

261. Gaskin, D. J., Richard, P. (2012). The economic costs of pain in the United States. *The Journal of Pain*, 13(8), 715–724. https://doi.org/10.1016/j.jpain.2012.03.009

262. Madden, K., Middleton, P., Cyna, A. M., Matthewson, M., Jones, L. (2016). Hypnosis for pain management during labour and childbirth. *The Cochrane Database of Systematic Reviews*, 2016(5), CD009356. https://doi.org/10.1002/14651858.CD009356.pub3

263. Bernardy, K., Füber, N., Klose, P., Häuser, W. (2011). Efficacy of hypnosis/guided imagery in fibromyalgia syndrome: A systematic review and meta-analysis of controlled trials. *BMC Musculoskeletal Disorders*, 12, 133. https://doi.org/10.1186/1471-2474-12-133

264. Montgomery, G. H., Bovbjerg, D. H., Schnur, J. B., David, D., Goldfarb, A., Weltz, C. R., Schechter, C., Graff-Zivin, J., Tatrow, K., Price, D. D., Silverstein, J. H. (2007). A randomized clinical trial of a brief hypnosis intervention to control side effects in breast surgery patients. *Journal of the National Cancer Institute*, 99(17), 1304–1312. https://doi.org/10.1093/jnci/djm106

265. Montgomery, G. H., Schnur, J. B., David, D. (2011). The impact of hypnotic suggestibility in clinical care settings. *The International Journal of Clinical and Experimental Hypnosis*, 59(3), 294–309. https://doi.org/10.1080/00207144.2011.570656

266. Montgomery, G. H., DuHamel, K. N., Redd, W. H. (2000). A meta-analysis of hypnotically induced analgesia: How effective is hypnosis? *The International Journal of Clinical and Experimental Hypnosis*, 48(2), 138–153. https://doi.org/10.1080/00207140008410045

267. Einstein, G. O., McDaniel, M. A. (1996). Retrieval processes in prospective memory: Theoretical approaches and some new empirical findings. In: M. Brandimonte, G. O. Einstein, and M. A. McDaniel (Eds.). *Prospective Memory: Theory and Applications* (pp. 115–141). Mahwah: Lawrence Erlbaum Associates Publishers.

268. Li, C. L., Ahlberg, D., Lansdell, H., Gravitz, M. A., Chen, T. C., Ting, C. Y., Bak, A. F., Blessing, D. (1975). Acupuncture and hypnosis: Effects on induced pain. *Experimental Neurology*, 49(1, Pt 1), 272–280. https://doi.org/10.1016/0014-4886(75)90210-1

269. Abshire, L. (2022). What does "triggered" actually mean? https://sageclinic.org/blog/what-does-triggered-actuallymean

270. Anbar, R. D. (2021). Understanding the subconscious: Its central role in hypnosis. https://www.psychologytoday.com/us/blog/understandinghypnosis/202112/understanding-the-subconscious-its-central-role-in-hypnosis

271. Du Plessis, C., Nel, L., Taylor, H. (2021). You're going to do what? Patients' myths regarding hypnotherapy as described by South African psychologists. *Integrative Psychological & Behavioral Science,* 55(3), 527–540. https://doi.org/10.1007/s12124-021-09604-0

272. Hammer, A. G., Orne, M. T. (2022). Hypnosis. *Encyclopedia Britannica.* https://www.britannica.com/science/hypnosis

273. He, J., Yan, X., Wang, R., Zhao, J., Liu, J., Zhou, C., Zeng, Y. (2022). Does childhood adversity lead to drug addiction in adulthood? A study of serial mediators based on resilience and depression. *Frontiers in Psychiatry*, 13, 871459. https://doi.org/10.3389/fpsyt.2022.871459

274. Iorio, C., Brattico, E., Munk Larsen, F., Vuust, P., Bonetti, L. (2022). The effect of mental practice on music memorization. *Psychology of Music*, 50(1), 230–244. https://doi.org/10.1177/0305735621995234

275. Hamblen, J. L., Norman, S. B., Sonis, J. H., Phelps, A. J., Bisson, J. I., Nunes, V. D., Megnin-Viggars, O., Forbes, D., Riggs, D. S., Schnurr, P. P. (2019). A guide to guidelines for the treatment of posttraumatic stress disorder

in adults: An update. *Psychotherapy*, 56(3), 359–373. https://doi.org/10.1037/pst0000231

276. Schnurr, P. P., Lunney, C. A. (2019). Residual symptoms following prolonged exposure and present-centered therapy for PTSD in female veterans and soldiers. *Depression and Anxiety*, 36(2), 162–169. https://doi.org/10.1002/da.22871
277. Schottenbauer, M. A., Glass, C. R., Arnkoff, D. B., Tendick, V., Gray, S. H. (2008). Nonresponse and dropout rates in outcome studies on PTSD: Review and methodological considerations. *Psychiatry*, 71(2), 134–168. https://doi.org/10.1521/psyc.2008.71.2.134
278. Michael, T., Schanz, C. G., Mattheus, H. K., Issler, T., Frommberger, U., Kollner, V., Equit, M. (2019). Do adjuvant interventions improve treatment outcome in adult patients with posttraumatic stress disorder receiving trauma-focused psychotherapy? A systematic review. *European Journal of Psychotraumatology*, 10(1), 1634938. https://doi.org/10.1080/20008198.2019.1634938
279. Ehlers, A., Hackmann, A., Michael, T. (2004). Intrusive re-experiencing in post-traumatic stress disorder: Phenomenology, theory, and therapy. *Memory*, 12(4), 403–415. https://doi.org/10.1080/09658210444000025
280. Types of hypnotherapy. Hypnotherapy Directory. https://www.hypnotherapy-directory.org.uk/content/hypnotherapy-types.html#bwrt
281. Halsband, U., Wolf, T. G. (2015). Functional changes in brain activity after hypnosis in patients with dental phobia. *Journal of Physiology*, 109(4–6), 131–142. https://doi.org/10.1016/j.jphysparis.2016.10.001
282. Webster, A. (2024). Does Medicaid cover drug and alcohol rehab? https://americanaddictioncenters.org/insurance-coverage/usingmedicaid-to-pay-for-rehab
283. Dufresne, A., Rainville, P., Dodin, S., Barré, P., Masse, B., Verreault, R., Marc, I. (2010). Hypnotizability and opinions about hypnosis in a clinical trial for the hypnotic control of pain and anxiety during pregnancy termination. *The International Journal of Clinical and Experimental Hypnosis*, 58(1), 82–101. https://doi.org/10.1080/00207140903310865
284. De Pascalis, V., Scacchia, P. (2016). Hypnotizability and placebo analgesia in waking and hypnosis as modulators of auditory startle responses in healthy women: An ERP study. *PLoS One*, 11(8), e0159135. https://doi.org/10.1371/journal.pone.0159135

285. Jiang, H., White, M. P., Greicius, M. D., Waelde, L. C., Spiegel, D. (2017). Brain activity and functional connectivity associated with hypnosis. *Cerebral Cortex*, 27(8), 4083–4093. https://doi.org/10.1093/cercor/bhw220

286. Williams, R. M., Ehde, D. M., Day, M., Turner, A. P., Hakimian, S., Gertz, K., Ciol, M., McCall, A., Kincaid, C., Pettet, M. W., Patterson, D., Suri, P., Jensen, M. P. (2020). The chronic pain skills study: Protocol for a randomized controlled trial comparing hypnosis, mindfulness meditation and pain education in Veterans. *Contemporary Clinical Trials*, 90, 105935. https://doi.org/10.1016/j.cct.2020.105935

287. Santarcangelo, E. L., Carli, G. (2021). Individual traits and pain treatment: The case of hypnotizability. *Frontiers in Neuroscience*, 15, 683045. https://doi.org/10.3389/fnins.2021.683045

288. Hochberg, L. R., Serruya, M. D., Friehs, G. M., Mukand, J. A., Saleh, M., Caplan, A. H., Branner, A., Chen, D., Penn, R. D., Donoghue, J. P. (2006). Neuronal ensemble control of prosthetic devices by a human with tetraplegia. *Nature*, 442, 164–171. https://doi.org/10.1038/nature04970

289. DeSouza, D. D., Stimpson, K. H., Baltusis, L., Sacchet, M. D., Gu, M., Hurd, R., Wu, H., Yeomans, D. C., Williams, N., Spiegel, D. (2020). Association between anterior cingulate neurochemical concentration and individual differences in hypnotizability. *Cerebral Cortex*, 30(6), 3644–3654. https://doi.org/10.1093/cercor/bhz332

290. Valentine, K. E., Milling, L. S., Clark, L. J., Moriarty, C. L. (2019). The efficacy of hypnosis as a treatment for anxiety: A meta-analysis. *The International Journal of Clinical and Experimental Hypnosis*, 67(3), 336–363. https://doi.org/10.1080/00207144.2019.1613863

291. Trumm, A. (2018). Effectiveness of hypnotherapy in anxiety disorders: A systematic review. http://dx.doi.org/10.13140/RG.2.2.33559.32169

292. Hinze, J., Röder, A., Menzie, N., Müller, U., Domschke, K., Riemenschneider, M., Noll-Hussong, M. (2021). Spider phobia: Neural networks informing diagnosis and (virtual/augmented reality-based) cognitive behavioral psychotherapy: A narrative review. *Frontiers in Psychiatry*, 12, 704174. https://doi.org/10.3389/fpsyt.2021.704174

293. Fuhr, K., Bender, A., Wiegand, A., Janouch, P., Drujan, M., Cyrny, B., Schweizer, C., Kreifelts, B., Nieratschker, V., Batra, A. (2023). Hypnotherapy for agoraphobia-Feasibility and efficacy investigated in a pilot study. *Frontiers in Psychology*, 14, 1213792. https://doi.org/10.3389/fpsyg.2023.1213792

294. Ebrahimi, O. V., Pallesen, S., Kenter, R. M. F., Nordgreen, T. (2019). Psychological interventions for the fear of public speaking: A meta-analysis. *Frontiers in Psychology*, 10, 488. https://doi.org/10.3389/fpsyg.2019.00488

295. Harrison, N., Peacock, N. (2013). Cultural distance, mindfulness and passive xenophobia: Using Integrated Threat Theory to explore home higher education students' perspectives on "internationalisation at home." *British Educational Research Journal*, 36(6), 877–902. https://doi.org/10.1080/01411920903191047

296. Mc Intosh, I. (2007). Brief selective hypnotherapy in the treatment of flying phobia. *Vertex*, 18(74), 268–271. https://pubmed.ncbi.nlm.nih.gov/18219399/

297. Papagianni, V., Kotera, Y. (2022). The beneficial effects of a single hypnotherapy session using parts negotiation for specific phobias. *International Journal of Spa and Wellness*, 5(2), 167–184. https://doi.org/10.1080/24721735.2022.2049103

298. Khazraee, H., Bakhtiari, M., Kianimoghadam, A. S., Ghorbanikhah, E. (2022). The effectiveness of mindful hypnotherapy on depression, self-compassion, and psychological inflexibility in females with major depressive disorder: A single-blind, randomized clinical trial. *The International Journal of Clinical and Experimental Hypnosis*, 71(1), 63–78. https://doi.org/10.1080/00207144.2022.2160257

299. Fuhr, K., Meisner, C., Broch, A., Cyrny, B., Hinkel, J., Jaberg, J., Petrasch, M., Schweizer, C., Stiegler, A., Zeep, C., Batra, A. (2021). Efficacy of hypnotherapy compared to cognitive behavioral therapy for mild to moderate depression: Results of a randomized controlled rater-blind clinical trial. *Journal of Affective Disorders*, 286, 166–173. https://doi.org/10.1016/j.jad.2021.02.069

300. Haipt, A., Rosenbaum, D., Fuhr, K., Giese, M., Batra, A., Ehlis, A. C. (2022). The effects of hypnotherapy compared to cognitive behavioral therapy in depression: A NIRS-study using an emotional gait paradigm. *European Archives of Psychiatry and Clinical Neuroscience*, 272(4), 729–739. https://doi.org/10.1007/s00406-021-01348-7

301. Cottraux, J. (2002). Nonpharmacological treatments for anxiety disorders. *Dialogues in Clinical Neuroscience*, 4(3), 305–319. https://doi.org/10.31887/DCNS.2002.4.3/jcottraux

302. Lesmana, C. B. J., Suryani, L. K., Tiliopoulos, N. (2022). The biobehavioural effectiveness of spiritual-hypnosis-assisted therapy in PTSD with childhood trauma. *The Egyptian Journal of Neurology, Psychiatry & Neurosurgery*, 58(1). https://doi.org/10.1186/s41983-022-00475-9

303. Xian-Yu, C. Y., Deng, N. J., Zhang, J., Li, H. Y., Gao, T. Y., Zhang, C., Gong, Q. Q. (2022). Cognitive behavioral therapy for children and adolescents with post-traumatic stress disorder: Meta-analysis. *Journal of Affective Disorders*, 308, 502–511. https://doi.org/10.1016/j.jad.2022.04.111

304. Gardner, T., O'Hagan, E., Gilanyi, Y. L., McAuley, J. H., Jensen, M. P., Rizzo, R. R. (2024). Using hypnosis in clinical practice for the management of chronic pain: A qualitative study. *Patient Education and Counseling*, 119, 108097. https://doi.org/10.1016/j.pec.2023.108097

305. Moura, V. L., Faurot, K. R., Gaylord, S. A., Mann, J. D., Sill, M., Lynch, C., Lee, M. Y. (2012). Mind-body interventions for treatment of phantom limb pain in persons with amputation. *American Journal of Physical Medicine & Rehabilitation*, 91(8), 701–714. https://doi.org/10.1097/PHM.0b013e3182466034

306. Oakley, D. A., Whitman, L. G., Halligan, P. W. (2002). Hypnotic imagery as a treatment for phantom limb pain: Two case reports and a review. *Clinical Rehabilitation*, 16(4), 368–77. https://doi.org/10.1191/0269215502cr507oa

307. Connelly, C. (2023). Hypnosis based treatments for internet addiction disorder: Systematic review. *Journal of Addiction Research*, 7(1), 01–06. https://www.opastpublishers.com/open-access-articles/hypnosis-based-treatments-for-internet-addiction-disorder-systematic-review.pdf

308. O'Keefe, T. (2020). Clinical hypnotherapy for stopping drug and alcohol addiction: Building resilience in clients to reduce relapses and remain clean and sober. *Australian Journal of Clinical Hypnotherapy and Hypnosis*, 41(1), 16–26. https://link.gale.com/apps/doc/A663245333/AONE?u=anon~e04550d7&sid=googleScholar&xid=fe392210

309. Pickering, D., Keen, B., Entwistle, G., Blaszczynski, A. (2018). Measuring treatment outcomes in gambling disorders: A systematic review. *Addiction*, 113(3), 411–426. https://doi.org/10.1111/add.13968

310. Vasiliu, O. (2022). Therapeutic management of buying/shopping disorder: A systematic literature review and evidence-based recommendations. *Frontiers in Psychiatry*, 13, 1047280. https://doi.org/10.3389/fpsyt.2022.1047280

311. Delestre, F., Lehéricey, G., Estellat, C., Diallo, M. H., Hansel, B., Giral, P. (2022). Hypnosis reduces food impulsivity in patients with obesity and high levels of disinhibition: HYPNODIET randomized controlled clinical trial. *The American Journal of Clinical Nutrition*, 115(6), 1637–1645. https://doi.org/10.1093/ajcn/nqac046

312. Tomic, N. (2011). Treating nocturnal enuresis with direct and indirect suggestions by using hypnosis. *Australian Journal of Clinical Hypnotherapy and Hypnosis,* 32(1). https://link.gale.com/apps/doc/A311718884/AONE?u=anon~eee6c9a5&sid=googleScholar&xid=26f8a0d4

313. Li, Z., Li, S. X. (2022). The application of hypnosis in sports. *Frontiers in Psychology,* 12, 771162. https://doi.org/10.3389/fpsyg.2021.771162

314. Zotey, V., Andhale, A., Shegekar, T., Juganavar, A. (2023) Adaptive neuroplasticity in brain injury recovery: Strategies and insights. *Cureus,* 15(9), e45873. https://doi.org/10.7759/cureus.45873

315. Tukaev, R. (2020). The integrative theory of hypnosis in the light of clinical hypnotherapy. In: *Hypnotherapy and Hypnosis.* IntechOpen. https://doi.org/10.5772/intechopen.92761

316. Fredette, C., El-Baalbaki, G., Neron, S., Palardy, V. (2013). Using hypnosis in the treatment of anxiety disorders: Pros and cons. In: *New Insights into Anxiety Disorders.* IntechOpen. https://doi.org/10.5772/53768

317. Eason, A. D., Parris, B. A. (2024). The importance of highlighting the role of the self in hypnotherapy and hypnosis. *Complementary Therapies in Clinical Practice,* 54, 101810. https://doi.org/10.1016/j.ctcp.2023.101810

318. Saucier, J. F. (2012). Neuro-hypnosis: Using self-hypnosis to activate the brain for change. *Journal of the Canadian Academy of Child and Adolescent Psychiatry,* 21(2), 154–155. https://www.ncbi.nlm.nih.gov/pmc/articles/PMC3338184/

319. Zimberoff, D., Hartman, D. (2014). Applications of hypnosis and hypnotherapy: A compilation of brief monographs. *Journal of Heart-Centered Therapies,* 16(2), and 17(1), 3–183. https://hypnotherapy.co.za/wp-content/uploads/2022/07/Review_Zimberhoff-Hartman-Training-material-Application-of-hypnosis-Zimberhoff-Hartman.pdf

320. Alman, B. (2001). Medical hypnosis: An underutilized treatment approach. *The Permanente Journal,* 5(4), 35–40. https://www.ncbi.nlm.nih.gov/pmc/articles/PMC6220618/

321. Xue, G., Chen, C., Lu, Z. L., Dong, Q. (2010). Brain imaging techniques and their applications in decision-making research. *Xin Li Xue Bao,* 42(1), 120–137. https://doi.org/10.3724/SP.J.1041.2010.00120

322. Geagea, D., Ogez, D., Kimble, R., Tyack, Z. (2024). Redefining hypnosis: A narrative review of theories to move towards an integrative model. *Complementary Therapies in Clinical Practice,* 54, 101826. https://doi.org/10.1016/j.ctcp.2023.101826

323. Gill, P., Fraser, E., Tran, T. T. D., De Sena Collier, G., Jago, A., Losinno, J., Ganci, M. (2023). Psychosocial treatments for nightmares in adults and children: A systematic review. *BMC Psychiatry*, 23(1), 283. https://doi.org/10.1186/s12888-023-04703-1

324. Reyna, V. F., Helm, R. K., Weldon, R. B., Shah, P. D., Turpin, A. G., Govindgari, S. (2018). Brain activation covaries with reported criminal behaviors when making risky choices: A fuzzy-trace theory approach. *Journal of Experimental Psychology. General*, 147(7), 1094–1109. https://doi.org/10.1037/xge0000434

325. Hsu, L. Chen, Y. J. (2020). Neuromarketing, subliminal advertising, and hotel selection: An EEG study. *Australasian Marketing Journal (AMJ)*, 28(4), 200–208. https://doi.org/10.1016/j.ausmj.2020.04.009

326. Peksa, J., Mamchur, D. (2023). State-of-the-art on brain-computer interface technology. *Sensors*, 23(13), 6001. https://doi.org/10.3390/s23136001

327. Thoma, M. V., La Marca, R., Brönnimann, R., Finkel, L., Ehlert, U., Nater, U. M. (2013). The effect of music on the human stress response. *PLoS One*, 8(8), e70156. https://doi.org/10.1371/journal.pone.0070156

328. University of Queensland. (2025). Timeline of brain development. https://stories.uq.edu.au/the-brain/2022/timeline-of-brain-development/index.html

Index

For Product Safety Concerns and Information please contact our EU representative GPSR@taylorandfrancis.com Taylor & Francis Verlag GmbH, Kaufingerstraße 24, 80331 München, Germany

Batch number: 10726507

Printed by Printforce, the Netherlands